Occupational Stress

Occupational Stress

A Handbook

Edited by

Rick Crandall, Ph.D.
Pamela L. Perrewé, Ph.D.

Taylor & Francis
Publishers since 1798

USA	Publishing Office:	Taylor & Francis
		1101 Vermont Avenue, N.W., Suite 200
		Washington, DC 20005-2531
		Tel: (202) 289-2174
		Fax: (202) 289-3665
	Distribution Center:	Taylor & Francis
		1900 Frost Road, Suite 101
		Bristol, PA 19007-1598
		Tel: (215) 785-5800
		Fax: (215) 785-5515
UK		Taylor & Francis, Ltd
		4 John Street
		London WC1N 2ET, UK
		Tel: 071 405 2237
		Fax: 071 831 2035

The chapters in this book are revisions of papers published in the Journal of Social Behavior and Personality, © 1988, 1989, 1991, 1992, 1993, 1994 by Select Press.

OCCUPATIONAL STRESS: A Handbook

2 3 4 5 6 7 8 9 0   B R B R   9 8

Cover design by Michelle Fleitz.

A CIP catalog record for this book is available from the British Library.

♾ The paper in this publication meets the requirements of the ANSI Standard Z39.48-1984 (Permanence of Paper)

Library of Congress Cataloging-in-Publication Data

Occupational stress: a handbook / edited by Rick Crandall, Pamela L. Perrewé
  p.   cm.—(The Series in health psychology and behavioral medicine)
  Includes bibliographical references and index.
  1. Job stress.   I. Crandall, Rick.   II. Perrewé, Pamela L.
  III. Series.
HF5548.85.O248   1995   94-36229
158.7—dc20   CIP
ISBN 1-56032-367-1

ISSN 8756-467X

Contents

Contributors

Stephen M. Allie
Hurst-Euless-Bedford
  Independent School District
Bedford, TX 76022

David F. Barone
Center for Psychological Studies
Nova Southeastern University
3301 College Avenue
Fort Lauderdale, FL 33314

Rabi S. Bhagat
Department of Management
202 Fogelman College of
  Business Economics
Memphis State University
Memphis, TN 38152

Arthur P. Brief
A.B. Freeman School of Business
Department of Psychology
Tulane University
New Orleans, LA 70118

Ronald J. Burke
Department of Organiational Behavior
  and Industrial Relations
Faculty of Administrative Studies
York University
4700 Keele Street
North York, ON M3J 1P3 Canada

Cary L. Cooper
Manchester School of Management
University of Manchester
Institute of Science and Technology
P.O. Box 88
Manchester, M60 1QD England

M. Lynne Cooper
Research Institute on Addictions
1021 Main Street
Buffalo, NY 14203

Kevin Corcoran
Graduate School of Social Work
Portland State University
P.O. Box 751
Portland, OR 97207

Angelo S. DeNisi
Institute of Management and
 Labor Relations
Rutgers University, Cook Campus
New Brunswick, NJ 08903

Thomas A. DeCotiis
T. A. DeCotiis and Associates
P.O. Box 3747
Rancho Santa Fe, CA 92067

Vincent DiSalvo
Department of Communication Studies
428 Oldfather Hall
University of Nebraska-Lincoln
Lincoln, NE 68588

Shimon L. Dolan
School of Industrial Relations
University of Montreal
P.O. Box 6128 "A"
Montreal, PQ H3C 3J7 Canada

Lisa Fiksenbaum
Department of Psychology
York University
4700 Keele Street
North York, ON M3J 1P3 Canada

David L. Ford, Jr.
University of Texas at Dallas
PO Box 830688
Richardson, TX 75083

Michael R. Frone
Research Institute on Addictions
1021 Main Street
Buffalo, NY 14203

Daniel C. Ganster
Department of Management
402 Business Administration Building
University of Arkansas
Little Rock, AR 72204

Anna-Maria Garden
GDS Consultants
25A Maresfield Gardens
Hampstead, London NW3 5SD England

Jennifer M. George
Department of Management
Texas A&M University
College Station, TX 77843

Esther R. Greenglass
Department of Psychology
York University
4700 Keele Street
North York, ON M3J 1P3 Canada

David M. Gudanowski
Department of Psychology
Central Michigan University
Mt. Pleasant, MI 48859

James R. Harris
Department of Business Administration
School of Business & Economics
North Carolina A&T State University
Greensboro, NC 27411

Stephen J. Havlovic
Faculty of Business Administration
Simon Fraser University
Burnaby, BC V5A 1S6 Canada

William H. Hendrix
Department of Management
Clemson University
Clemson, SC 29634

Wayne A. Hochwarter
Department of Management and MIS
Mississippi State University
P.O. Drawer MG
Mississippi State, MS 39762

Steve M. Jex
Department of Psychology
Central Michigan University
Mount Pleasant, MI 48859

John P. Keenan
Management Institute
University of Wisconsin
432 North Lake Street
Madison, WI 53706

Russell L. Kent
Department of Management
Georgia Southern University
College of Business Administration
Statesboro, GA 30460

Richard S. Lazarus
Professor Emeritus
Department of Psychology
University of California
Berkeley, CA 94720

Terry L. Leap
Department of Management
Clemson University
Clemson, SC 29634

James Lewis
Department of Communication
University of South Dakota
Vermillion, SD 57069

Charles Lubbers
Department of Communication Studies
428 Oldfather Hall
University of Nebraska-Lincoln
Lincoln, NE 68588

Ronald A. Newman
Department of Psychology
University of New Haven
West Haven, CT 06516

E. Dara Ogus
103-70 Elmsthorpe Avenue
Toronto, ON M5P 2L7 Canada

Pamela L. Perrewé
Department of Management
College of Business Administration
Florida State University
Tallahassee, FL 32306

Eric C. Reheiser
Center for Research in Behavioral
    Medicine and Health Psychology
University of South Florida
Tampa, FL 33620

Ana M. Rossi
Department of Communication Studies
428 Oldfather Hall
University of Nebraska-Lincoln
Lincoln, NE 68588

Marcia Russell
Research Institute on Addictions
1021 Main Street
Buffalo, NY 14203

Golnaz Sadri
Manchester School of Management
University o f Manchester
Institute of Science and Technology
P.O. Box 88
Manchester, M60 1QD England

John Schaubroeck
Department of Management
University of Nebraska-Lincoln
Lincoln, NE 68588

Paul E. Spector
Department of Psychology
University of South Florida
Tampa, FL 33620

Charles D. Spielberger
Department of Psychology
University of South Florida
4202 East Fowler Ave., BEH 339
Tampa, FL 33620

Robert P. Steel
Department of Communication and
    Organizational Sciences
School of Systems and Logistics
Air Force Institute of Technology
Wright-Patterson AFB, OH 45433

Timothy P. Summers
Department of Mangement
Clemson University
Clemson, SC 29634

Alan P. Wolfgang
Department of Pharmacy Practice
School of Pharmacy and Pharmacal
  Sciences
Purdue University
West Lafayette, IN 47907

Foreword

This handbook brings together a number of renowned scholars in the area of occupational stress. These experts describe their current theoretical perspectives as well as their empirical work in occupational stress.

Theoretical Approaches

The handbook is divided into five sections. The first section of this handbook is entitled, "Theoretical Perspectives in Job Stress Research." Who better to lead off this section, and the handbook, than Professor Emeritus Richard S. Lazarus? Dr. Lazarus examines the occupational stress field through his transactional approach highlighting the stress *process,* the stress *context,* and the individual *appraisal* of stress. The next three papers offer their comments and perspectives. First, Drs. Brief and George offer their own guidelines and suggestions for occupational stress research. Second, Dr. Harris examines the utility of the transactional approach. Third, Dr. Barone develops a transactional psychology for the study of occupational stress. The fifth paper is by Drs. DiSalvo, Lubbers, Rossi, and Lewis. They take an innovative approach to examining perceptions of occupational stress by measuring stress through unstructured means. The final paper in this section is by Drs. Charles D. Spielberger and Eric C. Reheiser. They provide a comprehensive overview of the research in occupational stress. Their paper also includes some very interesting findings relating gender and occupational level to experienced occupational stress and coping using the *Job Stress Survey,* a new measure developed over a number of years.

Model Testing

The second section of the handbook is entitled, "Sources and Consequences of Job Stress; Model Testing." The lead article in this section develops and tests a stress and health model. Drs. Hendrix, Summers, Leap, and Steel focus on individual and organizational characteristics, stress as an intervening variable, and organizational effectiveness and health outcomes. The second article examines a model of organizational stress as well as personal life stress. Drs. Bhagat, Allie, and Ford develop and test the effects of organizational stress and personal life stress on symptoms of life strain, while emphasizing the moderating role of problem-focused and emotion-focused coping styles. Drs. Summers, DeCotiis, and DeNisi examine both attitudinal and behavioral consequences of experienced stress. The final article in this section, by Drs. Frone, Russell, and Cooper, examines both work and family stressors on psychological distress. They develop and test their model focusing on the moderating influences of social support, mastery, active coping, and self-focused attention.

Coping

The next section of the handbook is, "The Roles of Coping and Dispositional Influences in Job Stress Research." Drs. Hochwarter, Perrewé, and Kent examine stressors, strain, and intentions to leave, focusing on the dispositional effects of persistence. The second article examines role stress and worker health. In this paper, Drs. Ganster and Schaubroeck demonstrate the moderating effects of self-esteem in the stressor-worker health relationship. Drs. Havlovic and Keenan examine a number of coping methods and their relationship to worker stress and strain. The final paper in this section, by Dr. Wolfgang, examines occupational stress, coping, and dissatisfaction of health professionals.

Burnout

The fourth section of the handbook includes five papers and is entitled, "An Examination of Burnout." The first article is a theoretical piece by Dr. Garden who discusses the purpose of burnout using a Jungian perspective. In the second article, Dr. Dolan uses a multivariate approach to examine the individual, organizational, and social determinants of managerial burnout. Drs. Greenglass, Fiksenbaum, and Burke take a longitudinal approach to examining the relationship between social support and burnout. Following this, Dr. Ogus looks at burnout and coping with health professionals. The final article in this section is by Dr. Corcoran who takes a hard look at the measurement of burnout.

Interventions

The final section of the handbook examines "Interventions Aimed at Job Strain Reduction." The first paper, by Drs. Cooper and Sadri, examines the usefulness of stress counseling at work. The final article examines exercise as a possible buffer to occupational stress. Using the information gathered from two studies, Drs. Jex, Spector, Gudanowski, and Newman examine the relations between exercise and psychological worker strain and health outcomes.

We hope this handbook will prove to be invaluable for practitioners and researchers in the occupational stress field. We believe it provides a comprehensive perspective on the field of occupational stress and offers suggestions for future directions in this area for the 1990s and beyond.

Pamela L. Perrewé
Rick Crandall

PART ONE

Theoretical Perspectives in Occupational Stress Research

Psychological Stress in the Workplace

Richard S. Lazarus

If one scans the research on work stress, one discovers that attention has been given mainly to the organizational arrangement of work as stressful, less to person variables, and almost none to the stress process, that is, actual stressful transactions that take place between workers and the environment, coping, and changes in stress from moment to moment and encounter to encounter. In the research tradition that has been dominant, antecedent variables of stress reactions, which include both environmental conditions and person characteristics, are treated as separate and static causes of behavioral and medical states, such as illness, distress and burnout, work dissatisfaction, performance, and absenteeism.

I present here a somewhat heretical outlook, one with which I have long been identified, namely, a transactional, process, contextual, and meaning-centered approach to stress (Lazarus, 1966; Lazarus & Folkman, 1984, 1987). It is my conviction that the traditional outlook toward work stress, while correctly identifying many normative antecedent and consequent environmental and personality variables, is not serviceable when we wish to deal with individual working people

Author's Note: Parts of this paper were originally published in the Proceedings of the International Conference on Industrial Health and the VIIIth International Symposium – Health Surveillance of Workers, University of Occupational and Environmental Health, Japan, March 1989.

and groups who are suffering from excessive stress and dysfunction. This sweeping conclusion needs to be defended.

Prodigious reviews, such as that of Holt (1982), divide research into stressful stimulus conditions such as work overload, role ambiguity and conflict, poor labor-management relations, monotony, and lack of control over job demands and possibilities; Holt lists 57 such variables. He also identifies 55 response variables that mitigate or exacerbate stress effects. The reader might also examine a series of edited books by Cooper and Kasl (e.g., Cooper & Payne, 1980), and one by Ivancevich and Matteson (1980), which is more theoretical in focus, to get the flavor of social science thinking about stress in the workplace. In short, industrial psychologists have recognized that stress at work is important, but, in the main, continue to do lip service to the most advanced theories about the stress process.

The ideas about person-environment fit presented by the Michigan group of French, Caplan, and Van Harrison (1982), which combine the separate causal variables of environment and person into an ongoing relationship characterized by fit or misfit, were, in my opinion, an important advance in thinking. Good person-environment fit is said to be associated with good adaptation, and misfit with bad adaptation. However, this approach falls short because the concept of fit between the person and the environment is static. It emphasizes stable relationships between the person and the workplace rather than flux or process in which stress constantly changes over time and varies with specific work-related contexts. A static or structural approach is indigenous to the field of industrial stress.

It is simplistic to carry over from medicine, clinical psychology, and personality psychology an emphasis on psychopathology or dysfunction, and to make the assumption that, as a result of personality traits, some people usually or always function badly whereas others usually or always function well. Although this assumption may have some probabilistic validity, sound workers not only experience stress at work, but they may also cope badly with certain stressful encounters; and, vice versa, unsound workers sometimes function well. In effect, even when there is, in general, a good stable fit between the work setting and person, stress can still be generated in particular encounters such as being evaluated, failure to be promoted or receive a raise in pay, dealing with difficult co-workers, and other difficulties to which all of us are subject in our working lives. A worker might deal very well with one work encounter yet experience major stress in other encounters.

It is also simplistic to carry over from social psychology and sociology the idea that some social environments generally result in healthy functioning and other environments generally result in dysfunction. Except for the most destructive and tyrannical environments, which are comparatively rare and whose pathogenic qualities are easily recognized, it is a half truth that most people respond similarly to the same conditions.

I do not mean by this to challenge the conclusion that uprooting, dehumanization at work, work changes, and environmental constraints are stressful institutional conditions. Of course they are. Nor do I dispute Levi, Frankenhaeuser, and

Gardell's (1982) claim that mass-production technology, including shift work, can generate stress. My point is simply that psychological stress and its damaging effects are quite an individual matter. Without knowing what is involved personally for individuals and particular collectivities, and the particular contexts in which they operate, we will be handicapped in our understanding and in our efforts to ameliorate or prevent stress in the workplace.

In the remainder of this paper, I focus on what I think are keys to understanding not only work stress but stress in any context of living. The analysis is centered on the concepts of transaction, process, and personal meaning.

METATHEORY

Two metatheoretical principles underlie the approach I am advocating. *Transaction* means not only that, in a particular adaptational encounter, the person influences the environment and vice-versa, but also that person-environment relationships transcend the separate interacting variables of person and environment, and are constantly subject to change. For a relationship to be stressful, there must, first, be some stake in the outcome. This is another way of saying that the person believes the transaction is relevant to personal goals of importance. Second, psychological stress occurs only when a person had made an evaluation (appraisal) that external or internal demands tax or exceed his or her resources. Stress is not a property of the person, or of the environment, but arises when there is a conjunction between a particular kind of environment and a particular kind of person that leads to a threat appraisal.

Since we usually attempt to change that which is undesirable or distressing, stress implies a *process* rather than a static arrangement. Process means that the psychological state changes over time and across diverse encounters. Person-environment relationships, and the goodness of fit between person and environment, are not constant over time, or from one work task or activity to another. This principle cannot be grasped and evaluated unless we employ research designs that are intra-individual, in addition to inter-individual or normative, in order to observe the degree of stability or instability of the reaction over time and across adaptational encounters.

THEORY

I said, above, that a transaction between the person and the environment is stressful only when it is evaluated by the person as a harm, threat or challenge to that person's well-being. The term I have used for this evaluation is *appraisal*. In a cognitive-motivational-relational theory, stress depends on the balance of power, as judged subjectively, between the environmental demands, constraints, and resources and the ability of the person to manage them.

Harm refers to damage that has already occurred, as in a loss of job, a poor job evaluation, a failure to be promoted, or disapproval by management or one's peers. Threat refers to a harm that has not yet happened, but is anticipated in the future. Challenge refers to a condition of high demand in which the emphasis is on mastering the demands, overcoming obstacles, and growing and expanding as an individual. In threat, the focus is on protecting against harm. In challenge, the emphasis is on the positive outcome possibilities. We like challenge, but dislike threat. The attitude of challenge allows us to feel enthused, engaged, and expansive, rather than endangered, defensive, and self-protective.

My early experimental and field research on appraisal was begun in the 1960s (Lazarus, 1966; 1968), and has continued to the present (Lazarus, Averill, & Opton, 1970, 1974; Lazarus & Folkman, 1984, 1987). It demonstrated that the way people evaluate what is happening with respect to their well-being, and the way they cope with it, influences whether or not psychological stress will result, and its intensity.

There are two basic kinds of appraisal. *Primary appraisal* concerns whether or not there is any personal stake in the encounter. *Secondary appraisal* concerns the available coping options for dealing with harm, threat, or challenge.

The task of appraisal is to integrate two sets of forces operating in every adaptational transaction, namely, personal agendas (e.g., goals and beliefs) brought to the transaction by the person, and the environmental realities that affect the outcome. To overemphasize personal agendas is autism, and to overemphasize the environmental realities is to abandon one's personal identity. Survival would not be possible if appraisals were constantly in bad fit with the environmental realities. It would be equally in jeopardy if we failed to take into account our personal agendas. It is essential to consider how the individual person appraises what is happening to understand his or her emotional reactions.

Since personal agendas vary from person to person, and even within the person from moment to moment, and since the environment is often quite complex and ambiguous, we attend selectively to what is happening, and evaluate it in diverse ways. This results in great variation in the appraisals people make in the same environmental context. Appraisal is usually not fixed, but we are constantly trying to evaluate what is happening in a way that is both realistic and also allows us to see the conditions of our lives in as favorable a way as possible. A sound theory of psychological stress must be capable of helping us understand the variations in the ways individuals appraise adaptational transactions with their environments.

I define *coping* as the cognitive and behavioral efforts a person makes to manage demands that tax or exceed his or her personal resources (Lazarus & Folkman, 1984). This is a process because the relationship with the environment is constantly changing (Folkman & Lazarus, 1978), although there are, of course, stable patterns too. Many of the changes are the result of coping itself, which provides new information that feeds back to the person and alters subsequent appraisals. Coping has a profound effect on psychological stress and emotional states (Folkman & Lazarus, 1988a, 1988b), as a result of what it does to the appraisal process, which is always the proximal cause of the stressful or emotional reaction.

Coping affects the stress reaction in two fundamental ways: *Problem-focused* coping consists of efforts to alter the actual relationship, as when we seek information about what needs to be done and change either our own behavior or take action on the environment. When coping actions alter the person-environment relationship for the better, they eliminate or reduce the psychological grounds for harm or threat and result, thereby, in a changed appraisal, which in turn changes the emotional reaction.

Emotion-focused coping consists of efforts to regulate the emotional distress caused by harm or threat. One basic strategy is to deploy one's attention by avoiding thoughts about sources of distress. The other consists of a variety of strategies that change the meaning of what is happening, or what will happen, such as denial, positive thinking, and distancing. Emotion-focused coping does not change the objective terms of the person-environment relationship, but only how these terms are attended to or interpreted.

My colleagues and I developed the *Ways of Coping Scales* to assess the main ways people cope with stress (Folkman & Lazarus, 1988c; Lazarus & Folkman, 1984). The approach to coping is contextual and process-oriented in contrast with the tradition of treating coping as broad personality traits or styles of relating to the world. Although people do have such styles, their coping response is always, to some extent, governed by the adaptational requirements of the particular encounter.

For example, if an encounter is appraised by the person as subject to control by his or her actions, problem-focused strategies predominate. If, on the other hand, the person judges that nothing can be done to change the situation, emotion-focused coping predominates. Our research has shown that patterns of coping vary from one stressful encounter to another, and over time (Folkman & Lazarus, 1978; Folkman, Lazarus, Dunkel-Schetter, DeLongis, & Gruen, 1986; Folkman, Lazarus, Gruen, & DeLongis, 1986). Some coping strategies, such as positive thinking, are relatively stable across encounters. They are obviously influenced by personality. Others, such as seeking social support, are quite unstable. They are obviously heavily influenced by the environmental conditions of the stressful encounter.

Coping is central to the stress process and its adaptational outcomes. It is influential, via the appraisal process, in immediate emotional reactions, in how the person acts, and in the person-environment relationship. It also most probably affects long-term adaptational outcomes such as subjective well being, social functioning, and somatic health.

Coping is also extremely important in the stressful *transitions* that take place in people's lives over the life course. Societies change within a lifetime; the social and work roles people occupy also change, as when companies fail, when there is organizational change, or workers become ill or handicapped, or decide to retire. It is important to consider how the sources of stress and the coping process change as society changes, and as adults grow older, as well as in personal crises (Baltes,

1987; Cantor, Norem, Brower, Niedenthal, & Langston, 1987; Caspi, 1987; Felner, Farber, & Primavera, 1983; Labouvie-Vief, Hakim-Larson, & Hobart, 1987). It is probable, for example, that the sources of stress change considerably over the life course and coping changes with them. Patterns of coping change with age, too. Older adults depend more than younger ones on distancing and less on active confrontation with an often unresponsive world (Morse & Weiss, 1955; Whitbourne, 1985). Life course issues are also relevant to the workplace, but space does not permit me to elaborate on this (see also Dewe, 1993).

One of the major problems in research on coping concerns its measurement. Conflict exists over whether coping should be measured as a *process*—that is, as a changing pattern of reactions to particular kinds of stress, at a particular time as coping changes, and in particular individuals over time and across stressful situations—or as a personality *style* that has some consistency over time and across stressful situations.

My current view of this conflict is that these approaches to measurement ask different questions, each of which is valuable. Both approaches should be followed depending on what is being asked. I have recently discussed the advantages and limitations of each approach in a review of the history of theory and research on coping and made suggestions for resolving the conflicts between both approaches (Lazarus, 1993a).

IMPLICATIONS FOR STRESS IN THE WORKPLACE

A transactional, process, and appraisal- (or meaning) centered approach to stress offers a very different perspective on work stress, and stress in general, than what has been traditional. It has a number of profound implications for (a) the way research is conducted and (b) efforts to ameliorate stress by interventions aimed at prevention and treatment. I shall address some of the most important of these implications below.

The Conduct of Research

It should come as no surprise that many conditions of work such as time pressure, noise, work overload, lack of decisional control, role ambiguity, conflicts with superiors and subordinates, and so on, are stressful for large numbers of workers. Likewise, certain types of persons, for example, rigid personalities, those addicted to drugs, those with neuroticism, with a tendency to depression, etc., are likely to react with stress more often or more intensely than others. Recurrent and intense stress will, of course, increase the incidence and severity of symptoms such as emotional distress, disruptions of performance, illness, absenteeism, and so on.

Although this kind of knowledge has value, it misses the central point that the sources of stress are always, to some extent, individual, as are the ways people cope with stress. One worker is upset primarily about his relationships with others in the shop. For that worker, being accepted and liked is of the greatest importance. Another worker is upset primarily about evaluations by his supervisors, or about whether he will advance in position and salary. For him, success in the job is of the greatest importance. To describe and understand stress in the workplace requires that these individual patterns be studied to generate knowledge about the kinds of persons who are more or less vulnerable to divergent sources of stress.

In their excellent discussion of stress, coping, and the meaning of work, organizational psychologists Locke and Taylor (1990) have written the following about individual meanings of work:

> ... the meaning of work varies widely among individuals in our society. While each of us can observe this variation from personal experience, the research literature provides even stronger evidence of such differences. In one of the earliest studies of the meaning of work, Morse and Weiss (1955) examined a sample of 401 American males, and found that individuals in middle class and farming occupations tended to emphasize the intrinsic interest and significance of their work, whereas those in lower class occupations viewed work simply as an activity that kept them busy. Similarly, Near, Rice, and Hunt (1980) reviewed the literature on work and nonwork domains and concluded that the importance of work varied by occupation; people in higher skill jobs saw their work as more important than those in lower skill jobs. Finally, Buchholz (1978) examined the work beliefs of over one thousand individuals including employees, union leaders, and managers and found individual differences according to age, education, and occupation. Young people displayed a stronger pro-work ethic than did older people, and those with graduate education expected more intrinsic outcomes (e.g., interest, challenge) from their work than those with less education.

Locke and Taylor then identify five values that people want to fulfill at work. These include material values, achievement-related values, a sense of purpose, social relationships, and enhancement or maintenance of self. Harm and threat at work stem from the thwarting of values or goals that are important to the person. To some extent, what is important varies from person to person and collectivity to collectivity.

To a large extent, too, stress and distress vary within the person from occasion to occasion, and setting to setting. A person's emotional state and coping efforts will also change over the course of a stressful transaction. If stress is indeed a process, there is no way to study it effectively by doing single assessments. Stress research must be conducted, at least in part, intra-individually, that is, by studying the same persons in different work contexts and over time, as well as inter-individually.

Researchers are also likely to be misled about the stress process when complex situations that change systematically over time are treated as a single stressful event. This was shown in a study of three stages of examination stress in college students: just before the exam, just after the exam but before grades were announced, and just following the announcement of grades (Folkman & Lazarus, 1978). Ignoring individual differences, which were substantial, the normative emotional state and the coping process changed dramatically from stage to stage.

For example, before the exam, students coped by making heavy use of information-seeking, whereas distancing was virtually absent. However, after the exam, distancing became predominant as students could do nothing but wait for the results. Furthermore, when grades were announced, seeking emotional support increased greatly, and distancing virtually disappeared as a coping strategy. Aggregating all this change as if there were no discrete stages would have led us to miss what is actually happening and the reasons for it, and would have produced misleading or uninterpretable results.

Similarly, we studied stress and coping in the field with a sample of 35-45 year old men and women and obtained data through careful interviews on the most stressful experiences each month for five months. We found that patterns of coping varied in the same person from one stressful experience to another (Folkman, Lazarus, Dunkel-Schetter, DeLongis, & Gruen, 1986; Folkman, Lazarus, Gruen, & DeLongis, 1986). In short, each stressful experience was, to a high degree, an independent event that had to be understood in its own right. Stressful experiences cannot be lumped together for the same person, or for any given population, without endangering clarity of understanding about the stress process.

We need to have a road map of the type of encounters that are stressful at the job for different individuals and collectivities. To this end, we created the *Hassles Scale* (Lazarus, 1984; Lazarus & Folkman, 1989). In the U.S., this scale has been analyzed into eight factors, including concerns about *future security* (e.g., retirement), *time pressures* (e.g., too many things to do), *work* (e.g., problems getting along with fellow workers), *household* (e.g., planning meals), *health* (e.g., difficulties seeing or hearing), *inner problems* (e.g., being lonely), *financial responsibilities* (e.g., owing money), and *neighborhood and environment* (e.g., troublesome neighbors).

Patterns of daily stress vary with age, socio-economic variables, type of job, and personality. Measuring tools like the Hassles Scale can provide a diagnostic picture of the patterns of stress over time for individuals, social groups, different organizations and industries, and different cultures. They provide a tool in the study of individual and group patterns of vulnerability to stress. At present, the scales we have created are, for my taste, too sociological and not psychological enough. That is, they do not address the psychodynamics underlying stressful encounters, for example, being evaluated when one has low self-esteem, being accepted socially, being shamed or guilty, being insulted and therefore angry, being threatened by uncertainty, and so on.

Stress Management

On the basis of knowledge gained from traditional research methods, there are three main strategies for reducing stress in the workplace. One is to alter the conditions of work so that they are either less stressful or less counterproductive for effective coping (Hay & Oken, 1972). When conditions of work are unnecessarily and severely stressful, and destructive of coping, especially for large numbers of workers, this is a useful strategy.

However, to the extent that the sources of stress vary from worker to worker, or group to group, relieving the problem for one worker or group of workers may exacerbate it for others. For example, providing a less competitive or evaluative atmosphere may be just what some workers need, but others will experience even more distress. Or, if we provide more control over the work process, the change might be favorable for some but unfavorable for others for whom having control is itself stressful. This principle has been pointed up clearly by Locke and Taylor (1990):

> For a relatively large segment of the population, work seems to be of value primarily because of its association with material outcomes, especially money. This meaning has been termed the economic function by Morse and Weiss (1955) and the instrumental function by Locke, Sirota, and Wolfson (1976). In an explanation of clerical workers' indifferent reactions to job enrichment, Locke et al. offered the following observations about the relative importance of material and achievement values:

> "The workers' greatest concern was to get good ratings so that they could get promoted and get more pay. Many had given up more interesting jobs in order to take their present ones. They were quite willing, if anxious, to have more interesting tasks but only on the condition that some practical benefit would result" (p. 710).

The strategy of changing the work environment works mainly, as I noted above, when there is a widespread or common reaction to that environment. However, as work environments become less harrowing, individual and group differences in sources of stress, and in the resources for coping with it, become greater, making it more difficult to produce favorable environmental change.

The second strategy is to try to help those who are having difficulty adapting to conditions that are impossible or difficult to change to cope more effectively. Most stress management programs fail because they are superficial and treat people as if they were all alike. It is very costly to focus on individuals and to address the person-environment relationship and how it is appraised by individuals with markedly different goal commitments, belief systems, and coping resources and patterns. It is less costly to deal with groups of individuals sharing common stressful relationships with the environment.

The third strategy is transactional and requires that we identify the individual or group *relationships with the work setting* that are stressful, and try to change them for the individual or group on the basis of relational findings. The emphasis is on the person or group *and* the work environment taken as a *single analytic unit,* rather than as separate sets of variables to be manipulated independently. This strategy may entail shifts in worker assignments to create a better person-environment fit, keeping in mind that what might be a good fit in one work context or for one individual might not be in another. Or it could involve coping skills training for persons sharing common coping deficits in the jobs they are performing.

A transactional, process, appraisal- and coping-centered approach offers great potential by providing a more sophisticated matching of person and environment. It eschews treating everyone as though they were alike, and work environments as though they had common effects on everyone. It also recognizes that the adaptational encounter is the basic unit of stress analysis. Every person in the work context has many such encounters and each is often quite different in the way the stress process operates.

When I referred to groups or collectivities earlier, I meant a collection of workers who share common goal commitments and beliefs, or stress dynamics. The basis of grouping can be any set of person-environment relationships that transactional research finds to be important in the generation and amelioration of stress.

This is what happens when people join voluntary, self-help groups whose members share a common life crisis and who benefit from sharing the same distressing experiences. Examples include people who have had a child killed in an automobile accident, women with alcoholic husbands, people with spinal cord injuries, and so on. By getting together with others who share a common stress, some of the problems inherent in individual differences in stress dynamics are minimized.

In the organizational setting, suitable bases for grouping workers must be found for this kind of stress management to work. In my opinion, we must get away from solely normative studies that tell us about the work environment or person variables as separate causal antecedents of stress and distress, and apply the principles of transaction, process, appraisal and coping, both in the conduct of research and in stress management.

Cross-Cultural Considerations

What I have been saying has been, to some extent, predicated on western values and social norms. Looking across cultures, what is stressful in Japan, for instance, might not be stressful in Europe or the U.S., and vice versa. Similarly, because of differences in social values, Japanese may cope in different ways than Westerners, though this still remains to be demonstrated (Lazarus, Tomita, Opton, & Kodama, 1966). I have left consideration of this for last because I do not consider myself adequately qualified to examine the similarities and differences in sources of stress and ways of coping in different cultures. Analyses by Doi (1963, 1986) suggest that these similarities and differences might hinge on quite subtle meanings, which could be missed in superficial comparisons. This is a matter for programmatic research, and I am constantly dismayed by the absence of cross-cultural research on the processes of stress and coping and, indeed, on the entire emotion process.

Second, differences in detail about goal commitments, belief systems, and coping preferences do not obviate what I have been saying about transaction, process, appraisal and coping. These metatheoretical and theoretical principles apply cross-culturally just as they do in the U.S. or anywhere. I think they are relatively culture free. I also believe that systematic adoption of these principles would facilitate progress toward understanding stress in the workplace and toward discovering effective methods of stress management.

A problem with the focus on stress has long been the tendency to view stress as a unidimensional variable, and, even when several kinds of stress have been distinguished, such as harm, threat, and challenge (Lazarus & Folkman, 1984), the range of information available to those concerned with understanding and helping to manage an individual's stress as a result of adaptational struggles is still extremely limited.

I have recently developed the position that our emphasis should be on the *emotions generated* in adaptational struggles rather than on stress *per se* (Lazarus, 1991, 1993b, 1993c). There are at least 15 different emotions, and each results from a different kind of person-environment relationship, the personal meaning of which is constructed by the individual on the basis of the process of appraisal. Coping is also relevant to these emotions, whether or not they are positive or stress emotions. In effect, each emotion has what I refer to as a distinctive *core relational theme,* which defines the personal harm or benefit contained in the relationship.

The advantage of emphasizing the emotions produced by stressful encounters is that we learn much more about individuals and their adaptational struggles from knowing the emotion or emotions generated than we do from knowing that the individual is undergoing this or that amount of stress. Anger in the workplace tells us something different and important in a practical sense about how the individual construes what is happening than, say, anxiety, guilt, envy, or pride. So we would gain greatly from a focus on the emotions of work stress rather than stress alone.

REFERENCES

Baltes, P.B. (1987). Theoretical propositions of lifespan developmental psychology: On the dynamics between growth and decline. *Developmental Psychology, 23,* 611–626.

Buchholz, R.A. (1978). An empirical study of contemporary beliefs about work in American society. *Journal of Applied Psychology, 63,* 219–227.

Cantor, N., Norem, J.K., Brower, A.M., Niedenthal, P.M., & Langston, C.A. (1987). Life tasks, self-concept ideals, and cognitive strategies in a life transition. *Journal of Personality and Social Psychology, 53,* 1178–1191.

Caspi, A. (1987). Personality in the life course. *Journal of Personality and Social Psychology, 53,* 1201–1213.

Cooper, C.L., & Payne, R. (Eds.). (1980). *Current concerns in occupational stress.* Chichester, England: Wiley.

Dewe, P.J. (1993). Primary appraisal, secondary appraisal, and coping: Their role in stressful work encounters. *Journal of Occupational Psychology, 64,* 331–351.

Doi, T. (1963). Amae: A key concept for understanding Japanese personality structure. In R.J. Smith & R.K. Beardsley (Eds.), *Japanese culture: Its development and characteristics.* Chicago, IL: Aldine.

Doi, T. (1986). *The anatomy of self: The individual versus society.* Tokyo: Kodansha International.

Felner, R.D., Farber, S.S., & Primavera, J. (1983). Transitions and stressful life events: A model for primary prevention. In R.D. Felner, L.A. Jason, J.N. Moritaugu, & S.S. Farber (Eds.), *Preventive psychology: Theory, research, and practice* (pp. 199–215). New York: Pergamon.

Folkman, S., & Lazarus, R.S. (1985). If it changes it must be a process: Study of emotion and coping during three stages of a college examination. *Journal of Personality and Social Psychology, 48,* 150–170.

Folkman, S., & Lazarus, R.S. (1988a). Coping as a mediator of emotion. *Journal of Personality and Social Psychology, 54,* 466–475.

Folkman, S., & Lazarus, R.S. (1988b). The relationship between coping and emotion: Implications for theory and research. *Social Science in Medicine, 26,* 309–317.

Folkman, S., & Lazarus, R.S. (1988c). *Manual for the Ways of Coping Scales.* Palo Alto, CA: Consulting Psychologists Press.

Folkman, S., Lazarus, R.S., Dunkel-Schetter, C., DeLongis, A., & Gruen, R.J. (1986). The dynamics of a stressful encounter: Cognitive appraisal, coping, and encounter outcomes. *Journal of Personality and Social Psychology, 50,* 992–1003.

Folkman, S., Lazarus, R.S., Gruen, R.J., & DeLongis, A. (1986). Appraisal, coping, health status, and psychological symptoms. *Journal of Personality and Social Psychology, 50,* 571–579.

French, J.R.P., Jr., Caplan, R.D., & Van Harrison, R. (1982). *The mechanisms of job stress and strain.* Chichester, England: Wiley.

Hay, D., & Oken, S. (1972). The psychological stresses of intensive care unit nursing. *Psychosomatic Medicine, 34,* 109–118.

Holt, R.R. (1982). Occupational stress. In L. Goldberger & S. Breznitz (Eds.), *Handbook of stress: Theoretical and clinical aspects* (pp. 419–444). New York: Free Press.

Ivancevich, J.M., & Matteson, M.T. (1980). *Stress and work: A managerial perspective.* Glenview, IL: Scott, Foresman.

Labouvie-Vief, G., Hakim-Larson, J., & Hobart, C.J. (1987). Age, ego level, and the lifespan development of coping and defense processes. *Psychology and Aging, 2,* 286–293.

Lazarus, R.S. (1966). *Psychological stress and the coping process.* New York: McGraw-Hill.

Lazarus, R.S. (1968). Emotions and adaptation: Conceptual and empirical relations. In W.J. Arnold (Ed.), *Nebraska Symposium on Motivation* (pp. 175–266). Lincoln, NE: University of Nebraska Press.

Lazarus, R.S. (1984). Puzzles in the study of daily hassles. *Journal of Behavioral Medicine, 7,* 375–389.

Lazarus, R.S. (1991). *Emotion and adaptation.* New York: Oxford University Press.

Lazarus, R.S. (1993a). Coping theory and research: Past, present, and future. *Psychosomatic Medicine, 55,* 234–247.

Lazarus, R.S. (1993b). From psychological stress to the emotions: A history of changing outlooks. In *Annual Review of Psychology* (pp. 1–21). Palo Alto, CA: Annual Reviews.

Lazarus, R.S. (1993c). Book review essays by Shweder, R.A., Trabasso, T., & Stein, N., and by Panksepp, J., and the author's response. *Psychological Inquiry, 4,* 322–342.

Lazarus, R.S., Averill, J.R., & Opton, E.M., Jr. (1970). Towards a cognitive theory of emotion. In M. Arnold (Ed.), *Feelings and emotions* (pp. 207–232). New York: Academic Press.

Lazarus, R.S., Averill, J.R., & Opton, E.M., Jr. (1974). The psychology of coping: Issues of research and assessment. In G.V. Coelho, D.A. Hamburg, & J.F. Adena (Eds.), *Coping and adaptation* (pp. 249–315). New York: Basic Books.

Lazarus, R.S., & Folkman, S. (1984). *Stress, appraisal, and coping.* New York: Springer.

Lazarus, R.S., & Folkman, S. (1987). Transactional theory and research on emotions and coping. In L. Laux & G. Vossel (Eds.), Personality in biographical stress and coping research. *European Journal of Personality, 1,* 141–169.

Lazarus, R.S., & Folkman, S. (1989). *Manual for the Hassles and Uplifts scales.* Palo Alto, CA: Consulting Psychologists Press.

Lazarus, R.S., Tomita, M., Opton, E.M., Jr., & Kodama, M. (1966). A cross-cultural study of stress reaction patterns in Japan. *Journal of Personality and Social Psychology, 4,* 622–633.

Levi, L., Frankenhaeuser, M., & Gardell, B. (1982). Report on work stress related to social structures and processes. In G.R. Elliot & C. Eisdorfer (Eds.), *Stress and human health: Analysis and implications of research* (Chap. 6). New York: Springer.

Locke, E.A., Sirota, D., & Wolfson, A.D. (1976). An experimental case study of the successes and failures of job enrichment in a government agency. *Journal of Applied Psychology, 61,* 701–711.

Locke, E.A., & Taylor, M.S. (1990). Stress, coping, and the meaning of work. In W. Nord & A.P. Brief (Eds.), *The meaning of work* (pp. 135–170). New York: Heath.

Morse, N.C., & Weiss, R.S. (1955). The function and meaning of work and the job. *American Sociological Review, 20,* 191–198.

Near, J.P., Rice, R.W., & Hunt, R.G. (1980). The relationship between work and nonwork domains: A review of the empirical research. *Academy of Management Review, 5,* 415–429.

Whitbourne, S.K. (1985). The psychological construction of the lifespan. In J.E. Birren & K.W. Schaie (Eds.), *Handbook of the psychology of aging* (2nd ed., pp. 594–618). New York: Van Nostrand.

Psychological Stress and the Workplace: A Brief Comment on Lazarus' Outlook

Arthur P. Brief
Jennifer M. George

Lazarus' theorizing (e.g., Lazarus, 1966, 1991a; Lazarus & Folkman, 1984) has had a major impact on our understanding of psychological stress. His focus on the transaction between the person and the environment, primary and secondary appraisal, and coping has provided many valuable insights. Furthermore, his attention to the stress process and how it unfolds and changes over time is noteworthy. The contribution of his theoretical perspective to the study of psychological stress is evidenced by the number of empirical studies his theoretical work has driven (e.g., Folkman, Bernstein, & Lazarus, 1987; Folkman & Lazarus, 1980, 1985, 1986, 1988; Folkman, Lazarus, Dunkel-Schetter, DeLongis, & Gruen, 1986; Folkman, Lazarus, Pimley, & Novacek, 1987). Indeed, organizational stress researchers have much to gain by familiarizing themselves with Lazarus' theoretical and empirical contributions.

While we have great respect for Lazarus' work on psychological stress per se, we would like to present an alternative perspective to some of his views on the study and management of stress in the *workplace*. While we agree with Lazarus that it is important to study and understand *individual patterns* of, and reactions to, stress over time and across situations, we also propose that there are other important items on the agenda for the study of stress in the workplace. While a

basic understanding of the stress process is invaluable to organizational researchers, theory and research on job-related stress needs to actively take into account the
unique context that is focused on the workplace. The uniqueness of the workplace
as a social setting and its potential importance in people's lives is well known (cf.
Brief & Nord, 1990). Below, therefore, we do not dwell on arguments regarding
the importance of work per se. We focus on the simple idea that the challenge to
organizational researchers is the development of theory to guide one to identify
those conditions of employment likely to affect adversely the psychological well-
being of most persons exposed to them.

Lazarus suggests that it is *not* that useful to try to identify conditions of work
which adversely affect *most* workers because stress is ultimately an individual
phenomenon. For example, Lazarus (1994, p. 9) states that "to describe and
understand stress in the workplace requires that…individual patterns be studied to
generate knowledge about the kinds of persons who are more or less vulnerable to
divergent sources of stress." While we agree that stress essentially occurs at the
individual level, we believe that it is useful to try to discover working conditions
which are likely to adversely affect *most* workers exposed to them. That is, our
disagreement with Lazarus stems from his assertion that description and understanding "requires" a focus on "individual patterns." While he does go on to say
that "we need to have a road-map of the types of encounters that are stressful at the
job for different individuals and collectivities" (p. 10), his use of the term "different" brings us back to a focus on "individual patterns," a focus perfectly consistent
with his phenomenological perspective.

As is true of any collection of investigators, the work stress research community has limited resources with responsibilities to society as well as to its science.
By searching for those conditions of work likely to adversely affect most workers
exposed to them (rather than more fully attending to intraindividual processes)
and, thereby, seeking to construct a taxonomy of stressful job conditions, we better
fulfill our societal responsibilities. For example, we know that conditions threatening to the employment security of workers have profound, negative effects on most
people exposed to them (Kahn, 1981). By discovering other conditions having
such pervasive effects and ways to mitigate against them, we may better serve the
masses of workers. As we will show later, this position in no way dictates that
theory building be ignored.

Lazarus' apparent discomfort with such an approach may stem, in part, from
his observations of what he calls the traditional outlook towards work stress. We
too, find fault with traditional approaches (e.g., Kahn, Wolfe, Quinn, Snoek, &
Rosenthal, 1964; Rizzo, House, & Lirtzman, 1979; Van Sell, Brief, & Schuler,
1981; Jackson & Schuler, 1985) but for different reasons. While Lazarus appears
to believe that adequate attention has been paid to job conditions that are stressful
to most workers, we disagree. This is so for several reasons which we have
discussed in more detail elsewhere (e.g., Brief & Atieh, 1987; Brief, Burke,
George, Robinson, & Webster, 1988; Burke, Brief, & George, 1993). For example,

job satisfaction, in traditional approaches, often has been used as an indicator of strain (e.g., Jackson & Schuler, 1985). But, considerable evidence indicates that job satisfaction is not substantially related to more general measures of psychological well-being (e.g., life satisfaction) (e.g., Brief & Nord, 1990; Quinn, Staines, & McCullough, 1974; Rice, Near, & Hunt, 1980). Thus, while it may be observed that a given job condition adversely affects job satisfaction, such an observation does not imply that the job condition also adversely affects well-being in life. The reliance of traditionalists on such job-related indicators of strain as job satisfaction suggests that they may believe that affective disorders stemming from the conditions of employment stop at the workplace door. Our position is that, if a condition of employment adversely affects mental health, then one should be able to demonstrate an association between the presence of the condition and some general (not work-specific) indicator of psychological well-being. Thus, the traditionalists' focus on job-related affective reactions may tell us relatively little about job-related distress. Hence, while we may have some idea of the causes of job-related strain, considerably more attention needs to be focused on work conditions that adversely affect well-being in life.

While Lazarus suggests that we focus on intraindividual processes, we contend that certain job conditions adversely affect the well-being of most workers. In some sense, it is the duty of organizational researchers to try to identify these conditions and develop ways to alleviate them. Obviously, whether or not such conditions exist is an empirical question. To borrow from Lazarus' theoretical orientation, one could pose a critical question confronting work-stress researchers as follows: What sorts of job conditions do most people, upon exposure, cognitively appraise as threatening or harmful and have difficulty coping with to the extent that their well-being in life is impaired?

We previously have offered a tentative answer to this question (Brief & Atieh, 1987). Our response focused on the obvious economic meaning work has in people's lives (e.g., Brett, Cron, & Slocum, in press; Brief & Aldag, 1989; Brief & Nord, 1990; Doran, Stone, Brief, & George, 1991; George & Brief, 1990). This meaning influences the ways individuals emotionally respond to their jobs. The economic instrumentality of work represents a relatively unique aspect of the work context. It primarily is in the work context, and typically in no other, that people earn a living to support themselves and their loved ones. Moreover, these economic outcomes, more than other aspects of work, impact other domains of life central to psychological well-being (e.g., the family domain; Andrews & Withey, 1974). Given this centrality, therefore, it would seem that understanding how workers economically interpret their jobs would be critical to isolating those conditions of employment associated with experiencing distress.

Of course, other conditions of work likely have adverse affects on well-being. For example, working with hazardous materials and other physically dangerous work may adversely affect substantial proportions of the workforce exposed to these sorts of conditions. To the extent this is the case, in addition to trying to limit the potential for physical harm in such positions, we also have an obligation to try to find ways to lessen the psychological trauma such workers experience. For instance, research is needed which focuses on the effects of caring for AIDS patients on health care workers. While it is obvious that we need to do all we can to minimize the risks of infection, work-stress researchers need to attend to the concomitant psychological distress which may result from such work and try to uncover ways to alleviate it (George, Reed, Ballard, Colin, & Fielding, 1993).

More generally, work-stress research should seek to answer questions such as: What is it *about work* that may stimulate stress and coping processes? and, How might these processes unfold in the workplace? To guide us in seeking answers to these context-effect type questions, we of course, need theory. Indeed, Lazarus (1994, p. 5) urges us to use his theory which he suggests supplies the "keys to understanding not only work stress but stress in any context of living." Again, perhaps; but any theory we employ should be molded to capture the unique features of the work context. For instance, Lazarus identifies primary appraisal as one of two basic kinds of cognitive appraisal key to understanding stress. Primary appraisal addresses whether or not an individual sees a personal stake in an encounter. We, in the work stress area, desperately need conceptual guidance in isolating those *particular* job conditions which, when encountered, employees appraise as affecting their well-being. Theoretical orientations like Lazarus' most certainly help us frame the questions we need to ask but, by themselves, provide few clues to the mystery of the workplace.

In sum, while we have nothing but admiration for Lazarus' theorizing, we do take issue with some of his views of the workplace, or at least, how stress and coping processes should be studied in it. Those of us in the work-stress area need to recognize more fully how little may be known about the particular context in which our expertise purportedly lies. In seeking remedies to this deficiency, we should borrow from others the conceptual and methodological tools we may need. But, we must not lose sight of our contextual focus and our responsibility to the masses that toil in that context. The study of work stress differs from the study of psychological stress at home, at school, or in any particular social context because of these contextual boundaries.

REFERENCES

Andrews, F.M., & Withey, S.B. (1974). Developing measures of perceived life quality: Results from several national surveys. *Social Indicators Research, 1,* 1–26.

Brett, J.F., Cron, W.L., & Slocum, J.W., Jr. (in press). Economic dependency on work: A moderator of the relationship between organizational commitment and performance. *Academy of Management Journal.*

Brief, A.P., & Aldag, R.J. (1989). The economic functions of work. In G.R. Ferris & K.M. Rowland (Eds.), *Research in Personnel and Human Resources Management* (Vol. 7, pp. 1–24). Greenwich, CT: JAI Press.

Brief, A.P., & Atieh, J.M. (1987). Studying job stress: Are we making mountains out of molehills? *Journal of Occupational Behavior, 8,* 115–126.

Brief, A.P., Burke, M.J., George, J.M., Robinson, B.S., & Webster, J. (1988). Should negative affectivity remain an unmeasured variable in the study of job stress? *Journal of Applied Psychology, 73,* 199–207.

Brief, A.P., & Nord, W.R. (1990). *Meanings of occupational work: A collection of essays.* Lexington, MA: Lexington Books.

Burke, M.J., Brief, A.P., & George, J.M. (1993). The role of negative affectivity in understanding relationships between self-reports of stressors and strains: A comment on the applied psychology literature. *Journal of Applied Psychology, 78*(3), 402–412.

Doran, L.I., Stone, V.K., Brief, A.P., & George, J.M. (1991). Behavioral intentions as predictors of job attitudes: The role of economic choice. *Journal of Applied Psychology, 76,* 40–45.

Folkman, S., Bernstein, L., & Lazarus, R.S. (1987). Drug misuse and stress and coping processes in older adults. *Psychology and Aging, 2,* 366–374.

Folkman, S., & Lazarus, R.S. (1980). An analysis of coping in a middle aged community sample. *Journal of Health and Social Behavior, 21,* 219–239.

Folkman, S., & Lazarus, R.S. (1985). If it changes it must be a process: Study of emotion and coping during three stages of a college examination. *Journal of Personality and Social Psychology, 48,* 150–170.

Folkman, S., & Lazarus, R.S. (1986). Stress processes and depressive symptomology. *Journal of Abnormal Psychology, 95,* 107–113.

Folkman, S., & Lazarus, R.S. (1988). Coping as a mediator of emotion. *Journal of Personality and Social Psychology, 54,* 466–475.

Folkman, S., Lazarus, R.S., Dunkel-Schetter, C., DeLongis, A., & Gruen, R.J. (1986). The dynamics of a stressful encounter: Cognitive appraisal, coping, and encounter outcomes. *Journal of Personality and Social Psychology, 50,* 992–1003.

Folkman, S., Lazarus, R.S., Pimley, S., & Novacek, J. (1987). Age differences in stress and coping processes. *Psychology and Aging, 2,* 171–184.

George, J.M., & Brief, A.P. (1990). The economic instrumentality of work: An examination of the moderating effects of financial requirements and sex on the pay-life satisfaction relationship. *Journal of Vocational Behavior, 37,* 357–368.

George, J.M., Reed, T.F., Ballard, K.A., Colin, J., & Fielding, J. (1993). Contact with AIDS patients as a source of work-related distress: Moderating effects of support and estrangement. *Academy of Management Journal, 36,* 157–171.

Jackson, S.E., & Schuler, R.S. (1985). A meta-analysis and conceptual critique of research on role ambiguity and role conflict in work settings. *Organizational Behavior and Human Decision Processes, 36,* 16–78.

Kahn, R.L. (1981). *Work and health.* New York: Wiley.

Kahn, R.L., Wolfe, D.M., Quinn, R.P., Snoek, J.D., & Rosenthal, R.A. (1964). *Organizational stress: Studies in role conflict and ambiguity.* New York: Wiley.

Lazarus, R.S. (1966). *Psychological stress and the coping process.* New York: McGraw-Hill.

Lazarus, R.S. (1991a). *Emotion and adaptation.* New York: Oxford University Press.

Lazarus, R.S. (1994). Psychological stress in the workplace. In P.L. Perrewé & R. Crandall (Eds.), *Occupational stress: A handbook* (pp. 3–14). Washington, DC: Taylor & Francis. (Original work published 1991)

Lazarus, R.S., & Folkman, S. (1984). *Stress, appraisal, and coping.* New York: Springer.

Quinn, R.P., Staines, G.L., & McCullough, M.R. (1974). *Job satisfaction: Is there a trend?* Manpower Research Monograph No. 30. U.S. Department of Labor.

Rice, R.W., Near, J.P., & Hunt, R.G. (1980). The job satisfaction/life satisfaction relationship: A review of empirical research. *Basic and Applied Social Psychology, 1,* 37–64.

Rizzo, J.R., House, R.J., & Lirtzman, S.I. (1970). Role conflict and ambiguity in complex organizations. *Administrative Science Quarterly, 15,* 150–163.

Van Sell, M., Brief, A.P., & Schuler, R.S. (1981). Role conflict and role ambiguity: Integration of the literature and directions for future research. *Human Relations, 34,* 43–71.

An Examination of the Transaction Approach in Occupational Stress Research

James R. Harris

The literature on job stress continues to grow and, like other areas of psychology and occupational science, job stress research promotes a rich debate in the areas of theoretical systems, constructs, and relationships. One such debate concerns the understanding of exchange between the person and the environment. Currently, one approach, called the "traditional" approach, seeks to identify and measure discreet stress concepts on a broad sociological level (Barone, 1994). The traditional approach attends to the validity of the stress constructs, the predictability of dysfunction, and the generalizability of the empirical system findings. When examined collectively, the research generated from the traditional approach has led to substantial understanding of stress in the workplace.

A second approach, the transaction approach (Lazarus, 1994) examines stress as a unique process between the person and the environment. Transaction psychology suggests that stress arises from a single maladaptive encounter and occupational stress occurs when a particular kind of person is in a particular kind of occupational situation (Lazarus, 1994). When job stress is examined using transaction psychology, important questions surface concerning the basic philosophy, research agenda, and epistemology of the job stress research paradigm. A number of observations are central to his argument concerning the inadequacies of this field and the utility of the transaction process model for occupational stress research. Three important issues will be examined in this commentary: 1) the mediating characteristics of coping on the emotional process, 2) the generality of the transaction process to organizational settings, and 3) the adoption of an individual level unit of analysis in occupational stress research.

COPING

The transaction approach to occupational stress concentrates on the environment in which a stressful encounter takes place. This environmental emphasis surfaces clearly in the ideas of adaptation (Lazarus, 1968) and person/environment fit (Lazarus & Folkman, 1984). Transaction psychology predicts stress occurring when the environment is evaluated as either harmful, threatening, or challenging. This evaluation is conceptualized as a process of environmental appraisal and individual coping (Lazarus & Folkman, 1987). Each appraisal is based on the integration of the individual's personal agenda and the subjective realities of the situation (Lazarus, 1994). Coping strategies are selected via the amount of individual control one has over a particular situation.

Lazarus conceives coping and appraisal as transforming the relationship between the person/environment fit and the emotional response, thus making coping and appraisal mediating variables in the transaction process (Folkman & Lazarus, 1988). A mediating effect occurs when a variable changes or alters the relationship between independent and dependent variables within a given causal system (Folkman & Lazarus, 1988). In essence, mediators speak to how and why a causal system operates (Baron & Kenny, 1986). Because coping and appraisal are suggested to be mediators, each environmental/person fit is unique in its stress producing qualities due to the bidirectional relationship of appraisal and coping (Lazarus & Folkman, 1987).

The process approach fails, however, to fully acknowledge the potential for coping style, or a consistent reliance on a particular type of coping mechanism. The coping style concept

> ...acknowledges that (1) people may have a tendency to cope in a certain way over time, and that (2) this coping style may result either because the person tends to appraise events in a certain way, e.g., is more susceptible to feeling time pressured, or because they have a tendency to behave in a certain way, e.g., to avoid rather than approach stressors, and (3) that the pattern may be conditioned/socialized by particular environments, or even be largely a product of the existence in a certain type of environment, e.g., a very high demand environment (Newton, 1989, p. 454).

If coping style is acknowledged, then coping becomes a moderator variable, rather than a mediator as specified in the transaction process model. The moderating influence would change the causal structure of the transaction model because coping style would become an antecedent condition. As an antecedent, coping style would moderate the perception of the situation and how it is appraised, rather than, as Lazarus suggests, alter the "relationship between the original appraisal and its attendant emotion" (Folkman & Lazarus, 1988).

While the distinction between coping as a moderator or a mediator may, at first, seem trivial, the potential for the moderating effects of coping style change both the conceptual relationship of the variables within the transaction model and the analytical strategies for testing the model. Lazarus and his colleagues have also acknowledged the lack of causal inference for the mediating characteristics of coping behavior and often acknowledge the recursive nature of their theoretical arguments:

> In addition to appraisal influencing coping, coping may influence the person's reappraisal of what is at stake and what the coping options are (Folkman, Lazarus, Dunkel-Schetter, DeLongis, & Gruen, 1986, p. 1001).
>
> ...[often] theoretical models emphasize a unidirectional causal pattern in which emotion affects coping by both motivating it and impeding it. However, the relationship between emotion and coping in stressful encounters is bidirectional, with each affecting the other (Folkman & Lazarus, 1988, p. 466).

Given that coping may influence appraisal, or appraisal may influence coping, the current empirical evidence concerning the transaction process model does not specifically address one of the more important questions for occupational stress researchers: Does coping have stronger effects on stress or does stress have stronger effects on coping?

THE WORK SITUATION AND THE TRANSACTION PROCESS

One of the fundamental principles of transactional psychology argues that occupational stress research would benefit from viewing stress within the greater context of life stress. This observation serves as the primary logical argument for the adoption of Lazarus' transaction process approach model in the occupational stress field. However, the adoption of the transaction approach in occupational settings may require introducing a new variable in the emotion process: the organization.

The influence of the organization on employees may be profound. For example, Bowles (1989) suggests that organizations often provide meaning for symbols, images, and events for their members. Davis-Blake and Pfeffer (1989) argue that the relative strength of organizational culture (Schein, 1985; Van Maanen, 1975) determines, to a large extent, individual employees' job attitudes and behavior. In addition, job burnout (Cordes & Dougherty, 1993), organization member status (Bacharach, Bamberger, & Mundell, 1993), and economic instrumentality (Brief & George, 1994) point to a potentially unique organizational influence on cognitive processes of employees. Given the potential influence of organizational life, the process of emotion may be different within the occupational context and the organization may need to be represented in the transaction process model.

Lazarus (1994) suggests that the organization and its effects are best left at the beginning of the model. According to theory, the organization is part of the environment in the "fit" of the person and the environment. The worker interacts with occupational conditions and the organization becomes an antecedent and an appraised condition (Lazarus & Folkman, 1987). This conceptual treatment of work suggests a moderating effect on the transaction process. However, the pervasive influence of the organization on the cognitions of employees may require a different theoretical strategy.

For example, the organization effects may mediate the appraisal process. Folkman and Lazarus (1985) suggest that appraisal includes two processes: primary and secondary. Primary appraisal is the judgment of the individual regarding the encounter as irrelevant, benign-positive or stressful by determining the degree of personal "stake" in the encounter. Secondary appraisal determines the individual's coping options for the encounter (Folkman & Lazarus, 1985). In essence, the subject is defining an encounter as stressful through the process of primary appraisal, and the intensity of a stressful encounter is determined by the secondary appraisal process (Folkman & Lazarus, 1988).

In terms of primary appraisal, the organization may have influence over the "stakes" individuals have in particular encounters and, therefore, the organization may affect this part of the transaction model. Because the control of employee outcomes is often viewed as a motivational device in organizations, the personal stakes approach to primary appraisal may be subject to organizational influence. This influence would mediate, not, as Lazarus suggests, moderate, the relationship between the fit and the primary appraisal process.

During the secondary appraisal process, coping options are derived from the integration of individual's personal agenda and the demands of the situation (Lazarus, 1994). Again, as with the primary appraisal process, the organization setting may determine, to a large extent, the menu of potential coping options and, therefore, the secondary appraisal process may also be reflecting profound situational influences. In other words, respondents providing evaluations of coping options may be providing perceptions of coping strategies that are rewarded by the organization, such as acceptance, rather than those coping strategies derived from the person-environment transaction.

Lazarus and his colleagues typically measure secondary appraisal with a 4-item instrument assessing perception of coping options such as change, acceptance, information seeking, and reservation (Folkman & Lazarus, 1980; Folkman, Lazarus, Gruen et al., 1986). Unfortunately, when used in the occupational context, these items and their inherent transparency may be confounded by cultural attributes of the organization such as a rewarded decision making style (Kanter, 1983), perceptions of group norms (Schweiger, Sandberg, & Ragan, 1986), or social desirability response bias. For example, respondents in organizational settings may be more likely to respond to coping options such as "I knew what had to be done, so I doubled my efforts to make things work" rather than "I prayed" or "tried to make myself feel better by eating, smoking, using drugs or mediation, etc." (Folkman, Lazarus, Dunkel-Schetter et al., 1986).

In addition to the influences on the appraisal processes, the work context may also affect the way employees cope as well as the menu of coping strategies available to employees. Much of the research of Lazarus and his colleagues has focused on coping as a process (Lazarus & Folkman, 1987). Research evidence supports a strong contextual influence on the coping process (Folkman & Lazarus, 1980; Lazarus & Folkman, 1987). In addition, coping strategies vary with time and the age of the subjects (Folkman & Lazarus, 1985). Given that the coping process varies with context, time, and age, coupled with the observations by Davis-Blake and Pfeffer (1989) that organizations are strong situations, it is not unlikely that the organization could affect the coping process as well.

Lazarus measures coping with The Ways of Coping scale (Folkman & Lazarus, 1985), which has demonstrated a consistent eight-factor solution (e.g., Folkman, Lazarus, Dunkel-Schetter et al., 1986). Little is known about the generalizability of this factor result, particularly given the fact that the underlying dimensionality has been determined using middle income married couples in Contra Costa County (Folkman & Lazarus, 1988; Folkman, Lazarus, Dunkel-Schetter et al., 1986), elderly individuals residing in the San Francisco East Bay (Folkman & Lazarus, 1988), and undergraduates of the University of California–Berkeley (Folkman & Lazarus, 1985). While the analyses concerning these diverse samples suggest internal consistency in the Ways of Coping scale, little is known about the factor solution for a sample taken from unique contexts such as a specific organization, a single occupation, or an individual industry.

The effects of work contexts on specific variables and, indeed, on the entire transaction process model need to be examined carefully. While research evidence suggests that some overlap exists between working life and non-working life (Near, Rice, & Hunt, 1980), it is likely that the circumstance individuals find themselves in while working is different from the circumstance of non-working life. The adaptation of the transaction model to the occupational setting requires additional consideration of the effects of the organization. The question that naturally arises is where in the model should the organization and its effects be represented?

THE UNIT OF ANALYSIS

One of the more important outcomes in the development of the transaction process model has been the observation of daily stress, or hassles, on somatic health (Delongis, Coyne, Dakof, Folkman, & Lazarus, 1982) and the wide variation in individual differences in the within-subjects effects for hassles on same day mood and health (Lazarus & Folkman, 1987). Given these observations, it was logically concluded that stressful encounters were often very different for the same individuals.

On a general level, Lazarus is suggesting that knowledge of neither the person nor the environment alone is sufficient to understand the stress process (Lazarus & Folkman, 1987). It is knowledge of the transaction between the person and the environment that is the central issue (Lazarus, 1994). As a consequence, occupational stress becomes far more conceptually dynamic in that the outcomes derived from each situational encounter may change with each person-environment transaction.

Of the demands of process methodology, this one is probably the most constructive and, unfortunately, the most problematic. It can be argued that occupational stress researchers know something of people and something of environments, but have little knowledge of the dynamic interaction of people and environments. The problems of focusing on the interaction of person and environment can be seen on an empirical, philosophical, or paradigm level.

On the empirical level, representing "interaction" can be accomplished in a number of ways. One of the more popular approaches is to combine different variables, most often multiplicatively, to construct a composite variable, or interaction term. This strategy is often used in regression and when testing if the interaction term has a significant effect on a dependent variable. A second procedure is to choose a statistical model, such as ANOVA, which can partition variance in a number of different ways. This algorithm approach most often tests if a main effect of one variable is dependent on the interaction of that variable with another. Finally, research design strategy, such as repeated measures, can be used to observe within-subjects effects and interactions (Rusbult & Farrell, 1983). This strategy investigates whether main effects for, say, persons, vary across situations or time. In each case, the interaction, represented as a combined variable, partition variance, or observed over time is static as opposed to dynamic. As a consequence, these strategies fail to capture the richness of Lazarus' conception of the person/environment fit. Therefore, the adoption of Lazarus' transaction approach suggests a reduced emphasis on nomothetic approaches in favor of ideographic research designs.

While an ideographic research program may provide rich insights into the cognitive processes of stress and emotion for single subjects (Kessler, 1987), inferences into the cognitive process of others become very difficult to make. The limited generalizability of ideographic findings present a chronic problem in the verification and testing of the transaction process approach.

On the philosophical level, a unit of analysis which is a transaction between a worker and the occupational environment reduces occupational stress to an individual level phenomena. As a consequence, guidance to organizations in the management of stress may become quite limited. Because the transactions of persons and environments would likely become the central issue to testing Lazarus' theoretical model, the stress response and its potential physical and psychological effects are naturally de-emphasized. Unfortunately, understanding how single individuals interact with various occupational situations does not necessarily provide insight into acute or chronic environmental stressors or into ways of correcting the stressful circumstances.

Lazarus' (1994) suggestion of grouping individuals who share common stress is not particularly unique (e.g., Murphy, 1984) and therapeutic approaches (i.e., adapting employees as opposed to changing occupational situations) to stress management have been previously argued against (Ganster, Mayes, Sime, & Tharpe, 1982). As a consequence, the transaction process may not be of immediate use to researchers seeking stress reduction interventions.

On the paradigm level, reducing the concept of stress to an individual level phenomena with a unit of analysis which is an individual transaction may create a number of potential problems in understanding occupational stress. If we adopt the viewpoint that social science seeks general understanding of social patterns, rather than an explanation of individual events, doubt surfaces concerning the utility of the transaction approach. Knowledge of the generalizability of the stress response and the usefulness of particular interventions may be more important to occupational stress researchers and the community they serve. While Lazarus makes an important observation concerning the dynamics of the interaction of people and environments, the question that naturally arises is, how is the interaction of the person and environment measured, analyzed, and tested?

CONCLUSION

The issues discussed above are not intended to dismiss the contribution of transaction psychology to the understanding of human well-being, nor is it suggested that occupational stress research will not benefit from this new and exciting perspective. Indeed, job stress research employing transaction psychology has demonstrated much promise (cf. Ashford, 1988; Dewe, 1989). What is suggested here is only that care be taken in applying the transaction process approach to job stress.

Specifically, the influence of the work situation may change the relationship of the coping concept from a mediator to a moderator and, thus, significantly alter the transaction model. The occupational context may also affect the measurement and treatment of many of the variables in the model. Finally, questions concerning the analytical strategies which are focused on person/environmental fit must be considered. However, even with these potential problems, it is without doubt that the occupational stress field will benefit from careful and thoughtful application of the transaction process model.

REFERENCES

Ashford, S.J. (1988). Individual strategies for coping with stress during organizational transitions. *Journal of Applied Behavioral Science, 24*, 19–36.

Bacharach, S., Bamberger, P., & Mundell, B. (1993). Status inconsistency in organizations: From social hierarchy to stress. *Journal of Organizational Behavior, 14*, 21–36.

Baron, R., & Kenny, D.A. (1986). The moderator-mediator distinction in social psychological research: Conceptual, strategic, and statistical considerations. *Journal of Personality and Social Psychology, 51*, 1173–1182.

Barone, D.F. (1994). Work stress conceived and researched transactionally. In P.L. Perrewé & R. Crandall (Eds.), *Occupational stress: A Handbook* (pp. 29–37). Washington, DC: Taylor & Francis. (Original work published 1991)

Bowles, M. (1989). Mismeaning and work organizations. *Organizational Studies, 14*, 405–421.

Brief, A.P., & George, J.M. (1994). Psychological stress and the workplace: A brief comment on Lazarus' outlook. In P.L. Perrewé & R. Crandall (Eds.), *Occupational stress: A Handbook* (pp. 15–19). Washington, DC: Taylor & Francis. (Original work published 1991)

Cordes, C.L., & Dougherty, T.W. (1993). A review and an integration of research on job burnout. *Academy of Management Review, 18*(4), 621–656.

Davis-Blake, A., & Pfeffer, J. (1989). Just a mirage: The search for dispositional effects in organizational research. *Academy of Management Review, 14,* 385–400.

Delongis, A., Coyne, J.C., Dakof, G., Folkman, S., & Lazarus, R. (1982). Relationship of daily hassles, uplifts, and major life events to health status. *Health Psychology, 1,* 119–136.

Dewe, P.J. (1993). Primary appraisal, secondary appraisal, and coping: Their role in stressful work encounters. *Journal of Occupational Psychology, 64,* 331–351.

Folkman, S., & Lazarus, R.S. (1980). An analysis of coping in a middle aged community sample. *Journal of Health and Social Behavior, 21,* 219–239.

Folkman, S., & Lazarus, R.S. (1985). If it changes it must be a process: Study of emotion and coping during three stages of a college examination. *Journal of Personality and Social Psychology, 48,* 150–170.

Folkman, S., & Lazarus, R.S. (1988). Coping as a mediator of emotion. *Journal of Applied Psychology, 52,* 466–475.

Folkman, S., Lazarus, R.S., Dunkel-Schetter, C., DeLongis, A., & Gruen, R.J. (1986). The dynamics of a stressful encounter: Cognitive appraisal, coping, and encounter outcomes. *Journal of Personality and Social Psychology, 50,* 992–1003.

Folkman, S., Lazarus, R.S., Gruen, R.J., & DeLongis, A. (1986). Appraisal, coping, health status, and psychological symptoms. *Journal of Personality and Social Psychology, 50,* 571–579.

Ganster, D.C., Mayes, B., Sime, W.W., & Tharpe, G. (1982). Managing occupational stress: A field experiment. *Journal of Applied Psychology, 67,* 533–542.

Kanter, R.M. (1983). *The change masters: Innovations for productivity in the American corporation.* New York: Simon & Schuster.

Kessler, R.C. (1987). The interplay of research design strategies and data analysis procedures in evaluating the effects of stress on health. In S.V. Kasl & C.L. Cooper (Eds.), *Stress and health: Issues in research methodology.* New York: Wiley.

Lazarus, R.S. (1994). Psychological stress in the workplace. In P.L. Perrewé & R. Crandall (Eds.), *Occupational stress: A handbook* (pp. 3–14). Washington, DC: Taylor & Francis. (Original work published 1991)

Lazarus, R.S., Averill, J.R., & Opton, E.M., Jr. (1974) Towards a cognitive theory of emotion: Issues of research and assessment. In G.V. Coelho, D.A. Hamburg, & J.F. Adena (Eds.), *Adaptation.* New York: Basic Books.

Lazarus, R.S., & Folkman, S. (1984). *Stress, appraisal, and coping.* New York: Springer.

Lazarus, R.S., & Folkman, S. (1987). Transactional theory and research on emotions and coping. In L. Laux & G. Vossel (Eds.), *Personality in biographical stress and coping research. European Journal of Personality, 1,* 141–169.

Murphy, L.R. (1984). Occupational stress management: A review and appraisal. *Journal of Occupational Psychology, 57,* 1–15.

Near, J.P., Rice, R.W., & Hunt, R.G. (1980). The relationship between work and nonwork domains: A review of the empirical research. *Academy of Management Review, 5,* 415–429.

Newton, T.J. (1989). Occupational stress and coping with stress: A critique. *Human Relations, 42,* 441–461.

Rusbult, C., & Farrell, D. (1983). A longitudinal test of the investment model: The impact on job satisfaction, job commitment, and turnover of variations in rewards, costs, alternatives, and investments. *Journal of Applied Psychology, 68,* 429–438.

Schein, E.H. (1985). *Organizational culture and leadership.* San Francisco: Jossey-Bass.

Schweiger, D., Sandberg, W., & Ragan, J. (1986). Group approaches for improving strategic decision making: A comparative analysis of dialectical inquiry, devil's advocacy, and consensus. *Academy of Management Journal, 29,* 51–71.

Shea, G., & Guzzo, R. (1987, Summer). Group effectiveness: What really matters? *Sloan Management Review,* p. 25.

Staw, B.M. (1986, Summer). Organizational psychology and the pursuit of the happy and productive worker. *California Management Review,* p. 49.

Van Maanen, J. (1975). Police socialization: A longitudinal examination of job attitudes in an urban police department. *Administrative Science Quarterly, 20,* 207–228.

Work Stress Conceived and Researched Transactionally

David F. Barone

A transactional view of stress is associated primarily with Richard Lazarus (1966, 1991, 1994; Lazarus & Folkman, 1984), but Dewey and Bentley (1989) are credited with the origin of the concept (Coyne & Lazarus, 1980). Although not widely known today in psychology, John Dewey had spent his whole career advocating a transactional, contextual, process-oriented view for adequately constructing the new behavioral sciences (Barone, Maddux, & Snyder, in press; Cahan, 1992; Cahan & White, 1992; Sarason, 1981). In the early days of psychology, Dewey (1969) objected to construing traits as pre-formed and static rather than as products of social transactions; further, he warned against an elemental, mechanistic psychology based on physiological reflexes rather than a holistic one based on ongoing emergent transactions (Dewey, 1972). In his presidential address to the American Psychological Association, Dewey (1976) argued for context-sensitive psychological findings from ongoing social transactions (like his Laboratory School) rather than from decontextualized laboratories; he insisted that such work would guide more effective practical innovations. These views may now be palatable with the renewed interest in contextualism today (Altman & Rogoff, 1987; McGuire, 1992; Rosnow & Georgoudi, 1986), but they were largely ignored in psychology's early rush to the laboratory, mental measurement, and the S-R formulation.

There has been a stream of development within psychology consistent with Dewey's vision. Some of the work in this tradition is particularly relevant to a transactional view of work stress. The Hawthorne studies (Mayo, 1933) moved industrial psychology beyond mental measurement to human relations. Consistent with Dewey's experience in his Laboratory School, they demonstrated that research on transactions in industrial settings could guide innovation. Kurt Lewin was another voice reaching conclusions similar to Dewey's: that social psychology take as its unit of study contextualized transactions (Lewin, 1943) and that experimentation take place in real-life settings within preexisting social groups (Lewin, 1944). The approach to work stress of the Michigan group, which Lazarus noted favorably, built explicitly on Lewin's theory (French & Kahn, 1962).

Lazarus and associates continue this tradition for the topic of stress (e.g., Coyne & Lazarus, 1980; Lazarus, 1990a, 1991, 1994). They argue the inadequacy of discrete inputs and outputs as constructs, the advantages of a transactional perspective, the concern about ecological validity, and the need for research in natural settings. Lazarus came to this view gradually. His early laboratory studies of reactions to industrial accidents (Lazarus, Averill, & Opton, 1970) have been supplanted in his recent work by repeated assessment of everyday stress via questionnaire and interview (Lazarus, 1994).

CURRENT EXEMPLARS OF WORK STRESS STUDIED TRANSACTIONALLY

The work setting offers special advantages to researching stress from the transactional perspective. Work stress is contextually defined. Situational factors appraised as stressful can be consensually defined, observed, and modified. Subjects can be observed coping and continue to be available for repeated assessment. Work stress can be related to different conditions between groups (departments, shifts, sites, etc.) or changed conditions within the whole organization, particular groups, or individuals (natural disaster, corporate takeover, reorganization, reassignment, etc.). Applied researchers have their own work stress—coping with political and social processes. They do not have the luxury of some psychologists' "misdirected" focus on the individual (Sarason, 1981).

It is just such contextual factors that a transactional view seeks to study. They are more easily observed in a work setting than in more private (family) or public (community) settings. As a consequence of their involvement in organizational process and politics, work stress researchers have direct influence (whether they acknowledge it or not) on the use and impact of their findings. The research below is limited to exemplars of how work stress is being studied transactionally in America and related Western cultures. Stress management and cross-cultural considerations are left to others.

Inventorying Transactions

Only a limited number of work stress inventories constructed along traditional psychometric lines are available (Jones & DuBois, 1989). They all evaluate either organizational stressors or personal strain. The Work Stress Inventory (Barone, Caddy, Katell, Roselione, & Hamilton, 1988; Barone, Katell, & Caddy, 1991) and the Job Stress Survey (Spielberger & Reheiser, 1994; Turnage & Spielberger, 1991) offer alternatives based on a transactional view of work stress. They have individuals appraise the frequency and intensity of stressfulness of work-based experiences. Many items reflect the psychological rather than sociological perspective that Lazarus calls for: not knowing what superiors expect of you, being injured as the result of the mistakes of others, having to make decisions that will dramatically affect other peoples' lives, having to do things on the job that are against your better judgment, and feeling that your work ability is underrated. Both inventories identified two large comparable factors across many occupations. One involves appraised stress from organizational policies and supervision; the other involves pressures and risks experienced on the job. Combining such an inventory with collateral independent assessment of stressful conditions and outcomes offers a more transactional view than relying solely on inventories of workers' stresses and strains, which are now known to be confounded in workers' thinking (Jex, Beehr, & Roberts, 1992).

An unresolved issue in inventory-reporting of work stress has to do with defensive responding. Research separating impression management and self-deception as sources of response bias (Linden, Paulhus, & Dobson, 1986) needs to be applied to the assessment of work stress. A transactional viewpoint would predict that, as a result of impression management, less work stress would be reported. Consistent with this prediction, it was found that human service workers signing their names reported significantly less stress on the WSI than those responding anonymously (Barone et al., 1988; Roselione & Barone, 1988). Research needs to establish public-private differentials for stress inventories, and research reports need to identify clearly under which condition responses were given.

The other component of this issue, self-deception, is a more difficult problem. To measure appraised stress in those who use denial or repression for coping is to confront a long-standing problem in personality assessment. For example, police officers reported relatively low levels of stress on the WSI, even under anonymous conditions, despite evidence of a relatively high occurrence in police of stress-related symptoms (Barone et al., 1988; Simons & Barone, 1994). It has been argued that such a repression/sensitization bias is not a response style but a significant personality trait (negative affectivity) that needs to be assessed, and that objective measures of stressors and stress outcomes free of its influence are also needed (Agho, Price, & Mueller, 1992; Ben-Porath & Tellegen, 1990; Burke, Brief, & George, 1993; Costa & McCrae, 1990; Paulhus, 1984; Watson, 1990).

Research on today's social-cognitive or cognitive-relational theories is leading back to affective and dispositional factors. Such factors cannot be ignored. However, in a transactional view, they are conceived to be coping resources (or liabilities) to emphasize their functional role. Thus, adequate research on the transaction and process of work stress needs to assess coping resources like personal control, defensiveness, and social support.

Simons and Barone (in press) tried to resolve a debate about whether social support moderates the relationship of work stressors to experienced strain in a high-risk group, police officers. Transactional appraisals were assessed with the Work Stress Inventory, the Social Support Questionnaire (Sarason, Sarason, Shearin, & Pierce, 1987), and the Maslach Burnout Inventory (Maslach & Jackson, 1981). It was found that emotional support as well as stressors accounted directly for significant amounts of variance in emotional exhaustion, depersonalization, and symptom frequency. In contrast to a previous study, a reverse buffering effect for social support (i.e., its increasing strain) made no significant contribution to predicting strain. Conflicted relationships, which are a source of stress as well as support, also contributed to strain. Half of supportive relationships were also conflicted, equally divided between ones on and off the job. These findings indicate that, to gauge the impact of stress, it is necessary to assess the workers' appraisals of work stressors, of social support, and of relationship stressors implicated in coping with work stress. Thus, this and other recent studies (Burke, 1993; Corrigan et al., 1994; Lee & Ashforth, 1993a, 1993b) demonstrate the value of such a transactional approach to understanding work stress and burnout. However, social support is not always associated with reduced job strain, as in a recent study of executives (Dolan, 1994).

Assessing Specific Transactions

Dewe (1989) provides an alternative to using a general-purpose inventory of work stress. He used a two-stage approach. The first was open-ended interviews to identify items reflecting local conditions and concerns. The second was a questionnaire on the appraisal of stressors and coping strategies. Such an assessment strategy is consistent with a transactional view in a number of ways. Rather than using a pre-formed, highly abstract, aggregating "ruler," he allowed his measure to emerge out of local transactions with a small sample. Then, with some understanding of stress-as-experienced in hand, he moved to issues of aggregation and accuracy by administering the questions to a much larger sample. Although he asked only for an aggregated appraisal of intensity, frequency, and impact of work stressors, he asked for frequency of usage of a large set of coping strategies. Unfortunately, he does not report whether responding on the questionnaire was public (named) or private (anonymous). Dewe has provided a partial model of how to combine the ipsative and normative styles, as discussed by Lazarus (1990b), in research on work stress.

Assessing work stress via more individualizing transactions would benefit from the development of structured interviews like the Life Event and Difficulties Schedule (Brown, 1990; Brown & Harris, 1989) or that used in the Berkeley Stress and Coping Project (Folkman, Lazarus, Gruen, & DeLongis, 1986). As Murphy and Hurrell (1987) pointed out, interviews allow flexibility missing in self-administered questionnaires and can better uncover sources of work stress. Another alternative method is the use of stress diaries in the work setting (Weber & Laux, 1990). As Dewey (1976) argued long ago, and Lazarus (1994) reasserts, applied research does not seek decontextualized abstract truths. Its method should aid its action objectives, which pertain to specific individuals in specific settings. Work settings pay for intensive evaluation of specific machines and operations to reduce errors and assure quality. Rather than relying on cost-saving deindividuating and nontransactional assessment, the same level of evaluative specificity and thoroughness needs to be devoted to identifying work stress in our human resources.

Evaluating the Transitional Process

An even more adequate exemplar of transactional research on stress is provided by Ashford (1988), who studied stress and coping during the Bell System's divestiture. She satisfies most of Lazarus' (1990a, 1994) recommendations in her repeated measurement across the acute stress of a organizational transition and her multifaceted measurement of appraised stressors, coping resources, coping responses, and stressful symptoms. She developed an inventory of stressors specific to transition and had subjects respond anonymously. An oversight from a transactional point of view is the omission of social support as a coping resource. The complex design paid off. It revealed that specific coping resources and coping responses buffered transitional stressors, that others had a reverse buffering effect, and that these effects were in some cases limited to stressful symptoms prior to or after the transition. The findings of this study are consistent with a large body of research on stress and demonstrate how applied research can be based on ongoing historical events (cf. Hobfoll, 1988). Another recent example studied the strain introduced when new technologies were implemented (Korunka, Weiss, & Karetta, 1993).

The final exemplar is the most transactionally complex. It involved studying work stress in collaboration with a labor union during hotel contract negotiations (Neale, Singer, & Schwartz, 1987; Singer, Neale, & Schwartz, 1989). A systems perspective is used to organize the various levels of transactions affecting the work stress process. Extended interviews were conducted to gauge local stressors, including ones specific to different departments in the hotel. Despite the large number of potential stressors assessed, only management policies and practices were related to negative emotions. The authors relate their findings to the specific transitional context of contract negotiations.

To psychologists accustomed to focusing only on technical issues of measurement accuracy, this study is unwieldy. To John Dewey, Seymour Sarason (1981), and advocates of a fully transactional psychology, this study is wonderful example of action research. It shows how research is embedded in the ongoing social process and how, in turn, the research becomes a contributor to that process. In this case, the research contributed to a contractual agreement for monthly meetings between shop stewards and upper management to discuss work practices and stress. The authors' use of systems theory demonstrates another line of fruitful development: To research work stress in its full complexity is to recognize and deal with the embedded social systems in which work stress is located.

CONCLUSION

Current research on work stress demonstrates that the vision of Dewey and Lewin is being actualized and is producing fruitful results. Such research begins by focusing on the specifics of time and place: what is happening in this organization now? If the particular stressful work experiences to be studied are not already inventoried or discussed in the literature, then interviews or focus groups are necessary to generate them. Inventories of coping resources and strategies also are needed (Dewe, 1993). If available ones are inadequate, they, too, need to be generated. An inventory of stress symptoms also can be included. It is desirable (if researching under named conditions) to supplement it with stress outcomes such as absences and medical reports, in which case a latent factor of stress reactivity can be derived. Finally, stressful organizational conditions also need to be included in some way in the research design. For example, a rating of jobs for stressfulness could be a collateral measure to workers' appraisal of stressfulness. Alternately, comparisons can be made before and after transitions, such as reassignment or restructuring, or between units or positions differing in job demands. Such research does not evaluate essences or stable traits but a transactional process; thus, it needs to be ongoing. However, work stress research has clearly favored cross-sectional over longitudinal research designs (Murphy & Hurrell, 1987). Such multivariate repeated-measures research requires sophisticated statistical analysis, including structural equation modeling (Kelloway & Barling, 1991; Lee & Ashforth, 1993b).

Part of the transaction of doing research on work stress is the reactivity of employees—their social and political awareness. A machine being tested for stress isn't defensive; but it also has no information to volunteer and no potential to be mobilized to solve problems. The challenge to psychologists and human-resource specialists is to engage employees constructively and to gauge their response biases given the conditions of the research. How can employees' cooperation in such intrusive research be elicited and the view be combatted that it represents monitoring to be used against them? Assessment by name (preferable for repeated measurement) may be possible in the context of training or team development if it is viewed as integral to goals being sought. Anonymous assessment can always be sabotaged by false reporting. Employees need to believe that the results are consequential in ways that will benefit them. From a transactional perspective, work stress research does not begin when political preparation is complete. Instead the research is an elaborate transactional process beginning with initial meetings (who attends is significant) and ending with final meetings (who endorses the findings and recommendations is significant). Work stress research pursued in this manner develops the transactional psychology advocated by Dewey, Lewin, Sarason, and Lazarus. It recognizes the involvement of such research in larger social systems; and it is consistent with organized psychology's current commitment to tackling the problem of work stress (Keita & Jones, 1990).

REFERENCES

Agho, A.O., Price, J.L., & Mueller, C.W. (1992). Discriminant validity of measures of job satisfaction, positive affectivity and negative affectivity. *Journal of Occupational and Organizational Psychology, 65,* 185–196.

Altman, I., & Rogoff, B. (1987). World views in psychology: Trait, interactional, organismic and transactional perxpectives. In D. Stokols & I. Altman (Eds.), *Handbook of environmental psychology* (Vol. 1, pp. 1–40). New York: Wiley.

Ashford, S.J. (1988). Individual strategies for coping with stress during organizational transitions. *Journal of Applied Behavioral Science, 24,* 19–36.

Barone, D.F., Caddy, G.R., Katell, A.D., Roselione, F.B., & Hamilton, R.A. (1988). The Work Stress Inventory: Organizational stress and job risk. *Educational and Psychological Measurement, 48,* 141–154.

Barone, D.F., Katell, A.D., & Caddy, G.R. (1991). The Work Stress Inventory. In *Tests in Microfiche* (Set P, #016638). Princeton, NJ: Educational Testing Service.

Barone, D.F., Maddux, J.E., & Snyder, C.R. (in press). *Social-clinical psychology: Cognitive, personal, and interpersonal processes.* New York: Plenum.

Ben-Porath, Y.S., & Tellegen, A. (1990). A place for traits in stress research. *Psychological Inquiry, 1,* 14–17.

Brown, G.W. (1990). What about the real world: Hassles and Richard Lazarus. *Psychological Inquiry, 1,* 19–22.

Brown, G.W., & Harris, T.O. (1989). *Life events and illness.* New York: Guilford.

Burke, R.J. (1993). Toward an understanding of psychological burnout among police officers. *Journal of Social Behavior and Personality, 8*(3), 425–438.

Burke, M.J., Brief, A.P., & George, J.M. (1993). The role of negative affectivity in understanding relationships between self-reports of stressors and strains: A comment on the applied psychology literature. *Journal of Applied Psychology, 78*(3), 402–412.

Cahan, E.D. (1992). John Dewey and human development. *Developmental Psychology, 28,* 17–24.

Cahan, E.D., & White, S.H. (1992). Proposals for a second psychology. *American Psychologist, 47,* 224.

Corrigan, P.W., Holmes, E.P., Luchins, D., Buican, B., Basit, A., & Parks, J.J. (1994). Staff burnout in a psychiatric hospital: A cross-lagged panel design. *Journal of Organizational Behavior, 15,* 65–74.

Costa, P.T., Jr., & McCrae, R.R. (1990). Personality: Another "hidden factor" in stress research. *Psychological Inquiry, 1,* 22–24.

Coyne, J.C., & Lazarus, R.S. (1980). Cognitive style, stress perception, and coping. In I.L. Kutash & L.B. Schlesinger (Eds.), *Handbook on stress and anxiety* (pp. 144–158). San Francisco: Jossey-Bass.

Dewe, P.J. (1989). Examining the nature of work stress: Individual evaluations of stressful experiences and coping. *Human Relations, 42,* 993–1013.

36   THEORETICAL PERSPECTIVES IN OCCUPATIONAL STRESS RESEARCH

Dewe, P.J. (1993). Primary appraisal, secondary appraisal, and coping: Their role in stressful work encounters. *Journal of Occupational Psychology, 64,* 331–351.

Dewey, J. (1969). Galton's statistical methods: Review of natural inheritance. In J.A. Boydston (Ed.), *The early works of John Dewey, 1882–1898* (Vol. 3, pp. 43–47). Carbondale, IL: Southern Illinois University Press. (Original work published 1889)

Dewey, J. (1972). The reflex arc concept in psychology. In J.A. Boydston (Ed.), *The early works of John Dewey, 1882–1898* (Vol. 5, pp. 96–110). Carbondale, IL: Southern Illinois University Press. (Original work published 1896)

Dewey, J. (1976). Psychology and social practice. In J.A. Boydston (Ed.), *The middle works of John Dewey, 1899–1924* (Vol. 1, pp. 131–150). Carbondale, IL: Southern Illinois University Press. (Original work published 1900)

Dewey, J., & Bentley, A.F. (1989). Knowing and the known. In J.A. Boydston (Ed.), *The later works of John Dewey, 1925–1953* (Vol. 16). Carbondale, IL: Southern Illinois University Press. (Original work published 1949)

Dolan, S.L. (1994). Individual, organizational and social determinants of managerial burnout: Theoretical and empirical update. In P.L. Perrewé & R. Crandall (Eds.), *Occupational stress: A handbook* (pp. 223–238). Washington, DC: Taylor & Francis. (Original work published 1992)

Folkman, S., Lazarus, R.S., Gruen, R.J., & DeLongis, A. (1986). Appraisal, coping, health status, and psychological symptoms. *Journal of Personality and Social Psychology, 50,* 571–579.

French, J.R.P., Jr., & Kahn, R.L. (1962). A programmatic approach to studying the industrial environment and mental health. *Journal of Social Issues, 18,* 1–47.

Hobfoll, S.E. (1988). *The ecology of stress.* New York: Hemisphere.

Jex, S.M., Beehr, T.A., & Roberts, C.K. (1992). The meaning of occupational stress items to survey respondents. *Journal of Applied Psychology, 77,* 623–628.

Jones, J.W., & DuBois, D. (1989). A review of organizational stress assessment instruments. In L.R. Murphy & T.F. Schoenborn (Eds.), *Stress management in work settings.* New York: Praeger.

Keita, G.P., & Jones, J.M. (1990). Reducing adverse reaction to stress in the workplace: Psychology's expanding role. *American Psychologist, 45,* 1137–1141.

Kelloway, E.K., & Barling, J. (1991). Job characteristics, role stress and mental health. *Journal of Occupational Psychology, 64,* 291–304.

Korunka, C., Weiss, A., & Karetta, B. (1993). Effects of new technologies with special regard for the implementation process per se. *Journal of Organizational Behavior, 14,* 331–348.

Lazarus, R.S. (1966). *Psychological stress and the coping process.* New York: McGraw-Hill.

Lazarus, R.S. (1990a). Theory-based stress measurement. *Psychological Inquiry, 1,* 3–13.

Lazarus, R.S. (1990b). Author's response. *Psychological Inquiry, 1,* 41–51.

Lazarus, R.S. (1991). *Emotion and adaptation.* New York: Oxrord University Press.

Lazarus, R.S. (1994). Psychological stress in the workplace. In P.L. Perrewé & R. Crandall (Eds.), *Occupational stress: A handbook* (pp. 3–14). Washington, DC: Taylor & Francis. (Original work published 1991)

Lazarus, R.S., Averill, J.R., & Opton, E.M., Jr. (1970). Toward a cognitive theory of emotion. In M. Arnold (Ed.), *Feelings and emotions* (pp. 207–232). New York: Academic Press.

Lazarus, R.S., & Folkman, S. (1984). *Stress, appraisal, and coping.* New York: Springer.

Lee, R.T., & Ashforth, B.E. (1993a). A further examination of managerial burnout: Toward an integrated model. *Journal of Organizational Behavior, 14,* 3–20.

Lee, R.T., & Ashforth, B.E. (1993b). A longitudinal study of burnout among supervisors and managers: Comparisons between the Leiter and Maslach (1988) and Golembiewski et al. (1986) models. *Organizational Behavior and Human Decision Processes, 54,* 369–398.

Lewin, K. (1943). Psychology and the process of group living. *Journal of Social Psychology, 17,* 119–129.

Lewin, K. (1944). Constructs in psychology and psychological ecology. In K. Lewin, C.E. Myers, J. Kalhorn, & J.R.P. French, Jr. (Eds.), Authority and frustration: Studies in topological and vector psychology III. *University of Iowa Studies in Child Welfare, 20,* 23–27.

Linden, W., Paulhus, D.L., & Dobson, K.S. (1986). Effects of response styles on the report of psychological and somatic distress. *Journal of Consulting and Clinical Psychology, 54,* 309–313.

Maslach, C., & Jackson, S.E. (1981). *Maslach Burnout Inventory Manual* (Research ed.). Palo Alto, CA: Consulting Psychologists Press.

Mayo, E. (1933). *The human problems of an industrial civilization.* New York: Macmillan.

McGuire, W.J. (1992). Toward social psychology's second century. In S. Koch & D.E. Leary (Eds.), *A century of psychology as science*. American Psychological Association. (Original work published 1985)

Murphy, L.R., & Hurrell, J.J. (1987). Stress measurement and management in organizations: Development and current status. In A.W. Riley & S.J. Zaccaro (Eds.), *Occupational stress and organizational effectiveness* (pp. 29–51). New York: Praeger.

Neale, M.S., Singer, J.A., & Schwartz, G.E. (1987). A systems assessment of occupational stress: Evaluating a hotel during contract negotiations. In A.W. Riley & S.J. Zaccaro (Eds.), *Occupational stress and organizational effectiveness* (pp. 167–203). New York: Praeger.

Paulhus, D.L. (1984). Two-component models of socially desirable responding. *Journal of Personality and Social Psychology, 46,* 598–609.

Roselione, F.B., & Barone, D.F. (1988, March). *A comparison of cognitive restructuring, relaxation training, and their combination in occupational stress reduction.* Paper presented at the meeting of the Southeastern Psychological Association, New Orleans.

Rosnow, R.L., & Georgoudi, M. (Eds.). (1986). *Contextualism and understanding in behavioral science: Implications for research and theory.* New York: Praeger.

Sarason, I.G., Sarason, B.R., Shearin, E.N., & Pierce, G.R. (1987). A brief measure of social support: Practical and theoretical implications. *Journal of Social and Personal Relationships, 4,* 497–510.

Sarason, S.B. (1981). *Psychology misdirected.* New York: Free Press.

Singer, J.A., Neale, M.S., & Schwartz, G.E. (1989). The nuts and bolts of assessing occupational stress: A collaborative effort with labor. In L.R. Murphy & T.F. Schoenborn (Eds.), *Stress management in work settings.* New York: Praeger.

Simons, Y., & Barone, D.F. (in press). The relationships of work stressors and emotional support to strain in police officers. *International Journal of Stress Management.*

Spielberger, C.D., & Reheiser, E.C. (1994). Job stress in university, corporate, and military personnel. *International Journal of Stress Management, 1,* 19–31.

Turnage, J.J., & Spielberger, C.D. (1991). Job stress in managers, professionals, and clerical workers. *Work & Stress, 5,* 165–176.

Watson, D. (1990). On the dispositional nature of stress measures: Stable and nonspecific influences on self-reported hassles. *Psychological Inquiry, 1,* 34–37.

Weber, H., & Laux, L. (1990). Bringing the person back into stress and coping measurement. *Psychological Inquiry, 1,* 37–40.

Unstructured Perceptions of Work-Related Stress: An Exploratory Qualitative Study

Vincent Di Salvo
Charles Lubbers
Ana M. Rossi
James Lewis

Stress has become recognized as one of the most serious occupational health hazards of our time (Cummins, 1990; Kreps, 1990). Work-related stress jeopardizes the organizational member's health, with 50 to 80% of all diseases being psychosomatic or of a stress-related nature (Pelletier, 1984). In addition, job-related stress results in the organizational problems of employee dissatisfaction, withdrawal, high turnover and absenteeism, and attendant low productivity, at an estimated annual cost of $50 to $75 billion (Wallis, 1983).

Several authors have concluded that coronary heart disease is related to the amount of stress an individual endures and his/her ability to cope with it (Fox, Dwyer, & Ganster, 1993). Still others have identified stress as causing somatic problems, including migraine, tension headache, nausea, muscular discomfort, pain (Hoiberg, 1982), burnout (Burke, 1993), and emotional exhaustion (Lee & Ashforth, 1993).

Psychological consequences of stress have been studied at the University of Michigan Institute for Social Research (Zander & Quinn, 1962). These studies associated stress with mental illness, psychosomatic illnesses, low self-esteem, anxiety, worry, tension, and impaired interpersonal relations.

Chronic stress on employees can have a direct impact on the organization itself, resulting in a variety of problems such as high absenteeism and turnover (Parker & DeCotiis, 1983), poor industrial relations, poor productivity, high accident rates, poor organizational climate, low morale, antagonism at work, and job dissatisfaction (Chen & Spector, 1992).

Although the majority of research has focused on the negative effects of job-related stress on both individuals and organizations, a much smaller percentage of stress research has focused on causes, and research has failed to ask employees their views of stress instead of having them respond to a list of causes developed by the researcher. Where causes are identified, they typically include areas of organizational activity where gender has been found to have a significant impact.

Causes of Work-Related Stress

Work-related stress has generally been viewed as resulting from an imbalance between environmental demands and individual capabilities (Fried, Rowland, & Ferris, 1984), but empirical probing of the specific causes of job-related stress has been more limited. Early research by Zander and Quinn (1962) correlated mental health problems to the organizational factors of role ambiguity, responsibility for managing people, shift work, little autonomy, rapid technological change, and threats to self-esteem inherent in methods of worker performance evaluation. These are suggested causes of work-related stress. Kahn, Wolfe, Quinn, Snoek, and Rosenthal (1964) presented one of the earliest category schemas for organizing the causes of stress. They noted that the most common stressors in organizations were role conflict, role ambiguity, unmet expectations, work overload, and interpersonal conflict among members.

Three general categories of stressors in organizational settings have been identified by Parasuraman and Alutto (1984): contextual, role-related, and personal stressors. Parker and DeCotiis (1983) developed a theoretical model of the causes and outcomes of stress. Their model identified six more specific causes: "(1) characteristics and conditions of the job itself, (2) conditions associated with the organization's structure, climate, and information flow, (3) role-related factors, (4) relationships at work, (5) perceived career development, and (6) external commitments and responsibilities" (p. 165).

More recently, Summers, DeNisi, and DeCotiis (1994) created a research model of causes of job stress consisting of the following areas: 1) personal characteristics; 2) structural organizational characteristics; 3) procedural organizational characteristics; and 4) role characteristics. Cummins' (1990) review of the literature suggests that the most common causes of job stress are: 1) role conflict and ambiguity; 2) work overload; 3) underutilization of skills; 4) resource inadequacy, and 5) lack of participation.

Previous attempts to identify causes of job-related stress have developed categorical schemas by examining the writing of other authors in the field of stress and organizational behavior or by using quantitative measures of worker attitudes. In contrast, the present investigation is based on a belief that it is important to ground research in the "real world" of day-to-day existence (Browning, 1978). It concentrates on what the organizational members say are their causes of stress, rather than measuring reactions to a list of potential stressors.

Harre and Secord (1973) suggested that behavioral scientists must move from mechanistic and rational methods of analysis to more human-oriented methods. They argue that theory-building regarding human interaction should begin with the subjects' perceptions of what is important and what causes the problems they experience in their day-to-day lives. Stano (1983) noted that the critical incident method and other qualitative methods have an advantage over some other methods because they rely on the self-reports of the individuals under examination. The qualitative approach "generates data which are based upon actual behavior rather than on a particular researcher's inferences, hunches, stereotypes and subjective estimates of what is important or necessary" (p. 2). Therefore, in contrast to existing research in the area of organizational stress, the present research employed a qualitative open-ended method for collecting data. The objective was to make the respondent's view of stress the focus rather than the respondent's response to a predetermined list of causes of stress.

Those utilizing qualitative methodologies begin with a very different type of warrant for their argument. Reality is viewed as a social phenomenon, embedded in the contexts in which it is observed, and understandable only in terms of the perceptions of the actors involved. Downs and Conrad (1981) stated the case when they commented, "First, critical incidents are self-reported by people who actually experienced the situations. Consequently, they stress what these people assess as having high priority. Second, the reports deal primarily with incidents which have significant impact upon the success or failure of an operation" (p. 31). According to Bryman (1988), "the most fundamental characteristic is to viewing events, action, norms, values, etc. from the perspective of the people who are being studied" (p. 61). Therefore, in contrast to existing quantitative research in the area of organizational stress, the present research employed a qualitative open-ended method of collecting data. The objective was to make the respondents' views of stress the focus.

PURPOSE

This research was undertaken to answer two research questions left unanswered by current stress research. First, how do organizational members perceive causes of stress when asked with a free-response format? Second, are there gender differences when comparing male and female perceptions of job-related stress?

METHODOLOGY

Sample

A sample of 220 professionals (110 men and 110 women) was selected from the directories of four professional organizations in a mid-sized Midwestern city. Every fifth name was selected to receive a cover letter and a copy of the instrument used. Ten days after the initial mailing, a follow-up postcard was sent to those who had not returned the survey. Fourteen days after the postcards were mailed, a personal letter was sent to those who had still not returned the form. This final letter was handwritten, personally addressed, and informal in tone; it also included an additional copy of the form.

Most respondents held managerial and supervisory positions in manufacturing, banking, human service agencies, state government agencies, hospitals, and educational institutions. A total of 85 females and 63 males returned the questionnaire. The age distribution of females ranged from 23 to 65, with a median of 37. The ages of the males were slightly higher, ranging from 28 to 68, with a median of 41.

Instrumentation

A questionnaire was developed, using critical incident methodology, as outlined by Flanagan (1955) and Herndon and Kreps (1993), to determine which situations on the job were most stressful for the subjects. In addition to providing demographic information, respondents were asked to "discuss the major causes of stress" that they "experience in the workplace." They were also asked to identify which of these causes of stress they generated were the "most devastating."

To ensure the face validity of the instrument, the questionnaire was pilot-tested by 20 individuals. Minor wording changes resulted from the pilot test.

Critical Incident Technique (CIT) is an epistemological process in which qualitative, descriptive data are collected regarding real-life accounts. The viability of this approach is supported by its successful application. Japp (1986) reviewed over 60 studies in a wide variety of disciplines in which this procedure had been used successfully. We have used it to investigate such topics as: conflict behaviors of difficult employees (Monroe, Di Salvo, & Borzi, 1989); supervisors and difficult employees (Monroe, Borzi, & Di Salvo, 1993); perceptions of dysfunctional work groups (Di Salvo, Nikkel, & Monroe, 1989); people problems in organizations (Di Salvo, Gill, & Monroe, 1989); employees' perceptions of communication problems at work (Di Salvo, Bartling, & Ulrich, 1989); and communication as a cause of work-related stress (Di Salvo, Lubbers, & Rossi, 1989).

Content Analysis

Categories based on similar stress causes were developed using the actual responses as the point of departure and employing the guidelines established by Flanagan (1955). Category membership was first determined by the consensus of three researchers (4.9% of the responses had to be dropped because consensus could not be obtained). To ensure the reliability of the coding process, category labels and descriptions were first given to three individuals unrelated to this research project, who coded a random sample of 75 responses into 14 categories. This pilot sort was used to test the adequacy of the method and to determine necessary revisions. Agreement with the researchers' consensus classification by these three coders was 70.6, 52.0, and 77.3%. This level of agreement was deemed sufficient to warrant further testing of the category descriptions.

In an effort to further check the reasonableness of the category system and to ensure that the categories reflected the answers of the respondents and not the background or gender of the coders, six coders (three males and three females, three study respondents and three others unrelated to the study in any other way) were trained and coded 75 responses into 14 categories. The ratings ranged from 68% to 73% agreement with the 14 category system devised by the researchers. The average agreement of these six coders with the researchers was 71 percent.

These results indicate nonsignificant differences between male and female coders or between those involved in the study and those not. They also show a level of agreement sufficiently high, especially considering the number of categories (14) raters had to sort responses into and the exploratory nature of the research dealing with qualitative responses. We felt the agreement was high enough to keep the category system intact and permit an exploratory analysis of the data.

RESULTS

Seventy percent (70%) of the sample returned completed forms. Seven of the original 220 forms were undeliverable because individuals had changed jobs and left no forwarding addresses. One form had to be discarded because it had not been completed adequately. A total of 148 usable responses was obtained (female N = 85, male N = 63).

Content Analysis

The 148 respondents listed 1,001 causes of stress in the workplace. Fifty of these were discarded because respondents did not provide enough information to prevent major ambiguities. In response to the first research question, how do organizational members perceive causes of stress when asked with a free-response format, the responses or causes were classified into 14 categories that fell into two broad clusters. The first cluster was called *Work Content* (345 responses). The five categories in this area deal with the demands that the job itself imposes on the worker, usually requiring the worker to adapt in order to perform his/her job. The categories are listed in Table 1 in descending order of frequency.

The second cluster that made up research question number one was *Work Context* (606 responses), which consisted of nine categories dealing with the physical and social context in which the work is performed, but they do not directly

TABLE 1   Work Content as a Cause of Stress

Percent of Total		Category
1.	32.46%	**Unpleasant Internal Task Duties:** Unpleasant work demands that must be carried out or faced with others within the organization as a part of one's job. For example, firing, reporting bad news, resolving disputes, mediating, disciplining, and traveling.
2.	21.74%	**Unpleasant External Task Duties:** Unpleasant work demands and/or situations that must be carried out or faced when dealing with others outside the organization as a part of one's job. For example, meeting client/customer demands, handling client/customer complaints, and interacting with clients/customers.
3.	16.81%	**Performance of Others:** The work performance of others that one feels must be dealt with as part of one's own work demands. For example, errors, poor quality, and poor performance.
4.	16.52%	**Work Load:** Variations in one's work load. For example, when work load demands are too high or too low, when the work demands are seasonal or cyclical, having an occasional swamped feeling, and having many clients/customers to deal with at once.
5.	12.46%	**Professional Risk:** Various types of risks associated with specific individual professions. For example, uncontrollable overhead costs, price war competition, no business in the shopping mall, and making monthly sales quotas.
Total	100%	

affect the actual job performance. These categories are listed in Table 2, also in descending order of frequency.

Sex Differences

Research question number two was, are there gender differences when comparing male and female perceptions of job-related stress? Responses within each cluster were cross-tabulated by category of response, gender of respondent, and devastation identification of the response. These tables were tested using hierarchical log-linear modeling techniques. The Work Content Cluster had a chi-square of 12.73 and the Work Context Cluster, 24.92; both were significant at .05. This indicates that some of the categories served as significantly greater stressors for one sex than for the other.

To determine the relationship between the five Work Content categories and gender, a two-dimensional cross table was constructed. Results are shown in Table 3. As this table illustrates, the most common cause of stress listed by both men and women was facing "Unpleasant Internal Task Duties" (feeling that resolving disputes and disciplining had to be carried out). About one-third of the responses fell into this category. Yet even though this is the category reported most frequently by

TABLE 2   Work Context as a Cause of Stress

Percent of Total	Category
1. 29.92%	**Other People's Attitudes/Behavior:** The attitudes and/or behavior of others toward work and the people at work. These actions come from the boss, co-workers, clients, or people in general, but in all cases, exclude any reference to actual work performance. For example, people who don't listen, no sense of urgency, uncommitted, dishonest, unwilling to compromise, negative thinking and dealing with boss' personality, manipulative style, and an absolute perfectionist.
2. 20.10%	**Time:** Time *itself.* For example, not enough time, too much work, deadlines, pressure, schedule preparation time, procrastination, and long hours.
3. 13.84%	**Work Inhibitors:** Those events and situations that block the completion of work. For example, interruptions, government regulations and red tape, family, meetings, bureaucracy, decision making processes, and additional or atypical duties.
4. 10.54%	**Power:** Power issues and/or political plays. For example, lack of power and exercising power.
5. 9.06%	**Task Communication:** Communication regarding the task. For example, insufficient communication, incomplete information, lack of feedback, and vagueness in job responsibility.
6. 8.90%	**Resources:** The resources available. For example, budget, staff, and equipment.
7. 7.41%	**My Personal Behavior:** One's own behavior/attitudes; the focus is on the self. For example, fear of failure, persistence, occasional boredom, and having one's views challenged.
8. 4.61%	**Interpersonal Relations:** Relationships between people and/or departments. For example, conflict, low morale, and lack of cooperation or teamwork.
9. 4.61%	**Physical:** The physical conditions in which one works. For example, noise and space.
Total 100%	

both sexes, it is significantly more likely to be reported by men than by women.

The only other category significantly related to gender was "Work Load" (this category included stress from too much or too little work, as well as situational factors such as experiencing heavy seasonal trade or even an afternoon when a store would be swamped with customers). Although this category represented 15% of the responses in this cluster, women were twice as likely as men to mention it. More than one in five of the female responses, but fewer than one in ten of the male, were in this category.

The relationship of gender to the nine Work Context categories was determined in a similar fashion. The cross-tabulated results are shown in Table 4. Again,

TABLE 3   Percent Work Content Category Codes by Gender[a]

Category	Male	Female	Total
Unpleasant Internal Task Duties	39.72%	27.45%	32.46%**
Unpleasant External Task Duties	21.28	22.05	21.74
Performance of Others	18.44	15.69	16.81
Work Load	9.22	21.57	16.52**
Professional Risk	11.35	13.24	12.46
Total	100	100	100

[a]$Chi\text{-}square = 12.20,\ df = 4, probability = .016;\ **p < .05.$

two of the categories were differentially related to gender, and the most frequently reported category was one of them. This category represented the influence of "Other People's Attitudes/Behavior" (description of attitudes of others—boss, co-workers, clients, and people in general—that cause stress but do not refer to actual work performance, such as, not listening, dishonest, boss' personality). It was almost twice as likely to be mentioned by males as by females.

The second gender-related category was "Power" (the feeling that power issues and power plays—such as lacking power or exercising power—are sources of stress). Power was mentioned only about half as often as the previously discussed category, but it was more than twice as likely to be mentioned by a female than by a male.

There were no significant effects using the ratings of severity. There were no observable differences in how "devastating" any listed stressor was by content, context or gender.

DISCUSSION

The results of this study empirically support the viability of generating grounded constructs for research in the organizational context recommended by Browning (1978) and Herndon and Kreps (1993). First, respondents provided open-ended replies when asked to describe perceived stressors. Categorical constructs then emerged from a data base representing job stressors perceived by those affected by the stressors. Theory based on such constructs should be particularly helpful for those who want to deal with the pragmatic problems of stress on the job as they exist in the work world.

It is noteworthy to call attention to Work Content (Table 1). The two categories of "Unpleasant Internal" and "Unpleasant External Task Duties" accounted for approximately 50% of the responses in this cluster. In both cases, the two categories focused on having to deal with other people in an unpleasant context. When looking at Work Context (Table 2), three categories, "Other People's Attitudes/Behavior," "Time," and "Work Inhibitors" accounted for approximately two-

TABLE 4   Percent Work Context Category Codes by Gender[a]

Category	Male	Female	Total
Other People's Attitudes/Behavior	28.18%	16.80%	29.92%**
Time	17.73	21.45	20.10
Work Inhibitors	16.82	12.14	13.84
Power	5.91	13.18	10.54**
Task Communication	8.18	9.56	9.06
Resources	7.27	9.82	8.90
My Personal Behavior	6.36	8.01	7.41
Interpersonal Relations	3.18	5.43	4.61
Physical	6.36	3.62	4.61
Total	100	100	100

[a]$Chi\text{-}square = 24.55$, $df = 8$, $probability = .002$; $**p < .05$.

thirds of the responses. It is interesting to note that four out of the top five dealt with people directly or indirectly.

A major finding that emerged from the analysis of the category responses was that the intensity level of a stressor did not appear to be more severe in any one category of stressors. Respondents indicated that the consequences of the stressor could be equally "devastating" for content or context stressors, for any category of stressor, and independent of gender. This would suggest that variables other than those examined in the present study must be explored if we wish to understand the "intensity" of the stressor.

Analysis of the responses also indicates that causes of work-related stress constitute two broad clusters. Those which are an inherent part of the job description (the content cluster of stressors) are mentioned only half as frequently as the more indirect sources of stress (the context cluster of stressors). These are in contrast to the three causes of job-related stress (contextual, role-related, and personal stressors) as identified by Parker and DeCotiis (1983). However, the categories within each cluster in the present study suggest that personal stressors exist in both contextual job elements and work content (role-related) elements of organizational activity.

When looking at the causes of stress discussed by Summers et al. (1994) and Cummins (1990), a number of our categories provide some support for their models. This would include work load, work inhibitors, task communication, power, and resources. What is not so obvious from their models is the strong and consistent support for problematic people and having to deal with and/or rely on these people. (McDonald & Korabik, 1991, did identify treatment by others and negative impact on relationships with others at work.) This is something that appears to be potentially lost in the quantitative correlational type research where scales have been constructed in an a priori fashion. Using a qualitative, naturalistic, open-ended response research design allows researchers to gather a more comprehensive and richer category system of stress by allowing the employees to generate the data rather than just respond to a questionnaire.

This study indicates that, from a broad perspective, men and women perceive stressors quite similarly. No gender differences were found in the overall clusters. For both men and women, more causes of work-related stress stem from the organizational activity of people working daily in interdependent roles than from the specific content of the jobs themselves.

The frequency of specific causes of stress for men and women did differ in four out of the 14 cases. Within the Work Content Cluster, "Unpleasant Internal Task Duties" created stress more frequently for men than for women.

The second Work Content category showing a gender difference is "Work Load," which creates stress more frequently for women than for men. In the stressors created by the Work Context, "Other People's Attitudes/Behavior" caused stress more frequently for males than for females. A second contextual work stressor, "Power," created stress more frequently in women than in men.

Using a grounded data base does not, in itself, ensure the usefulness of the categories. Problems of reliability and rater perspective are common with this type of data and must be addressed (Downy & Ireland, 1983). The category schema that emerged from this study showed particular strengths in these areas. Satisfactory levels of agreement were attained between researcher coding and that of naive coders. Additionally, equivalent levels of agreement were obtained when coding was done by respondents in the study. Further, equivalent agreement was also achieved independent of coder gender, even though responses showed a gender effect. Finally, the categorical schema of work-related stress identified in the present study, while supportive of previous research (i.e., Cummins, 1990; Kahn et al., 1964; Parasuraman & Alutto, 1984; Parker & DeCotiis, 1983; Summers et al., 1994), more clearly elucidate and delineate these causes. Therefore, the free-response, qualitative method and the categories developed in this study show promise as a basis for future research on the causes of stress in organizational activity.

Future research into the causes of stress may investigate these differences by examining the implications raised by the identification of the stressors. Additionally, researchers may replicate this research using a population of individuals who do not hold managerial or supervisory positions.

In sum, causes of work-related stress appear to stem from nine indirect contextual work-related factors and five direct work content factors. Of these causes, four are gender-differentiated.

REFERENCES

Baird, J.E., Jr., & Bradley, P.H. (1979, June). Styles of management and communication: A comparative study of mean and women. *Communication Monographs, 46,* 101–111.

Brannon, R. (1985). Dimensions of the male sex role in America. In A.G. Sargent (Ed.), *Beyond sex roles* (pp. 296–316). St. Paul, MN: West.

Browning, L.D. (1978). A grounded organizational communication theory derived from qualitative data. *Communication Monographs, 45,* 93–109.

Bryman, A. (1988). *Quantity and quality in social research.* Mass: Allen and Unwin.

Burke, M.J., Brief, A.P., & George, J.M. (1993). The role of negative affectivity in understanding relationships between self-reports of stressors and strains: A comment on the applied psychology literature. *Journal of Applied Psychology, 78*(3), 402–412.

Chen, P.Y., & Spector, P.E. (1992). Relationships of work stressors with aggression, withdrawal, theft and substance abuse. *Journal of Occupational and Organizational Psychology, 65,* 177–184.

Cooke, P. (1985). Assertive communication: Why don't we just say what we think? In A.G. Sargent (Ed.), *Beyond sex roles* (pp. 411–430). St. Paul, MN: West.

Cummins, R. (1990). Job stress and the buffering effect of supervisory support. *Group and Organizational Studies, 15,* 92–104.

Daniels, T.D., & Logan, L.L. (1983). Communication in women's career development relationships. In R.N. Bostrom (Ed.), *Communication Yearbook 7* (pp. 532–552). Beverly Hills: Sage.

Di Salvo, V., Bartling, H., & Ulrich, H. (1989). *A field investigation and identification of employees' perceptions of communication problems at work.* Paper presented at the Speech Communication Association, San Francisco, CA.

Di Salvo, V., Gill, M., & Monroe, C. (1989). *People problems in organizations: An exploratory field study.* Paper presented at the Speech Communication Association, San Francisco, CA.

Di Salvo, V., Lubbers, C., & Rossi, A. (1989). *Communication as a cause of work-related stress.* Paper presented at the American Business Communication Association Convention, Las Vegas.

Di Salvo, V., Monroe, C., & Nikkel, E. (1989). Theory and practice: A field investigation and identification of group members perception of problems facing natural work groups. *Journal of Small Group Behavior, 20,* 551–567.

Downs, C., & Conrad, C. (1981). "Effective subordinacy." *Journal of Business Communication, 14,* 27–38.

Downy, H.K., & Ireland, R.D. (1983). Quantitative vs. qualitative: Environmental assessment in organizational studies. In J. Van Maanen (Ed.), *Qualitative methodology.* Beverly Hills: Sage.

Flanagan, J.C. (1955). The critical incident technique. *Psychological Bulletin, 51,* 327–358.

Fox, M.L., Dwyer, D.J., & Ganster, D.C. (1993). Effects of stressful job demands and control on physiological and attitudinal outcomes in a hospital setting. *Academy of Management Journal, 36*(2), 289–318.

Fried, Y., Rowland, K.M., & Ferris, G.R. (1984). The physiological measurement of work stress: A critique. *Personnel Psychology, 37,* 583–615.

Grandjean, B.D., & Bernal, H.H. (1979). Sex and centralization in a semiprofession. *Sociology of Work and Occupations, 6,* 84–102.

Hagen, R., & Kahn, A. (1975). Discrimination against women. *Journal of Applied Social Psychology, 5*(4), 362–376.

Harre, R., & Secord, P. (1973). *The explanation of social behavior.* New Jersey: Littlefield and Adams.

Henley, N., Hamilton, M., & Thorne, B. (1985). Womanspeak and manspeak: Sex differences and sexism in communication, verbal, and nonverbal. In A.G. Sargent (Ed.), *Beyond sex roles* (pp. 168–187). St. Paul, MN: West.

Hennig, M., & Jardim, A. (1978). *The managerial woman.* New York: Pocket Books.

Herndon, S., & Kreps, G. (1993). *Qualitative research: Applications in organizational communication.* Virginia: Hampton.

Hoiberg, A. (1982). Occupational stress and illness incidence. *Journal of Occupational Medicine, 24,* 445–451.

Jablin, F.M. (1985). Assimilating new members into organizations. In R.N. Brostrom (Ed.), *Communication Yearbook 8* (pp. 594–626). Beverly Hills: Sage.

Jablin, F.M. (1980). Subordinate's sex and superior-subordinate status differentiation as moderators of the Pelz effect. In D. Nimmo (Ed.), *Communication Yearbook 4* (pp. 349–366). New Brunswick, NJ: Transaction Books.

Japp, D., & Di Salvo, V. (1986). *Secretaries and communication: A view from the bottom.* Paper presented at the Speech Communication Association. Boston, MA.

Josefowitz, N. (1985). Women and power: A new model. In A.G. Sargent (Ed.), *Beyond sex roles* (pp. 199–214). St. Paul, MN: West.

Kahn, R.L., Wolfe, D.M., Quinn, R.P., Snoek, J.D., & Rosenthal, R.A. (1964). *Organizational stress: Studies in role conflict and ambiguity.* New York: Wiley.

Kanter, R.M. (1977). *Men and women of the corporation.* New York: Basic Books.

Kreps, G. (1990). *Organizational communication.* New York: Longman.

Lee, R.T., & Ashforth, B.E. (1993). A longitudinal study of burnout among supervisors and managers. Comparisons between the Leiter and Maslach (1988) and Golembiewski et al. (1986) models. *Organizational Behavior and Human Decision Processes, 54,* 369–398.

McDonald, L.M., & Korabik, K. (1991). Sources of stress and ways of coping among male and female managers. In P.L. Perrewé (Ed.), Handbook on job stress [Special issue]. *Journal of Social Behavior and Personality, 6*(7), 185–198.

Monahan, L. (1983, February). The effects of sex differences and evaluations on task performance and aspiration. *Sex Roles, 9,* 205–216.

Monroe, C., Borzi, M., & Di Salvo, V. (1993). Management decisions and difficult subordinates. *Southern Journal of Speech Communication, 59,* 23–33.

Monroe, C., Di Salvo, V., & Borzi, M. (1989). Conflict behaviors of difficult subordinates. *Southern Journal of Speech Communication, 54,* 311–329.

Moore, L.L. (1985). Issues for women in organizations. In A.G. Sargent (Ed.), *Beyond sex roles* (pp. 215–225). St. Paul, MN: West.

Nadler, L.B., & Nadler, M.K. (1986, Spring). The role of sex in organizational negotiation ability. *Women's Studies in Communication, 9,* 1–11.

Parasuraman, S., & Alutto, J.A. (1984). Sources and outcomes of stress in organizational settings: Toward the development of a structural model. *Academy of Management Journal, 27,* 330–350.

Parker, D.F., & DeCotiis, T.A. (1983). Organizational determinants of job stress. *Organizational Behavior and Human Performance, 32,* 160–177.

Pelletier, K.R. (1984). *Healthy people in unhealthy places: Stress and fitness at work.* New York: Delacorte.

Rossi, A.M., & Wolesensky, B. (1983). Women in management: Different strategies for handling problematic interactions with subordinates. *American Business Communication Association Proceedings* (pp. 79–93).

Russo, N.F. (1985). Sex-role stereotyping, socialization, and sexism. In A.G. Sargent (Ed.), *Beyond sex roles* (pp. 150–167), St. Paul, MN: West.

Sargent, A.G. (Ed.). (1985). *Beyond sex roles* (2nd ed.). St. Paul, MN: West.

Schein, V.E. (1978, Summer). Sex role stereotyping, ability and performance: Prior research and new directions. *Personnel Psychology, 31,* 95–100.

Seashore, E.W., & Bunker, B.B. (1985). Power, collusion, intimacy-sexuality, support. In A.G. Sargent (Ed.), *Beyond sex roles* (pp. 462–476). St. Paul, MN: West.

Sereno, K., & Wheathers, J. (1981). Impact of communicator sex on receiver reactions to assertive, nonassertive, and aggressive communication. *Women's Studies in Communication, 4,* 1–17.

Shepard, H. (1985). Men and organizational cultures. In A.G. Sargent (Ed.), *Beyond sex roles* (pp. 374–383). St. Paul, MN: West.

Stano, M.E. (1983). *The critical incident method: A description of the method.* Paper presented at the Central States Speech Association Convention, Lincoln, NE.

Summers, T.P., DeNisi, A.S., & DeCotiis, T.A. (1994). A field study of some antecedents and consequences of felt job stress. In P.L. Perrewe & R. Crandall (Eds.), *Occupational stress: A handbook* (pp. 113–128). (Original work published 1989)

Thorkelson, A.E. (1985). Women under the law: Has equity been achieved? In A.G. Sargent (Ed.), *Beyond sex roles* (pp. 477–496). St. Paul, MN: West.

Todd-Mancillas, W.R., & Rossi, A.M. (1985, Spring). Gender differences in the management of personnel disputes. *Women's Studies in Communication, 8,* 25–33.

Trahey, J. (1977). *Women and power: Who's got it—how to get it.* New York: Avon.

U.S. Department of Labor. (1983). *Women's bureau. Twenty facts on women workers.*

Waetjen, W.B., Schuerger, J.M., & Schwartz, E.B. (1979). Male and female managers: Self-concept, success, and failure. *Journal of Psychology, 103,* 87–94.

Wallis, C. (1983, June 6). Stress: Can we cope? *Time.*

Weaver, C.N. (1978). Sex differences in the determinants of job satisfaction. *Academy of Management Journal, 21,* 265–274.

Wertheim, E.G., Widom, C.S., & Wortzel, L.H. (1978). Multivariate analysis of male and female professional career choice correlates. *Journal of Applied Psychology, 63,* 234–242.

Zamutto, R.F., London, M., & Rowland, K.M. (1979). Effects of sex on commitment and conflict resolution. *Journal of Applied Psychology, 64,* 227–231.

Zander, A., & Quinn, R. (1962). The social environment and mental health: A review of past research at the Institute for Social Research. *Journal of Social Issues, 18*(3), 48–66.

Measuring Occupational Stress: The Job Stress Survey

Charles D. Spielberger
Eric C. Reheiser

It is now well established that stress in the workplace adversely affects productivity, absenteeism, worker turnover, and employee health and well-being (e.g., Cooper & Payne, 1988; Kahn, Wolfe, Quinn, Snoek, & Rosenthal, 1964; Karasek & Theorell, 1990; Keita & Sauter, 1992; Levi, 1981; Matteson & Ivancevich, 1982; Perrewé, 1991; Quick, Murphy, & Hurrell, 1992). For example, in a recent nationwide study of occupational stress, the proportion of workers who reported "feeling highly stressed" had more than doubled from 1985 to 1990. Those reporting "having multiple stress-related illnesses" increased from 13% to 25% (Northwestern National Life, 1991). Moreover, 69% of the 600 workers surveyed in the Northwestern National Life study reported that their productivity was reduced because of high stress levels, and "one in three say job stress is the single greatest stress in their lives" (1991, p. 2).

The effects of occupational stress on absenteeism and worker turnover were also clearly reflected in the same study (Northwestern National Life, 1991). Of the

Authors' Note: The authors acknowledge the invaluable assistance of Karen Unger and Virginia Berch in the preparation of the manuscript for this paper, and Dr. Barbara Fritzsche of Psychological Assessment Resources, Inc. (PAR) for her assistance in the construction and validation of the Job Stress Survey (JSS) and the preparation of the JSS Manual.

participants, 17% reported missing one or more days of work each year due to high stress levels, and 14% indicated that stress had caused them to quit or change jobs in the preceding two years. In a similar study of more than 1200 full-time, private-sector employees, 40% reported that their jobs were "very" or "extremely" stressful (Northwestern National Life, 1992). Compared with workers reporting lower levels of job stress, the employees who perceived their jobs as highly stressful were twice as likely to: work overtime frequently (62% vs. 34%); think about quitting their job (59% vs. 26%); suffer stress-related medical problems (55% vs. 21%); and experience burnout on the job (50% vs. 19%).

Absenteeism, employee turnover, and stress-related medical problems are clear-cut direct costs to employers. In addition, reduced productivity and diminished customer services are hidden costs that often result from "exhausted or depressed employees (who) are not energetic, accurate, or innovative at work" (Karasek & Theorell, 1990, p. 167). According to Matteson and Ivancevich (1987), preventable costs in the 1987 U.S. economy relating to reduced productivity, absenteeism, and worker turnover is estimated to be approximately $2800 per year per employee. Many employees with stress-related workplace problems also expect compensation, and seek disability payments and early retirement benefits. Sauter (1992, p. 14) has observed that, each year in the U.S., "nearly 600,000 workers are disabled for reasons of psychological disorders," costing $5.5 billion in annual payments to individuals and their families.

Growing concerns over the consequences of job stress for both employees and organizations have stimulated efforts to understand the sources and consequences of stressors in the workplace. These concerns are dramatically reflected in the increasing number of studies of occupational stress that have appeared in the psychological, organizational, and medical literature over the past 20 years. Publications listed in *PsycLit* with titles that specifically included "job stress," "work stress," "occupational stress" or "family stress" are presented in Figure 1 for each three-year period from 1971 to 1992, the last full year for which relatively complete data were available. Investigations in all three workplace-related stress categories have increased more than fifty-fold over the past two decades. The total number of studies in 1990–92 (N = 169) was more than 8 times greater than during the entire decade of the seventies (N = 19). In contrast, research relating to family stress has increased to a much lesser degree.

Unfortunately, the explosive increase in research on stress in the workplace has not clarified the interpretations of the findings in these studies because of ambiguity in the conceptual definition of occupational stress, which often differs from study to study (Kasl, 1978; Schuler, 1980). As Schuler (1991) recently noted, a major source of confusion in occupational stress research stems from the fact that some investigators have focused on antecedent conditions and pressures associated with the characteristics of a particular job. Others have been primarily concerned with the consequences of work-related stress. Such differences in approaches to

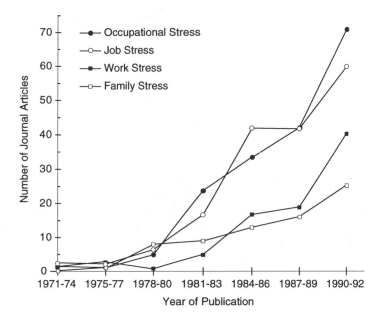

FIGURE 1   Publications Listed in *PsycLit* for each Three-Year Period from 1971 to 1992 in which Job Stress, Work Stress, Occupational Stress, or Family Stress Were Included in the Titles

stress in the workplace have greatly influenced the procedures that are used to measure occupational stress.

In this paper, influential conceptions of occupational stress are briefly reviewed and stress measures emanating from these conceptions are evaluated. Gender differences in occupational stress are also examined. Finally, a new psychometric measure for assessing job stress, the *Job Stress Survey* (Spielberger, 1994), is described in some detail. Research findings with this instrument that demonstrate substantial gender differences in occupational stress are then reported.

Person–Environment Fit Theory

Person–Environment Fit theory (French & Caplan, 1972; French, Caplan, & Harrison, 1982) is among the most utilized and widely accepted approaches to conceptualizing the nature of occupational stress (Chemers, Hays, Rhodewalt, & Wysocki, 1985). In the context of this theoretical orientation, occupational stress is defined in terms of job characteristics that pose a threat to the individual because of a poor match between the abilities of the employee and the demands of the job

(French & Caplan, 1972). The workplace stress that results from an incompatible person–environment fit produces psychological strain and stress-related physical disorders (French, Caplan, & Harrison, 1982).

In early studies guided by Person–Environment Fit theory, differences in sources of job stress, and in stressful work-related events were investigated for a variety of occupations (e.g., Caplan, Cobb, French, Harrison, & Pinneau, 1975). The concepts of role ambiguity and role conflict, and how these differed for various organizational settings and occupational groups were prominent in this research (Fisher & Gitelson, 1983; Jackson & Schuler, 1985). For example, in studies of sources of stress for different occupational levels, executives and managers tended to perceive more role ambiguity. Employees in less responsible positions experienced more role conflict (Hamner & Tosi, 1974; Kahn, Wolfe, Quinn, Snoek, & Rosenthal, 1964). It should be noted, however, that Jackson and Schuler (1985) found no relationship between occupational level and either role conflict or ambiguity, and little evidence that these variables influenced job performance or worker satisfaction.

Karasek (1979) proposed a variant of Person–Environment Fit theory in which interactions between level of control and job demands are emphasized as determinants of work-related psychological strain. For example, low control and high demand appear to contribute to lowered productivity and a greater risk of health problems. Sauter and Hurrell (1989) also recognized the importance of worker autonomy and control, noting that lack of control inhibits learning and undermines the motivation that is generally needed to overcome the stress associated with demanding work. However, they also pointed out that "fundamental questions remain concerning the conceptualization and operationalization of the construct [of job control]" (1989, p. XVI).

Other job conditions, such as workload and interpersonal conflict, also contribute to person–environment fit problems which often have adverse effects on job satisfaction and employee well-being. Spector (1987) reported significant positive correlations of excessive workload and work-related interpersonal conflicts with anxiety, frustration, job dissatisfaction, and health symptoms. Other investigators have reported positive correlations between work underload and dissatisfaction, health symptoms, and depression (Ganster, Fusilier, & Mayes, 1986). The concepts of work overload and underload are similar to work demands as this concept is related to job control.

Findings that support a person–environment fit model of occupational stress have been reported by Chemers et al. (1985). They also observed that interactions between person and environment variables predict job strain better than either person or environment variables separately (Caplan et al., 1975; Harrison, 1978; Locke, 1969; Porter, 1961). Chemers et al. (1985, p. 628) conclude that, although Person–Environment Fit theory provides a useful approach to conceptualizing and measuring occupational stress, "…[the] theory has not yielded a highly focused approach."

Measurement Issues

In research guided by Person–Environment Fit theory, measures of occupational stress have encompassed a wide range of contents. These have included the assessment of: job and organizational characteristics; employee skills; job satisfaction; individual differences in attitudes and personality traits; and health status (Beehr & Newman, 1978; Sharit & Salvendy, 1982). Control over how job demands are met (autonomy) is also emphasized in the assessment of occupational stress (Karasek & Theorell, 1990). Unfortunately, such diverse aspects of stress in the workplace are often confounded in the operational measures that are used to assess job stress, which makes it difficult to know exactly what is being measured.

The environmental antecedents of stress in the workplace have been measured primarily in terms of general role demands or expectations. Specific job pressures and task characteristics tend to be neglected. Within the context of his general theory of stress, Lazarus (1994) has recently explained his approach to conceptualizing occupational stress as a process involving a transaction between an individual and his/her work environment.

Lazarus' Transactional Process Theory

Lazarus' theory distinguishes between stressful antecedent conditions ("stressors"), how these are perceived and cognitively appraised by a particular person, and consequent emotional reactions when a stressor is perceived as threatening and the individual is not able to cope effectively with it. His theoretical approach requires a detailed analysis of the specific stressors that are associated with a particular job, and how different workers react to these stressors, taking into account the individual's coping skills and past experience. Thus, Lazarus conceptualizes stress in the workplace as essentially an individual phenomenon in which the effects of work-related stressor events on emotions and behavior are mediated by an employee's perceptions and appraisals of particular stressors and her/his coping skills for dealing with them.

Brief and George (1994) criticize Lazarus' emphasis on the idiographic nature of occupational stress. They argue that it is especially important to discover those working conditions that are likely to adversely affect groups of employees who are exposed to them. In a similar critique of Lazarus' model, Harris (1994) has noted that the occupational stressors associated with the climate and culture of an organization can have profound effects on employees, and that these may differ as a function of gender and individual differences in personality and coping skills.

In summary, Lazarus' conception of occupational stress and Person–Environment Fit theory both have merit and limitations, and can be construed as complimentary rather than contradictory in providing a meaningful conceptual framework for understanding stress in the workplace. A major difference between these perspectives is in the specificity and size of the unit of measurement for assessing the components of job stress. Whereas Person–Environment Fit theory identifies the general conditions that produce job strain (work demands), Lazarus' transactional process model focuses on how a particular stressor event is perceived and appraised, noting that these mediating processes are strongly influenced by the worker's coping skills and previous experience. A comprehensive assessment of occupational stress as a transactional process requires taking into account the nature of the stressor event, how it is perceived and appraised, and the emotional reactions of the worker.

Gender Differences in Occupational Stress

On the basis of a comprehensive review of stress in organizations, Beehr and Schuler (1980) concluded that there was little evidence that gender influences stress-related symptoms in the workplace. Similarly, Di Salvo, Lubbers, Rossi, and Lewis (1994), in a study of gender and work-related stress, observed that "from a broad perspective, men and women perceive stressors quite similarly. No gender differences were found in the overall clusters [of stressors]" (p. 48). Martocchio and O'Leary (1989) came to a similar conclusion based on the results of a meta-analytic evaluation of 15 studies which examined gender differences in occupational stress. In their words: "There are no sex differences in experienced and perceived work stress" (p. 495).

Jick and Mitz (1985) suggest that the difficulty in identifying gender-related differences in workplace stress may be due to sampling problems, noting that men are over-represented in managerial positions while more women hold clerical and service jobs. In their review of empirical evidence of sex differences in workplace stress, they identified numerous "gaps, ambiguity, and inconsistencies in the existing research" (1985, p. 408). Nevertheless, despite the oft quoted findings of no gender-related differences in occupational stress, they conclude that gender acts "not only as a direct predictor of the source of stress, but also as a moderator affecting how stress is perceived, what coping skills are called upon, and how stress is manifest" (p. 409). In supporting this conclusion, they call attention to evidence that women report more symptoms of psychological distress (e.g., emotional discomfort and depression), whereas men are more prone to develop severe stress-related physical illnesses, as reflected by a higher incidence of heart disease and cirrhosis of the liver due to alcohol abuse.

Nelson and Quick (1985) also reviewed research on gender differences in workplace stress, and concluded that women experience greater occupational stress than men because of the unique sources of job stress typically faced by women: for example, lower salaries, career blocks, discrimination and stereotyping, and the interface of marriage and work (cf. Goldsmith, 1988). Although no inherent gender differences have been found in research on job-related burnout, there may be differences in the causes of burnout for men and women and in their coping strategies for handling burnout symptoms, with men choosing more active and direct coping strategies (Lowman, 1993).

Although Di Salvo et al. (1994) found no overall gender differences in broad clusters of workplace stressors, they noted that the "frequency of specific causes of stress for men and women did differ in four out of the 14 cases" (p. 48). Work-load factors, such as too much or too little work, were twice as likely to be reported by women then men. Stresses relating to power such as lacking or expressing power were mentioned more than twice as often by men.

Measuring Occupational Stress

In addition to differences in gender representation at higher and lower occupational levels, and in broader versus narrower definitions of stressor events, job stress measures tend to confound the perceived severity of a stressful event with the frequency of its occurrence. The perceived severity of a stressor will greatly influence the intensity of an emotional reaction when that stressor occurs. However, even though a specific stressor event may be perceived as highly stressful, if it occurs infrequently it will have limited impact as a source of stress. Consequently, it is important to assess not only the perceived severity of a stressor, but also how often it occurs.

Jackson and Schuler (1985, p. 47) contended that research on occupational stress should focus on "...the development of good diagnostic tools for pinpointing *specific aspects about one's job* that are ambiguous or conflicting" [italics added]. In a similar vein, Beehr and Newman (1978) argued that work environment variables should be measured both objectively and subjectively in stress research, and Murphy and Hurrell (1987) called for the construction of a generic questionnaire or core set of questions to facilitate comparing stress levels in specific occupational groups. Barone, Caddy, Katell, Roselione, and Hamilton (1988) also affirmed the need to develop valid psychometric questionnaires for assessing occupational stress. In their words: "One would expect this burgeoning literature to be grounded in a sophisticated assessment of work stress; however, most of the literature has relied on answers to only one or a few questions about experienced stress" (pp. 141–142).

Noting such criticisms of existing job stress measures, Osipow and Spokane (1981) developed a promising generic measure to assess person–environment fit variables such as role overload, role ambiguity, and psychological strain across different occupational levels and work environments. Consistent with Lazarus' transactional process theory, Spielberger and his colleagues (Spielberger, Westberry, Grier, & Greenfield, 1981) developed the Police Stress Survey (PSS) to evaluate the perceived severity and frequency of occurrence of 60 specific stressors encountered by law enforcement officers. Most of these specific sources of stress in police work had been previously identified by Kroes (1976) and his colleagues (Kroes & Gould, 1979; Kroes, Margolis, & Hurrell, 1974).

The Police Stress Survey was field-tested with 50 Florida law enforcement officers from seven geographically diverse locations. Data were subsequently collected for a representative sample of 233 officers. Findings with the PSS are reported in a monograph by Spielberger et al. (1981), and in several brief reports

(Spielberger, Grier, & Pate, 1979, 1980). Since the representative sample of law enforcement officers only included 15 females, no gender differences were reported.

The Teacher Stress Survey (TSS) was designed to evaluate sources of stress relevant to secondary school teachers and to compare teacher stress with police stress (Grier, 1982). Of the 60 TSS items, 39 were selected from the PSS on the basis of their applicability to both teaching and police work. These TSS items were identical to the corresponding PSS items, except that "teacher" and "school" were routinely substituted for "police" and "department." Twenty-one additional items were generated from the teacher stress literature in consultation with experienced high school teachers. Research findings with the Teacher Stress Survey have been reported by Grier (1982) who found only one main effect of gender, but a number of sex-by-age and sex-by-experience interactions.

Brantly and Jones (1989) constructed a psychometric instrument designed to monitor and evaluate the frequency and impact of daily stressors. Commenting in general on the potential utility of self-report measures of occupational stress, Spector, Dwyer, and Jex (1988) noted that the validity of such instruments is supported by studies reporting findings of significant correlations between them and objective measures of stressful job conditions (Algera, 1983; Gerhart, 1986).

Barone (1994) and his colleagues (Barone et al., 1988) developed the Work Stress Inventory (WSI), a psychometric instrument similar to the PSS and TSS, to assess the frequency and intensity of stressors in the workplace. Designed with the intent of assessing stress associated with a wide range of circumstances for various occupations (Barone et al., 1988), the Work Stress Inventory consists of 40 items, each rated for both intensity and frequency. Components of occupational stress are assessed with two WSI 20-item subscales: Organizational Stress (OS), and Job Risk (JR). Females scored higher than males on the WSI Intensity indices, and lower on Frequency of Job Risk, which Barone et al. (1988) attributed to the over-representation of women in lower-risk jobs. No differences were found between men and women on the WSI frequency subscales.

The Job Stress Survey (JSS)

The *Job Stress Survey* (JSS) was designed to address the shortcomings that have been noted in existing measures of occupational stress (Spielberger, 1994; Spielberger & Reheiser, 1994; Turnage & Spielberger, 1991). The JSS was adapted from the earlier surveys that evaluated sources of stress specific to law enforcement officers (Spielberger et al., 1981) and high school teachers (Grier, 1982). This 30-item psychometric instrument was designed to assess the perceived intensity (severity) and frequency of occurrence of working conditions that are likely to adversely affect the psychological well-being of employees who are exposed to them (Spielberger, 1994). Items describing general sources of stress commonly experienced by managerial, professional, and clerical employees in a variety of occupational settings were selected to form a generic job stress measure.

The format for responding to the JSS Severity Scale is similar to the procedure employed by Holmes and Rahe (1967) in the *Social Readjustment Rating Scale* for rating stressful life events. Subjects first rate, on a 9-point scale, the relative amount (severity) of stress that they perceive to be associated with each of the 30 JSS job stressors (e.g., "Excessive paperwork," "Working overtime"), as compared to a standard stressor event, "Assignment of disagreeable duties," which was assigned a value of "5." In research with the PSS and TSS, this item was rated near the middle of the range in stress severity by both police officers and teachers.

In addition to rating the perceived severity of each stressor as compared to the standard, the JSS takes into account the state–trait distinction that has proved important in the assessment of anxiety (Spielberger, 1972, 1983) by requiring respondents to indicate how frequently each stressor event was encountered. Respondents are asked to report, on a scale from 0 to 9+ days, the number of days on which each workplace stressor was experienced during the preceding six months. Thus, ratings of the individual JSS items provide useful information regarding the perceived severity of each of the 30 stressor situations, and how often a particular person experiences each stressor event.

Summing the ratings for each individual JSS item yields overall Severity (JSS–S) and Frequency (JSS–F) scores based on all 30 items, and an overall Job Stress Index (JSS–X), which is based on the sum of the cross-products of the Severity and Frequency scores. Stress Severity and Frequency scores are also computed for the 10-item Job Pressure and Organizational Support subscales, which were derived in factor analyses of the 30 JSS items (Spielberger, 1994).

Turnage and Spielberger (1991) administered the JSS to white-collar employees of a large manufacturing firm to investigate differences in specific sources of job stress experienced by managerial, professional, and clerical personnel. They found that the professionals rated job pressures as more intense than the other groups, and had higher overall Job Stress Index scores than clerical employees. Managers reported experiencing a greater number of job pressures more frequently than professionals, who in turn reported more frequent job pressures than clerical personnel. Specific sources of stress, such as meeting deadlines, periods of inactivity, and frequent changes from boring to demanding activities, were reported significantly more often by females then by males. Spielberger and Reheiser (1994) also identified a number of significant gender differences relating to the perceived severity and frequency of occurrence of specific stressors.

Gender Differences in JSS Severity, Frequency and Index Scores

Gender differences in scores on the Job Stress Survey were evaluated for a sample of 1781 working adults (922 females, 859 males) employed in university and corporate work settings. These groups included 1387 faculty and staff from a large state university (579 males, 808 females); and 450 managerial, professional and clerical employees (280 males, 114 females) located at the corporate headquarters of two large industrial companies.

All participants responded anonymously to the JSS, which was distributed to them by company or university administrative personnel. They were encouraged to carefully read the instructions for both parts of the JSS, and were informed that their responses would be confidential and would contribute to the development of a university or company stress management program. In addition to responding to the JSS, they were asked to report their age, gender, and professional classification and the number of years they had been employed in their present work settings. The rate of return was approximately 60% for the corporate employees and 70% for the university personnel.

The means and standard deviations for the perceived Severity and Frequency scores of females and males are reported in Table 1 for each of the 30 JSS items, except for item 1, for which only the assigned score of 5.00 is reported. The items are listed in descending order of the mean Severity scores for the females. It is interesting to note that item 1, "Assignment of disagreeable duties," the standard on the basis of which the perceived severity of the other 29 items was rated, ranked 13th in severity for females and 16th for males. This finding provides additional evidence that the standard is about "average" in perceived severity as compared to the other JSS items.

Means and standard deviations of the scores for each of the Frequency items for males and females are also reported in Table 1. In addition, rankings of the mean scores for the Frequency items for both sexes, and the rank order of the Severity scores for the males, are reported in this Table. Gender differences in the JSS item means were evaluated in one-way analyses of variance (ANOVAs), and the resulting F tests and levels of significance for these analyses are reported in Table 1.

The most striking finding in the results of the ANOVAs for the item data is the number of items for which significant gender differences were found. Differences were found for almost half of the JSS Severity items (14 of 29), and for 60% of the Frequency items (18 of 30), as can be seen in Table 1. Moreover, the probability level for 19 of the 32 items for which significant gender differences were found was less than $p < .001$. It is interesting to note that both men and women rated the same 6 items as highest in perceived severity, and that both sexes rated the same 5 items as lowest in severity.

For both men and women, "Inadequate salary" and "Lack of opportunity for advancement" ranked highest in severity. The former also ranked relatively high in frequency of occurrence. These findings probably reflect the fact that more than three-fourths of the subjects were employed in a state university, where salaries for both faculty and staff are relatively low, with minimum raises during the several years before this study was conducted.

"Working overtime" was ranked very low in perceived severity by both men and women, whose mean scores for this item were essentially the same. The relative rank given to this stressor event in terms of its frequency of occurrence was much higher, especially for the males. Gender differences in Frequency scores for "Working Overtime" appear to be greater than for any other stressor event ($F = 70.28$, $p < .001$), males ranking this item 3rd in frequency of occurrence and females 15th.

TABLE 1   JSS Item Severity and Frequency Means and SDs, for Females and Males

JSS ITEMS		Severity				Frequency				
		Female	Male	Rank	F/Sig	Female	Rank	Male	Rank	F/Sig
19/49 Inadequate salary	M	6.68	6.16	01	25.99***	5.27	03	4.53	08	14.69***
	SD	2.14	2.18			4.02		4.07		
03/33 Lack of opportunity for advancement	M	6.22	6.01	02	4.02*	3.31	20	3.05	22	1.95
	SD	2.25	2.25			3.80		3.79		
05/35 Fellow workers not doing their job	M	5.74	5.63	06	1.13	4.16	07	4.21	10	0.09
	SD	2.29	2.12			3.55		3.53		
06/36 Inadequate support by supervisor	M	5.69	5.78	04	0.61	2.47	24	2.46	24	0.01
	SD	2.64	2.29			3.25		3.18		
15/45 Insufficient personnel to handle assignment	M	5.62	5.86	03	5.68*	4.07	10	4.70	07	13.40***
	SD	2.24	2.10			3.69		3.63		
08/38 Lack of recognition for good work	M	5.55	5.67	05	1.42	3.16	21	3.57	17	6.02**
	SD	2.25	2.18			3.47		3.52		
23/53 Frequent interruptions	M	5.44	5.08	13	11.43***	6.52	01	6.28	01	2.37
	SD	2.33	2.11			3.28		3.45		
07/37 Dealing with crisis situations	M	5.41	5.43	08	0.02	4.16	06	4.72	06	12.27***
	SD	2.15	2.04			3.29		3.43		
17/47 Personal insult from customer/colleague	M	5.32	5.28	10	0.19	2.18	25	1.83	28	7.12**
	SD	2.35	2.26			2.83		2.54		

(continued)

61

TABLE 1   JSS Item Severity and Frequency Means and SDs, for Females and Males (Continued)

JSS ITEMS		Severity				Frequency				
		Female	Male	Rank	F/Sig	Female	Rank	Male	Rank	F/Sig
29/59 Poorly motivated co-workers	M	5.21	5.14	12	0.38	3.88	13	3.69	14	1.28
	SD	2.35	2.22			3.59		3.48		
18/48 Lack participation in policy decisions	M	5.12	5.51	07	15.52***	3.39	18	3.89	12	9.19**
	SD	2.18	2.08			3.48		3.41		
13/43 Difficulty getting along with supervisor	M	5.07	5.21	11	1.14	1.41	30	1.43	30	0.03
	SD	2.89	2.63			2.52		2.51		
01/31 Assignment of disagreeable duties	M	5.00	5.00	16	0.01	2.80	23	3.01	23	2.21
	SD	—	—			2.96		2.90		
10/40 Inadequate or poor quality equipment	M	4.98	5.08	14	0.88	3.60	16	3.87	13	2.40
	SD	2.31	2.21			3.66		3.68		
20/50 Competition for advancement	M	4.97	4.59	20	13.23***	2.00	27	2.10	27	0.38
	SD	2.24	2.08			3.05		3.13		
14/44 Experience neg. attitude toward organization	M	4.95	5.34	09	13.10***	3.49	17	4.29	09	23.81***
	SD	2.36	2.21			3.47		3.47		
25/55 Excessive paperwork	M	4.91	5.07	15	2.44	5.02	04	4.97	04	0.08
	SD	2.27	2.15			3.58		3.65		
26/56 Meeting deadlines	M	4.85	4.95	17	0.95	5.49	02	5.87	02	5.29*
	SD	2.17	2.07			3.43		3.40		
16/46 Critical on-the-spot decisions	M	4.68	4.36	21	12.14***	3.79	14	4.78	05	39.52***
	SD	2.00	1.97			3.21		3.45		

Code	Item		M/SD 1	M/SD 2	Rank	F	M/SD 3	Rank	M/SD 4	Rank	F
22/52	Noisy work area	M	4.60	4.28	24	8.54**	3.98	12	3.52	19	6.63**
		SD	2.45	2.26			3.81		3.72		
21/51	Poor or inadequate supervision	M	4.54	4.82	19	5.99**	2.08	26	2.39	25	4.11*
		SD	2.48	2.38			3.13		3.23		
28/58	Covering work for another employee	M	4.48	4.12	25	12.12***	4.11	09	3.61	16	9.44**
		SD	2.28	2.13			3.48		3.38		
11/41	Assignment of increased responsibility	M	4.44	4.30	22	2.82	4.18	05	3.65	15	11.55***
		SD	2.06	1.94			3.32		3.22		
30/60	Conflict with other departments	M	4.44	4.83	18	13.16***	1.98	28	3.06	21	57.61***
		SD	2.31	2.18			2.81		3.21		
04/34	Assignment of new or unfamiliar duties	M	4.28	4.30	23	0.05	3.38	19	3.11	20	3.41
		SD	1.97	1.97			3.13		3.12		
09/39	Performing tasks not in job description	M	4.21	3.80	29	16.21***	4.12	08	4.08	11	0.06
		SD	2.20	2.09			3.62		3.72		
24/54	Frequent changes boring/demanding activities	M	4.04	3.86	26	3.44	3.99	11	3.54	18	7.01**
		SD	2.08	1.91			3.57		3.61		
27/57	Insufficient personal time	M	4.03	3.45	30	30.93***	3.04	22	2.38	26	16.05***
		SD	2.33	2.05			3.66		3.43		
12/42	Periods of inactivity	M	3.88	3.83	28	0.16	1.85	29	1.56	29	5.24*
		SD	2.30	2.19			2.70		2.59		
02/32	Working overtime	M	3.83	3.85	27	0.03	3.61	15	5.14	03	70.28***
		SD	2.19	2.09			3.81		3.93		

*p < .05; **p < .01; ***p < .001

Gender Differences in Job Pressures and Organizational Support

On the basis of the information provided by each respondent, employees were classified as either managerial, professional, clerical, or maintenance personnel. University administrators, deans, and department chairs were classified as managers; professors and mid-level administrators were assigned to the professional group. However, in order to have a sufficient number of subjects for more stable analyses of the data in terms of gender and occupational level, the participants assigned to the two higher levels (managers and professionals) and the two lowest levels (clerical and maintenance) were combined to form two occupational levels.

For the total sample, the number of men and women was relatively equal, but there were nearly twice as many men in the higher occupational level group, and more than twice as many women in the lower group. As was previously noted, such differences in the proportion of females to males in the managerial/professional and clerical groups are generally consistent with the prevailing gender representation in the occupational levels that were surveyed in this study.

Means, standard deviations and alpha coefficients for the JSS Severity, Frequency and Index scales are reported in Table 2 for the total sample. The alpha coefficients for all three JSS scales were remarkably high, especially for the relatively brief 10-item JSS Severity and Frequency scales. The mean JSS Severity and Index scores for men and women were quite similar for the total sample, and for the subsamples of managerial/professional and clerical/maintenance personnel, as can be noted in Table 2. Differences in these means were evaluated in 2 by 2 ANOVAs, for which the results are also reported in Table 2.

The results of the ANOVAs indicated no main effects for gender, nor any gender-by-occupational-level interactions for the total sample. Moreover, no significant main effects or interactions were found for the JSS Severity and Index scores for the two occupational groups. However, the main effect for occupational level was highly significant ($p < .001$), indicating that both female and male managerial/professional employees reported experiencing the occurrence of the 30 JSS stressor events much more frequently than the clerical/maintenance workers. Thus, in marked contrast to the numerous differences that were found in both the perceived severity and the frequency of occurrence of specific JSS stressor events, no differences were found in the JSS scale scores as a function of gender, nor any interactions of gender with occupational level.

The finding of numerous differences in the perceived severity and frequency of occurrence of specific JSS stressors, and the failure to find any gender-related differences in JSS Severity, Frequency and Stress Index scores, at first seems paradoxical. However, these results are actually quite consistent with, and help to clarify, the results reported by other investigators. Gender-related differences in occupational stress appear to be determined by differences in the perceived severity of *specific* stressors, and in the frequency that these stressors are experienced by men and women.

TABLE 2 Means, Standard Deviations, and Alpha Coefficients for JSS Stress Severity, Frequency and Index Scales for Managerial/Professional, Clerical/Maintenance Personnel and for the Combined Sample for Females and Males

JSS Stress Scales	Total Sample		Managerial/ Professional		Clerical/ Maintenance		F-test/Significance		
	F	M	F	M	F	M	Gender	Occup. Level	Inter-action
Severity									
Mean	148.06	146.69	148.92	147.62	147.57	143.86	0.67	1.71	0.43
SD	37.63	32.41	33.81	30.01	39.66	38.74			
N	922	859	333	646	589	213			
Alpha	.92	.90	.91	.89	.93	.93			
Frequency									
Mean	107.25	109.39	113.28	112.09	103.79	101.31	0.75	13.98***	0.06
SD	52.20	50.82	49.70	48.43	53.32	56.74			
N	898	838	327	628	571	210			
Alpha	.90	.90	.89	.89	.91	.92			
Index									
Mean	61.31	60.81	62.70	61.05	60.52	60.15	0.09	0.82	0.12
SD	34.93	32.32	31.26	29.80	36.84	39.00			
N	874	825	315	619	559	206			
Alpha	.90	.89	.88	.87	.91	.92			

***$p < .001$

Of the 14 JSS stressor events reported in Table 1 for which significant differences in perceived severity were found, women rated 9 of these stressors as significantly more severe than men, whereas men perceived 5 of them as more severe. Of the 18 JSS stressor events for which significant differences in frequency of occurrence were found, men reported experiencing 10 of these stressors more frequently than women, and women had higher Frequency scores for 8 of the JSS stressors. When scores on the specific JSS stressor events for which significant differences were found are summed to form Severity or Frequency scales, these differences tend to cancel out. If only the scale scores are considered, this would result in the erroneous conclusion that gender does not affect the perceived severity of workplace stressors and that the frequency of occurrence of these stressors was unrelated to gender.

SUMMARY AND CONCLUSIONS

Evidence of the adverse effects of stress in the workplace on employee productivity, absenteeism, worker turnover, and stress-related medical problems was examined. Direct and indirect costs to both workers and employers were noted. Person–Environment Fit theory and other current conceptions of occupational stress were briefly reviewed, and a number of measures that have been used to assess stress in the workplace were critically evaluated.

It was noted that ambiguity and inconsistencies in prevailing theories of stress in the workplace are reflected in the wide range of heterogeneous content that is assessed by measures of occupational stress. Confusion in occupational stress research has also resulted from the fact that some investigators have focused on job pressures, whereas others have been concerned primarily with the consequences of work-related stress. The importance of developing good diagnostic tools for assessing the specific job pressures and organizational factors that contribute to stress in the workplace is now recognized as a major priority by occupational stress researchers.

Inconsistencies in research on gender differences in occupational stress were also noted and attributed to sampling problems and limitations in the instruments that have been used to measure stress in the workplace. The Job Stress Survey assesses the perceived severity of specific stressor events and how frequently they occur. Research with this measure in studies of gender differences in the workplace was described. Numerous gender differences in the severity and frequency of occurrence of specific JSS stressors were found, along with consistencies in perceived severity and frequency for specific stressors. It was concluded that gender is extremely important in determining how different workplace stressors are perceived, and that men and women experience different stressors more or less often, depending to some extent on their occupational level.

REFERENCES

Algera, J.A. (1983). "Objective" and perceived task characteristics as a determinant of reactions by task performers. *Journal of Occupational Psychology, 56*, 95–107.

Barone, D.F. (1994). Work stress conceived and researched transactionally. In P.L. Perrewé & R. Crandall (Eds.), *Occupational stress: A handbook* (pp. 29–37). Washington, DC: Taylor & Francis. (Original work published 1991)

Barone, D.F., Caddy, G.R., Katell, A.D., Roselione, F.B., & Hamilton, R.A. (1988). The Work Stress Inventory: Organizational stress and job risk. *Educational and Psychological Measurement, 48*, 141–154.

Beehr, T.A., & Newman, J.E. (1978). Job stress, employee health, and organizational effectiveness: A facet analysis, model, and literature review. *Personnel Psychology, 31*, 665–699.

Beehr, T.A., & Schuler, R. (1980). Stress in organizations. In K. Rowland & G. Ferris (Eds.), *Personnel management* (pp. 390–419). Boston: Allyn and Bacon.

Brantly, P.J., & Jones, G.N. (1989). *Daily Stress Inventory professional manual.* Odessa, FL: Psychological Assessment Resources.

Brief, A.P., & George, J.M. (1994). Psychological stress and the workplace: A brief comment on Lazarus' outlook. In P.L. Perrewé & R. Crandall (Eds.), *Occupational stress: A handbook* (pp. 15–19). Washington, DC: Taylor & Francis. (Original work published 1991)

Caplan, R.D., Cobb, S., French, J.R.P., Jr., Van Harrison, R., & Pinneau, S.R., Jr. (1975). *Job demands and worker health: Main effects and occupational differences* (HEW NIOSH No. 75–160). Washington, DC: U.S. Government Printing Office.

Chemers, M.M., Hays, R.B., Rhodewalt, F., & Wysocki, J. (1985). A person–environment analysis of job stress: A contingency model explanation. *Journal of Personality and Social Psychology, 49*, 628–635.

Cooper, C.L., & Payne R. (Eds.). (1988). *Causes, coping and consequences of stress at work.* Chichester, England: Wiley.

Di Salvo, V., Lubbers, C., Rossi, A.M., & Lewis, J. (1994). The impact of gender on work-related stress. In P.L. Perrewé & R. Crandall (Eds.), *Occupational stress: A handbook* (pp. 39–50). Washington, DC: Taylor & Francis. (Original work published 1988)

Fisher, C.D., & Gitelson, R. (1983). A meta-analysis of the correlates of role conflict and ambiguity. *Journal of Applied Psychology, 68*, 320–333.

French, J.R.P., Jr., & Caplan, R.D. (1972). Organizational stress and individual strain. In A.J. Marrow (Ed.), *The failure of success* (pp. 30–66). New York: Amacom.

French, J.R.P., Jr., Caplan, R.D., & Van Harrison, R. (1982). *The mechanisms of job stress and strain.* Chichester, England: Wiley.

Ganster, D.C., Fusilier, M.R., & Mayes, B.T. (1986). Role of social support in the experience of stress at work. *Journal of Applied Psychology, 71*, 102–110.

Gerhart, B. (1986, August). *Sources of variance in incumbent perceptions of job complexity.* Paper presented at the annual meeting of the Academy of Management, Chicago.

Goldsmith, E.B. (Ed.). (1988). Work and family: Theory, research and applications [Special issue]. *Journal of Social Behavior and Personality, 3*(4).

Grier, K.S. (1982). A comparison of job stress in law enforcement and teaching (Doctoral dissertation, University of South Florida, 1981). *Dissertation Abstracts International, 43*, 870B.

Hamner, W.C., & Tosi, H.L. (1974). Relationship of role conflict and role ambiguity to job involvement measures. *Journal of Applied Psychology, 59*, 497–499.

Harris, J.R. (1994). An examination of the transaction approach in occupational stress research. In P.L. Perrewé & R. Crandall (Eds.), *Occupational stress: A handbook* (pp. 21–28). Washington, DC: Taylor & Francis. (Original work published 1991)

Harrison, R.V. (1978). Person–environment fit and job stress. In C.L. Cooper & R. Payne (Eds.), *Stress at work* (pp. 175–205). New York: Wiley.

Holmes, T.H., & Rahe, R.H. (1967). The Social Readjustment Rating Scale: A cross-cultural study of Western Europeans and Americans. *Journal of Psychosomatic Research, 14*, 391–400.

Jackson, S.E., & Schuler, R.S. (1985). A meta-analysis and conceptual critique of research on role ambiguity and role conflict in work settings. *Organizational Behavior and Human Decision Processes, 36*, 16–78.

Jick, T.D., & Mitz, L.E. (1985). Sex differences in work stress. *Academy of Management Review, 10*, 408–420.

Kahn, R.L., Wolfe, D.M., Quinn, R.P., Snoek, J.D., & Rosenthal, R.A. (1964). *Organizational stress: Studies in role conflict and ambiguity.* New York: Wiley.

Karasek, R.A., Jr. (1979). Job demands, job decision latitude and mental strain: Implications for job redesign. *Administrative Science Quarterly, 24,* 285–308.

Karasek, R.A., Jr., & Theorell, T. (1990). *Healthy work: Stress, productivity, and the reconstruction of working life.* New York: Basic Books.

Kasl, S.V. (1978). Epidemiological contributions to the study of work stress. In C.L. Cooper & R.L. Payne (Eds.), *Stress at work* (pp. 3–38). New York: Wiley.

Keita, G.P., & Sauter, S.L. (Eds.). (1992). *Work and well-being: An agenda for the 1990s.* Washington, DC: American Psychological Association.

Kroes, W.H. (1976). *Society's victim—the policeman: An analysis of job stress in policing.* Springfield: Charles C. Thomas.

Kroes, W.H., & Gould, S. (1979). Job stress in policemen: An empirical study. *Police Stress, 1,* 9–10, 44.

Kroes, W.H., Margolis, B., & Hurrell, J.J., Jr. (1974). Job stress in policemen. *Journal of Police Science and Administration, 2,* 145–155.

Lazarus, R.S. (1994). Psychological stress in the workplace. In P.L. Perrewé & R. Crandall (Eds.), *Occupational stress: A handbook* (pp. 3–14). Washington, DC: Taylor & Francis. (Original work published 1991)

Levi, L. (1981). Preventing work stress. Reading, MA: Addison-Wesley.

Locke, E.A. (1969). What is job satisfaction? *Organizational Behavior and Human Performance, 4,* 309–336.

Lowman, R.L. (1993). *Counseling and psychotherapy of work dysfunctions.* Washington, DC: American Psychological Association.

Martocchio, J.J., & O'Leary, A.M. (1989). Sex differences in occupational stress: A meta-analytic review. *Journal of Applied Psychology, 74,* 495–501.

Matteson, M.T., & Ivancevich, J.M. (1982). *Managing job stress and health: The intelligent person's guide.* New York: Free Press.

Matteson, M.T., & Ivancevich, J.M. (Eds.). (1987). *Controlling work stress: Effective human resource and management strategies.* San Francisco: Jossey-Bass.

Murphy, L.R., & Hurell, J.J. (1987). Stress measurement and management in organizations: Development and current status. In A.W. Riley & S.J. Zaccaro (Eds.), *Occupational stress and organizational effectiveness* (pp. 29–51). New York: Praeger.

Nelson, D.L., & Quick, J.C. (1985). Professional women: Are distress and disease inevitable? *Academy of Management Review, 10,* 206–213.

Northwestern National Life. (1991). *Employee burnout: America's newest epidemic.* Minneapolis, MN: Northwestern National Life Insurance Company.

Northwestern National Life. (1992). *Employee burnout: Causes and cures.* Minneapolis, MN: Northwestern National Life Insurance Company.

Osipow, S.J., & Spokane, A.R. (1981). *Occupational Stress Inventory manual research version.* Odessa, FL: Psychological Assessment Resources.

Perrewé, P.L. (Ed.). (1991). Handbook on job stress [Special issue]. *Journal of Social Behavior and Personality, 6*(7).

Porter, L.W. (1961). A study of perceived need satisfactions in bottom and middle management jobs. *Journal of Applied Psychology, 45,* 1–10.

Quick, J.C., Murphy, L.R., & Hurrell, J.J., Jr. (Eds.). (1992). *Stress & well-being at work: Assessments and interventions for occupational mental health.* Washington, DC: American Psychological Association.

Sauter, S.L. (1992). Introduction to the NIOSH Proposed National Strategy. In G.P. Keita & S.L. Sauter (Eds.), *Work and well-being: An agenda for the 1990s* (pp. 11–16). Washington, DC: American Psychological Association.

Sauter, S.L., & Hurrell, J.J., Jr. (1989). Introduction. In S.L. Sauter, J.J. Hurrell, Jr., & C.L. Cooper (Eds.), *Job control and worker health* (pp. XIII–XX). Chichester, England: Wiley.

Schuler, R.S. (1980). Definition and conceptualization of stress in organizations. *Organizational Behavior and Human Performance, 25,* 184–215.

Schuler, R.S. (1991). Foreword. In P.L. Perrewé (Ed.), Handbook on job stress [Special issue]. *Journal of Social Behavior and Personality, 6*(7), v–vi.

Sharit, J., & Salvendy, G. (1982). Occupational stress: Review and reappraisal. *Human Factors, 24,* 129–162.

Spector, P.E. (1987). Interactive effects of perceived control and job stressors on affective reactions and health outcomes for clerical workers. *Work and Stress, 1,* 155–162.

Spector, P.E., Dwyer, D.J., & Jex, S.M. (1988). Relation of job stressors to affective, health, and performance outcomes: A comparison of multiple data sources. *Journal of Applied Psychology, 73,* 11–19.

Spielberger, C.D. (1972). Anxiety as an emotional state. In C.D. Spielberger (Ed.), *Anxiety: Current trends in theory and research* (Vol. 1, pp. 23–49). New York: Academic Press.

Spielberger, C.D. (1983). *Manual for the State–Trait Anxiety Inventory (Form Y).* Palo Alto, CA: Consulting Psychologists Press.

Spielberger, C.D. (1994). *Professional Manual for the Job Stress Survey (JSS).* Odessa, FL: Psychological Assessment Resources.

Spielberger, C.D., Grier, K.S., & Greenfield, G. (1982, Spring). Major dimensions of stress in law enforcement. *Florida Fraternal Order of Police Journal,* pp. 10–12.

Spielberger, C.D., Grier, K.S., & Pate, J.M. (1979, Aug./Sept.). Sources of stress in police work. *Florida Fraternal Order of Police Journal.*

Spielberger, C.D., Grier, K.S., & Pate, J.M. (1980, Winter). The Police Stress Survey. *Florida Fraternal Order of Police Journal,* pp. 66–67.

Spielberger, C.D., & Reheiser, E.C. (1994). Job stress in university, corporate, and military personnel. *International Journal of Stress Management, 1,* 19–31.

Spielberger, C.D., Westberry, L.G., Grier, K.S., & Greenfield, G. (1981). *The Police Stress Survey: Sources of stress in law enforcement* (Human Resources Institute Monograph Series Three, No. 6). Tampa, FL: University of South Florida, College of Social and Behavioral Sciences.

Turnage, J.J., & Spielberger, C.D. (1991). Job stress in managers, professionals, and clerical workers. *Work & Stress, 5,* 165–176.

PART TWO

Sources and Consequences of
Occupational Stress: Model Testing

Antecedents and Organizational Effectiveness Outcomes of Employee Stress and Health

William H. Hendrix
Timothy P. Summers
Terry L. Leap
Robert P. Steel

Review of the literature reveals an exponential increase in interest of worksite health promotion (Gebhardt & Crump, 1990) and related interventions such as stress management (Ivancevich, Matteson, Freedman, & Phillips, 1990) and physical fitness (Gebhardt & Crump, 1990). Stress and stress-related factors are central to most of these health promotion models and a crucial aspect of intervention efforts to improve employee health.

Organizations have become increasingly interested in health promotion and stress management programs as medical costs for their employees have risen (Edwards & Gettman, 1980). On a national level, health and stress-related problems are a major cost for organizations, both in financial terms and through the loss of valued employees. For example, approximately ten years ago stress was estimated to cost approximately $75 to $90 billion annually nationally (Ivancevich & Matteson, 1980). More recently, Niehouse (1987) reported that the annual cost to U.S. businesses from stress and burnout is estimated to be $100 billion, excluding the cost of poor job performance and the cost to replace employees who die, who are ill, or who quit. In this country, about 75 million individuals have back pain, another stress related problem, that has been estimated to cost $1 billion in lost output and over $250 million in worker's compensation claims (Myers, 1986, pp.

666–667). Employers during 1983 paid $77 billion for the health benefits for 68 million employees for an average of about $1,100 per employee (Myers, 1986, p. 667). Evidence is beginning to emerge that indicates that health promotion programs are worthwhile ventures at some companies. For example, a recent case study of the DuPont Company indicated that, for each dollar invested in workplace health education, the yield is $1.42 over two years from reduced absences from illness (Bertera, 1990).

Although there is a great deal of health and stress research being performed, as well as a large number of health promotion-related intervention programs, few are being performed within a comprehensive health promotion model that has stress as a central focus. In part, the lack of sound research data has resulted from the absence of a multidisciplinary approach within a comprehensive framework that can guide the research. Instead, there have been two major approaches to stress research: the behavioral approach and the medical approach. To make substantial progress we need to incorporate these into a comprehensive systems model.

A number of theoretical health-related models have been proposed that have implications for a comprehensive health promotion model with stress as a central focus (Bhagat, 1983; Brief, Schuler, & Van Sell, 1981; Cooper & Marshall, 1976; Hendrix, 1985; Ivancevich, & Matteson, 1980). The major components suggested by these models have included: stress and health-related determinants (i.e., intra-organizational, extra-organizational and individual characteristics), job and life stress, psychological and physical outcomes, and organizational outcome factors. Therefore, these components should also be included when developing and testing a more comprehensive stress-based health promotion model.

The purpose of this research is to develop, through exploratory path analysis, a structural model of health promotion that has stress as a central focus. The model is to serve as a guide for future research in health promotion and stress management interventions. This model will incorporate both the medical and behavioral approaches to health promotion. In developing the model, this research will build on current health promotion and stress related research and will extend the work on a model proposed earlier by Hendrix (1985) and research that partially tested it (Hendrix, Ovalle, & Troxler, 1985).

Programs dealing with health promotion and wellness should have a clear vision of their purpose. It would appear reasonable that health promotion programs would be designed to reduce illness and premature death, and increase the general well-being of employees. A health promotion model, therefore, should include variables related to these desired goals. Figure 1 depicts the hypothesized path analytic model guiding this research which incorporates the variables noted above. This model was developed based on recent research literature and research models (e.g., Beehr & Newman, 1978; Gebhardt & Crump, 1990; Ivancevich & Matteson, 1980, p. 44; Ivancevich, Matteson, Freedman, & Phillips, 1990; Matteson & Ivancevich, 1979, p. 350; Zedeck & Mosier, 1990). The direction of an arrow indicates the hypothesized path, whereas the plus or minus sign indicates the direction of the hypothesized relationship between two variables. The absence of an arrow indicates a hypothesis of no relationship between two variables. The curved, double-headed arrows normally provided between exogenous variables have been omitted because including them would result in an extremely complicated path diagram.

Stress has been defined as a stimulus variable and as a response variable (Ivancevich & Matteson, 1980, pp. 6–8). When defined as a stimulus, stress is considered to be a force acting on a person that causes discomfort or strain. The response definition, on the other hand, considers stress to be a physiological or psychological response or outcome to a stressor. Adoption of one of these definitions influences how one operationalizes stress. If a stimulus definition is chosen, then variables such as quantitative workload and role conflict would be used to measure stress. On the other hand, if the response definition is used, psychological and physiological response outcomes such as anxiety, anger, headaches, blood pressure, and cortisol might be used as measures of stress. A third choice, the one adopted for this research, views stress as one's perception of being stressed (i.e., felt stress), not simple exposure to what others have labeled as a stressor (e.g., quantitative workload) or the effects of experiencing stress as indicated by a stress outcome (e.g., anxiety). Instead, perceptual or felt stress is viewed as an intervening variable located between stressors and strain outcomes. This concept of felt or perceived stress is consistent with that proposed by Summers, DeNisi, and DeCotiis (1994), who defined stress as the manifestly uncomfortable feeling that an individual experiences when he or she is forced to deviate from normal or desired patterns of functioning. Specifically, the stress definition used for this research is defined as an *uncomfortable cognitive state resulting from exposure to a stressor that can result in psychological and physiological strain.* This definition implies that one's perceived stress can vary in intensity. It also implies that exposure to a potential stressor does not necessary result in the perception or feeling of stress.

Figure 1 contains variables that can be categorized as *exogenous* or *endogenous.* Exogenous variables are those which are not influenced by other variables in the model and are depicted to the far left or preceding the model's endogenous variables in a path diagram. At first glance, alcohol consumption appears to be an endogenous variable, perhaps influenced by worker characteristics, e.g., demographic variables, or intraorganizational variables. However, based on their recent review of literature on alcohol and drug use in the workplace, Harris and Heft (1992) concluded that the relationship between intraorganizational variables and alcohol consumption is quite small. Moreover, their review showed weak or nonexistent relationships between alcohol use and life stress, and between depression (included among somatic symptoms) and alcohol consumption.

Endogenous variables are those that serve as dependent variables or criterion variables and are considered determined by some combination of the variables in the system. These are located to the right of the exogenous variables in a path diagram. The endogenous variables at the right in Figure 1 are the path model's

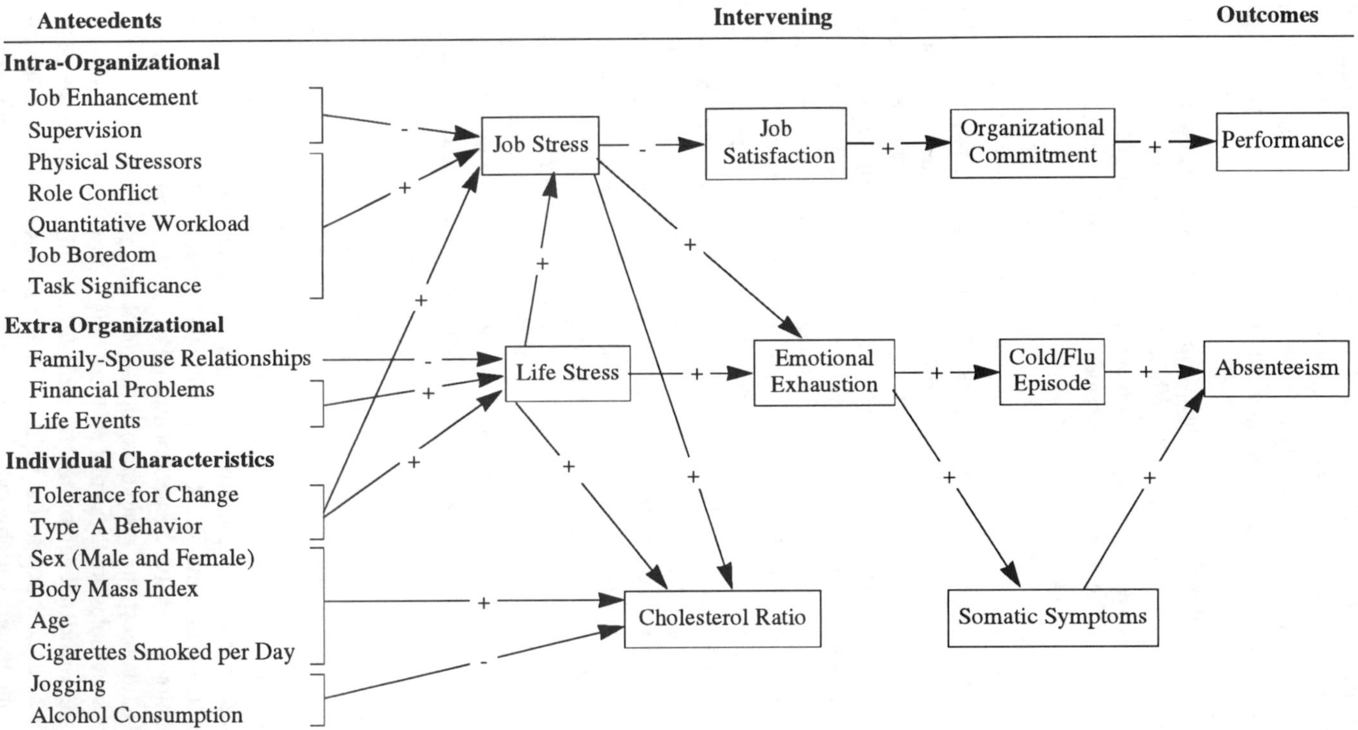
Antecedents
Intervening
Outcomes
Intra-Organizational
Job Enhancement
Supervision
Physical Stressors
Role Conflict
Quantitative Workload
Job Boredom
Task Significance
Extra Organizational
Family-Spouse Relationships
Financial Problems
Life Events
Individual Characteristics
Tolerance for Change
Type A Behavior
Sex (Male and Female)
Body Mass Index
Age
Cigarettes Smoked per Day
Jogging
Alcohol Consumption
Job Stress
Job Satisfaction
Organizational Commitment
Performance
Life Stress
Emotional Exhaustion
Cold/Flu Episode
Absenteeism
Cholesterol Ratio
Somatic Symptoms

FIGURE 1 Hypothetical Model

outcome variables. Three outcome variables are included in this research model: the cholesterol ratio, performance, and absenteeism. The cholesterol ratio in this research serves as a proxy for coronary heart disease (CHD). The remaining two outcome variables (performance and absenteeism) included in the hypothesized model (Figure 1) have been suggested as relevant outcome variables by previous stress and health promotion research (Israel, House, Schurman, Heaney, & Mero, 1989; Motowidlo, Packard, & Manning, 1986).

Exogenous Variables

The exogenous variables included in Figure 1 can be divided into three categories: intra-organizational, extra-organizational and individual characteristics. *Intra-organizational* variables are depicted in Figure 1 as having direct effects on job stress. Prior research suggests a major role for intra-organizational or job-related variables as contributors to job stress. Some of the intra-organizational variables linked to distress include role conflict, role ambiguity, role overload and time pressure, low job autonomy, low utilization of abilities, low participation and low control (Burke & Richardson, 1990; Caplan, Cobb, French, Van Harrison, & Pinneau, 1975; Hendrix, 1985), management-supervision, organizational climate, and group conflict (Hendrix, Ovalle, Troxler, 1985; Matteson & Ivancevich, 1979; Ivancevich & Matteson, 1980). More recently, Fox, Dwyer, and Ganster (1993) found that work load and job demands were significantly associated with blood pressure and cortisol levels.

The *extra-organizational* variables depicted in Figure 1 are depicted as directly affecting life stress and have been implicated as major contributors to perceived stress and especially to life stress. Extra-organizational variables are those variables external to the work environment other than individual or personal factors. They include economic factors, family-spouse relations, political uncertainty, life crisis, and lack of social support (Hendrix et al., 1985; Ivancevich & Matteson, 1980; Marshall & Cooper, 1979; Matteson & Ivancevich, 1979).

In Figure 1, the variables listed under the *individual characteristics* category are presented as directly affecting either job stress, life stress, or the cholesterol ratio. The cholesterol ratio, one of three outcome variables in this research, serves as a surrogate measure of coronary heart disease. Coronary heart disease (CHD) is the number one killer of individuals in the United States (Matarazzo, 1984). It has been estimated that about 30 million people in the United States have cardiovascular disease, and nearly one million die from it annually (DeBakey, Gotto, Scott, & Foreyt, 1984). Therefore, a major component of a comprehensive stress and health promotion model should be coronary heart disease. A number of variables have been related to CHD . These include stress (House, 1974; Houston, Smith, & Cates, 1989; Schnall, Devereaux, Pickering, & Schwartz, 1992), Type A behavior (Appels, Mulder, van't Hof, Jenkins, van Houtem, & Tan, 1987; Brief, Schuler, & Van Sell, 1981), age (Hendrix, 1985; Nanas, Pan, & Stamler, 1987), alcohol consumption (Hendrix, 1985, Miller, Bolton, & Hayes, 1988), Body Mass Index (BMI) (Hendrix,

1985), smoking (Lauer, Lee, & Clarke, 1988), gender (Barnett, Biener, & Baruch, 1987; Hendrix, 1985), and lack of exercise (Fripp & Hodgson, 1987; Gebhardt & Crump, 1990). For a more detailed review of the effects of Type A behavior see Strube (1990). The hypothesized model (Figure 1) incorporates these variables.

Measuring CHD has been accomplished by a variety of techniques including coronary angiography and the cholesterol ratio. Coronary angiography is the most accurate and expensive (Kaltenbach, Lichten, & Friesinger, 1973) and serves as a standard for other evaluation methods. Hendrix, Ovalle, and Troxler (1985) evaluated a series of factors that were predictive of CHD. These included family history, age, total cholesterol, HDL cholesterol, smoking level, blood pressure, and the cholesterol ratio. Their results indicated that the cholesterol ratio was the most predictive of all the variables evaluated. Other research studies have also found the cholesterol ratio to be one of the best single laboratory predictors of coronary artery disease (CAD) potential (Malaspina, Bussiere, & LeCalve, 1981; Uhl, Troxler, Hickman, & Clark, 1981). Therefore, the cholesterol ratio was included in the hypothesized model as a measure of CHD potential.

Numerous research studies have found that acute stress and life-style factors can result in increased cholesterol levels (e.g., Bryant, Story, & Yim, 1988; Friedman, Rosenman, & Carroll, 1958; Hartman, 1983; Quick & Quick, 1984, p. 104; Rahe, Rubin, Arthur, 1974). In fact, most medical textbooks attribute from 50% to 70% of illnesses to stress-related sources (Ivancevich & Matteson, 1980). Few studies, however, have looked at the relationship of stress and life-style to the cholesterol ratio (for one exception, see Steffy & Jones, 1988). Because research results, as noted before, indicate that the cholesterol ratio is a more sensitive predictor of future heart disease than total cholesterol alone (Hendrix et al., 1985), this research will focus on the effects of job stress, life stress, and life-style on the cholesterol ratio instead of total cholesterol.

In addition to affecting the cholesterol ratio, individual characteristics have been found to play a primary role in individuals' stress reactions. These have included tolerance for change, sex, weight, diet, age, exercise, smoking level (Epstein, 1965; Hendrix et al, 1985; Jenkins, 1976; Johnson, 1978; Matteson & Ivancevich, 1979; Christensen & Jensen, 1994), and Type A behavior (Hendrix, et al., 1985; Jenkins, 1976; Strube, 1990). However, Izraeli (1993) did not support hypotheses of sex differences in work/family conflict for similarly situated men and women.

Endogenous Variables

The endogenous variables can be further divided into intervening and out-come variables. The *intervening variables* in Figure 1 include three variables that are hypothesized to lead to or affect the *outcome variable of performance:* Job stress, job satisfaction, and commitment. Inclusion of these variables is based on research that has indicated that job and life stress (Friend, 1982; Motowidlo, Packard, & Manning, 1986; Brief, Schuler, & Van Sell, 1981), job satisfaction

(Hendrix, Steel, & Schultz, 1987), and organizational commitment (DeCotiis & Summers, 1987) affect job performance. A second major path included in Figure 1 incorporates a series of intervening variables which affect *absenteeism.* Specifically, absenteeism has been predicted in the literature by stress, cold/flu episode rates, somatic symptoms, and emotional exhaustion (Brief, Schuler, & Van Sell, 1981, p. 56; Hendrix, et al., 1987; Jackson, Schwab, Schuler, 1986; Quick, & Quick, 1984). The paths from job and life stress through emotional exhaustion to somatic symptoms were suggested by Bromet, Dew, Parkinson, Cohen, and Schwartz (1992) who studied 552 female blue collar employees. They found that a surrogate measure of job stress, job-related conflict, was associated with depressive symptomatology, severe headaches, light-headedness, fatigue, rashes, and presence of multiple symptoms. Similarly, Verbrugge (1993) found that stress poses health risks for young women. Emotional exhaustion, identified as a component of burnout, was linked to anxiety in a study by Corrigan et al. (1994).

This body of literature forms the basis for the hypothesized model depicted in Figure 1 and serves as the basis for this research.

METHOD

Subjects

Subjects consisted of 463 civilian employees working for the Department of Defense (DOD) at DOD facilities in three states: Colorado, Florida, and Texas. Approximately 55% were female, 3% black, 87% white, 5% Hispanic, and 5% were listed as Indian, Asian, or did not provide their race. The ages of the participants ranged from 18 to 70 years with an approximate mean of 44 years. Approximately 75% of the participants were classified as non-supervisors, while 24% were classified as supervisors. In addition, the percentage breakout for marital status was: 9% single, 12% divorced, 2% widowed/widowers, and 77% married. This demographic information is consistent with separate demographic information collected by the organization reflecting the total population at each data collection site. The population demographics were: age ranged from 18 to 74 years, with a mean of 43.2 years. Approximately 49% were females, 79% white, 8% Hispanic, 9% black, and 4% were listed as being of another race. Participation in the research program was voluntary.

Measures

The measures used in this research can be grouped as exogenous (or antecedent) variables and endogenous (or intervening and outcome) variables. The exogenous variables can be further divided into three categories labeled as intra-organizational, extra-organizational, and individual characteristics. The endogenous variables consist of intervening and outcome variables. Figure 1 provides these variables within the path analytic framework used for this research.

TABLE 1   Means, Standard Deviations, and Coefficient Alpha Reliabilities

Variable	Mean	SD	Overall Alpha
Job Enhancement	4.88	1.30	.86
Supervision	4.67	1.42	.94
Physical Stressors	3.70	1.38	.69
Role Conflict	3.14	1.27	.82
Quantitative Workload	5.56	1.26	.91
Boredom	3.01	1.47	.89
Task Significance	5.46	1.37	.91
Family-Spouse Relationships	5.17	1.34	.93
Financial Problems	2.97	1.61	NA
Life Events	191.73	120.74	NC
Tolerance for Change	4.05	1.16	.79
Type A Behavior	5.02	1.29	.73
Body Mass Index	.035	.005	NA
Age	49.93	9.65	NA
Cigarettes Smoked	1.71	1.53	NA
Jogging	1.71	1.53	NA
Alcohol Consumption	9.66	8.18	NA
Cholesterol Ratio	4.49	1.84	NA
Job Stress	4.10	1.37	.71
Life Stress	3.34	1.41	.86
Job Satisfaction	5.41	1.53	.90
Emotional Exhaustion	2.36	1.06	.88
Somatic Symptoms	2.62	1.20	.89
Cold/Flu Episodes	2.52	1.36	NA
Absenteeism	3.16	1.68	NA
Organizational Commitment	5.56	1.39	.72
Performance	3.43	.61	NA

Note: NC = Not Computed; NA = Single Item.

The measures used in this research except for the three major outcome variables of performance, absenteeism, and the cholesterol ratio were collected using a health-assessment survey. The scales in the health promotion survey, in the main, consisted of 7-point attitudinal scales. The multiple- and single-item measures used in this study are provided in Table 1 along with their means, standard deviations, and coefficient alpha reliabilities.

Outcome Variables. Performance was operationalized as an individual's last merit performance appraisal and was obtained from the employee's records. This measure was on a five-point scale (1 = low, 5 = high). *Absenteeism* was operationalized as the number of days of sick leave taken over a six-month period.

Operationalizing absenteeism as sick leave was selected since sick leave appeared to be a relevant operationalization for a health promotion model. The single physiological measure, the *Cholesterol Ratio,* was obtained from blood samples taken from each participant. Specifically, the cholesterol ratio was computed by dividing total serum cholesterol by high density lipoprotein (HDL) cholesterol. High levels of total serum cholesterol have been associated with increased risk of CAD while high levels of HDL cholesterol have been associated with decreased risk of CAD. As noted earlier, the cholesterol ratio has been found to be one of the best single laboratory predictors of coronary artery disease (CAD) potential (Malaspina, Bussiere, & LeCalve, 1981; Uhl, Troxler, Hickman, & Clark, 1981).

Antecedent and Intervening Variables. The measures of job enhancement, supervision, tolerance for change, family-spouse relations, job stress, life stress, job satisfaction, and Type A behavior were taken from the research reported by Hendrix (1985), and task significance from scale items reported by Hendrix (1984). The *Job Enhancement* scale consisted of 3 items which measured the extent that a person's work allowed him or her to use his/her talents and training to accomplish a worthwhile job. *Supervision* was measured by a 10-item scale which assessed the extent that one's supervisor was perceived as a good planner, represented the group consistently, established good work procedures, made responsibilities clear to the group, performed well under pressure, helped employees to perform well, and provided feedback on individual performance. *Tolerance for Change* was measured by a six-item scale that assessed the extent that individuals were resistant to change, were uncomfortable with change, or were made angry by change. The scale was reversed scored so that a high score on this scale indicated a high tolerance for change. Type A behavior was measured by a three-item scale that assessed the Speed and Impatience construct proposed by Friedman and Rosenman (1959), with a high score indicating Type A behavior. *Sex* or gender was coded as 1 if male and 0 if female. *Age* was measured simply as the person's chronological age as of his or her last birthday. The *Family-Spouse Relations* scale consisted of four items which assessed the extent that things were going well with one's wife/husband, the extent one was satisfied with his/her family life, had a good time together, had a good relationship, and felt his/her relationship with his/her spouse was good. The *Job Stress* scale consisted of three items which assessed the perception of stress felt by an individual. The scale reflected the extent to which individuals felt their jobs overall to be stressful and the degree that the jobs produced stress by thwarting personal growth. For example, one of the items included was, "All in all, I feel I have a great deal of stress on the job." The *Life Stress* scale consisted of three items which assessed the perception of felt stress outside of the job environment. One of the items in the scale is, "Overall, how stressful is your life when you are not at work?" The scale *Life Events* was measured by The Social Readjustment Scale (Holmes & Rahe, 1967). *Job Satisfaction* (Hendrix, 1984) was measured with a 2-item scale that assessed the extent that individuals were satisfied with the work itself and the job as a whole.

The scales of role conflict, quantitative workload, alcohol consumption, and organizational commitment were taken from Hendrix and Spencer (1989). *Role Conflict* was measured by a 4-item scale which assessed the extent that one received incompatible (conflicting) requests from two or more people, and that superiors assigned tasks that were in conflict with each other. *Quantitative Workload* was measured by a 4-item scale which assessed the extent to which one had a great deal to do in their job or had to work very hard. *Alcohol Consumption* measured the average number of alcoholic drinks consumed by each participant per week. *Organizational Commitment* was measured using a 4-item scale that assessed the intent to continue to work for the organization, willingness to exert effort on the job for the organization, and the extent that the goals of the organization are compatible with the individuals' goals. *Emotional Exhaustion* was measured with a nine-item frequency scale taken from the Maslach Burnout Inventory (Maslach, 1982).

The scales of physical stressors, job boredom, somatic symptoms, and financial problems were developed as a part of this research: The *Physical Stressors* measure was made up of 3-item scale that measured the extent one's work area was noisy, temperature was too hot or cold, and lighting was too bright or not bright enough. The *Job Boredom* measure consisted of a 2-item scale that measured the extent one considered his/her job boring or interesting. The *Somatic Symptoms* measure consisted of a 4-item scale that assessed the extent that an individual experienced headaches, had trouble falling asleep, had restless sleep, and experienced fatigue. *Financial Problems* measured the extent to which individuals felt that they had financial problems.

Single-item measures developed as a part of this research included cigarette smoking, cold/flu episodes, and jogging. *Cigarette Smoking* was operationalized as the average number of cigarettes smoked per day. *Cold/flu Episodes* consisted of self-reports of the number of cold or flu episodes an individual had over a one-year period. *Jogging* was the average number of miles jogged per week. The *Body Mass Index* (BMI) is a measure of the relative weight of an individual at a given height. It is computed as weight divided by height squared.

Procedure

A health assessment survey was administered en masse at each research site during a one-day health promotion seminar. Surveys were administered at the beginning of the seminar before the participants were exposed to information that could influence their responses on the survey. It took participants approximately 1 hour and 15 minutes to complete the survey. Individuals were motivated to participate in the seminar because they would receive health information designed to help them assess their risk of developing coronary artery disease, individual health survey feedback, and blood analysis results. After completing the survey, subjects provided blood samples to establish their potential for developing coronary artery disease using the cholesterol ratio (i.e., total serum cholesterol divided by HDL cholesterol). Performance measures and absenteeism rates were obtained from employees' records.

Analysis

Exploratory path analysis was performed to establish which paths in the hypothesized causal chain (see Figure 1) were statistically significant and to develop a revised, stress-based health promotion model. The first stage of this analysis involved regressing each variable in the hypothesized model, other than the exogenous variables, against all preceding variables. The beta weights served as the path coefficients and indicated the relative strength of the variables in the model.

The second stage involved eliminating variables not having statistically significant (p < .05) path coefficients and rerunning the regression equations to derive a revised model.

RESULTS

Zero-Order Correlations

Table 2 contains the zero-order correlations between the exogenous variables and endogenous variables used in this research. The first column contains the hypothesized antecedents (exogenous variables) of stress. The remaining columns contain the zero-order correlations between the exogenous variables and the endogenous variables (intervening and outcome variables). The table has been divided into two parts: one part contains the multiple-item scales while the second part contains the single-item measures.

Path Analysis

Exploratory path analysis was conducted to assist in establishing a revised stress-based health promotion model using regression analysis. This analysis resulted in the revised model depicted in Figure 2. All paths within the model are significant at or beyond the .05 level.

Examination of the revised model (Figure 2) reveals that 24 of the 30 hypothesized relationships in Figure 1 were supported. Overall, the major paths in the hypothesized model (Figure 1) were supported. For example, the path leading from the intra-organizational variables to job stress, job satisfaction, and to organizational commitment was supported, but not the relationship from commitment to performance. Also, the major path involving life stress was supported with the notable exception of the hypothesized link between life stress and emotional exhaustion. The remaining major paths leading to the cholesterol ratio were supported for the hypothesized individual characteristic variables but not for job or life stress. In addition to the relationships noted above that were hypothesized but not found to be significant, two other non-significant relationships are worth noting. These were life events to life stress and Type A behavior to life stress.

There were 13 significant paths included in the revised stress-based heath promotion model that were not originally hypothesized: specifically, the three separate paths from supervision, quantitative workload, and job satisfaction to emotional exhaustion, job boredom and job enhancement to job satisfaction, age to cold/flu episodes, supervision and role conflict to organizational commitment, organizational commitment to absenteeism, life events to absenteeism, life stress to somatic symptoms, task significance to performance, and jogging to somatic symptoms.

Table 3 contains, in the left column, the series of criterion variables used when performing the path analyses. The R^2 values associated with each full and restricted model predictor sets are provided in the remaining columns.

DISCUSSION

The purpose of the present research was to develop a multivariate stress-based health promotion model to serve as a basis for guiding stress and health promotion research and as a basis for health related interventions. There is a plethora of stress-related studies, yet few of these have been cast within the framework of a major health promotional model which incorporates the organizational setting. This is not surprising as few empirically derived comprehensive stress-based health promotion models have appeared in the literature to help guide research and intervention efforts. Therefore, the results of stress research have frequently focused on narrowly defined aspects of stress including only one, or a few variables (e.g., Type A behavior), and the results across these studies have, in a number of areas, been inconclusive and contradictory (e.g., effects of stress on heart attacks, absenteeism, and productivity). The development of a stress-based model should, therefore, be helpful in directing future research.

As predicted by the hypothesized model (Figure 1), job stress was affected mainly by intra-organizational variables, while life stress was affected by extra-organizational variables. The only individual characteristic affecting stress was tolerance for change, which affected job stress. Greater tolerance was associated with lower experienced job stress. The major variables affecting job stress were job boredom, job enhancement, and quantitative workload. The higher one scored on job boredom and quantitative workload the higher the job stress the person reported experiencing. The higher one scored on job enhancement the lower the reported stress the person experienced. Life stress was affected primarily by family-spouse relations and, to a lesser extent, by financial problems.

The lack of any sex differences in life or job stress, or stress-linked outcomes is consistent with the literature review by Korabik, McDonald, and Rosin (1993). They concluded that, despite being subjected to more stressors than their male counterparts, women managers do not appear to feel more stress, nor do women experience more negative consequences of stress. Greenglass (1993) proposed that sex differences in coping mechanisms and resources accounts for the lack of sex differences in stress. Future stress-health models would benefit by the inclusion of coping variables.

TABLE 2　Intervening and Outcome Variables

Antecedents	Life Stress	Job Stress	Job Satis.	Emotional Exhaustion	Organ. Commit.	Cold/Flu	Somatic Symptoms	Perform. Appraisal	Absenteeism	Cholesterol Ratio
Job Enhancement	-.03	-.41	.66	-.23	.47	-.11	-.11	.20	-.20	.11
Supervision	-.16	-.41	.39	-.37	.45	-.13	-.23	.13	-.20	.02
Physical Stressors	.11	.34	-.17	.33	-.20	.17	.22	-.05	.17	.21
Role Conflict	.12	.37	-.13	.36	-.26	.12	.23	.03	.09	.01
Quan. Workload	-.05	.09	.39	.18	.21	.06	.02	.05	-.05	-.07
Job Boredom	.06	.39	-.74	.27	-.51	.08	.14	.09	.24	.06
Task Significance	.06	.02	.36	.10	.21	.05	.02	.16	.02	-.01
Family-Spouse Rel.	-.64	-.12	.10	-.17	.14	-.14	-.22	-.06	-.12	-.01
Financial Problems	.38	.19	-.10	.22	-.16	.12	.25	-.08	.28	-.01
Life Events	.41	.14	-.11	.18	-.07	.16	.35	.02	.37	-.06
Tol. for Change	-.21	-.16	.08	-.14	.11	-.03	-.11	.03	.03	-.03
Type A Behavior	.19	.05	-.03	.13	-.02	.11	.10	.13	-.02	.06
Male	-.03	-.11	.02	-.17	.03	-.05	-.18	-.04	-.14	.31
BMI	.21	.08	.12	.02	.10	.04	.15	.10	.05	.44
Age	-.16	-.07	.13	-.08	.14	-.19	-.12	-.02	-.15	.23
Cig. Smoking	.17	.04	.03	.06	-.00	.01	.08	.04	.13	.09
Jogging	-.01	-.07	.02	-.01	-.00	-.10	-.10	-.03	-.10	-.09
Alcohol Consumption	.09	-.05	.02	-.04	-.04	-.05	-.12	-.02	-.04	-.09

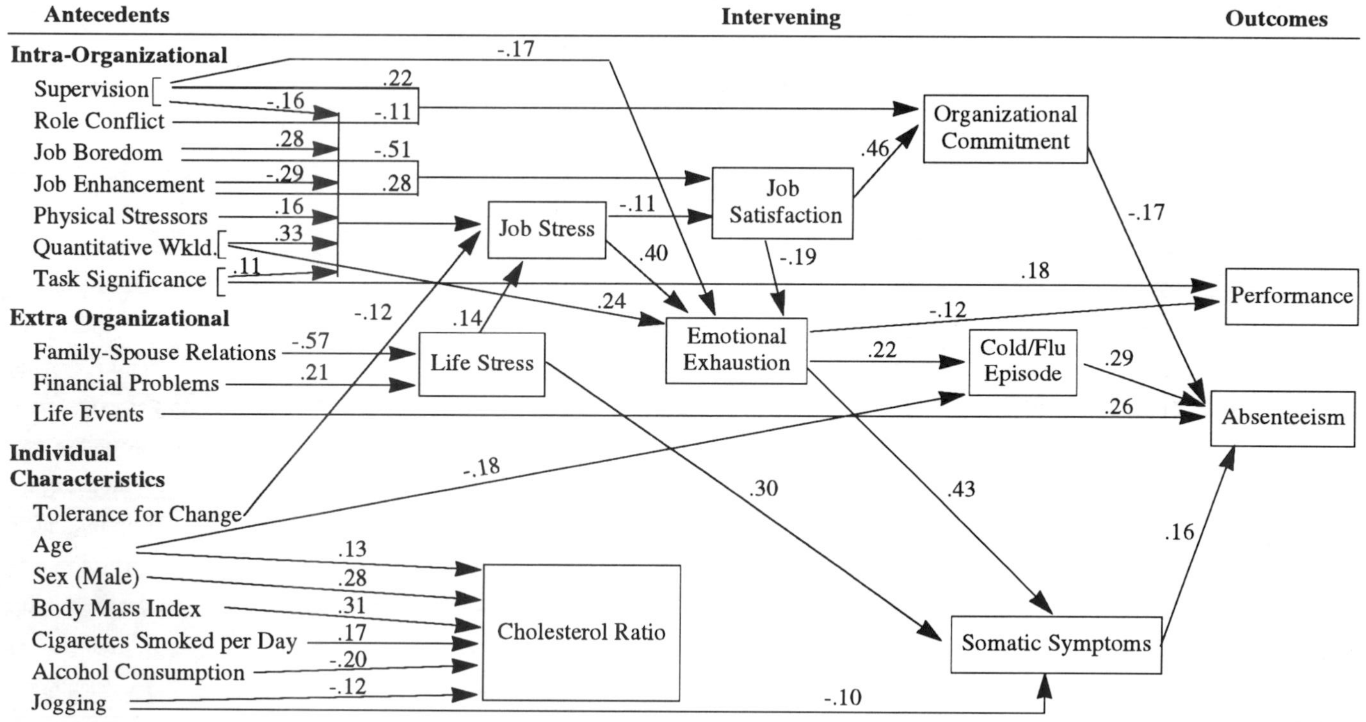

FIGURE 2   Revised Model

Antecedents

Intra-Organizational

Supervision
Role Conflict
Job Boredom
Job Enhancement
Physical Stressors
Quantitative Wkld.
Task Significance

.22
-.16
-.11
.28
-.51
.28
-.29
.16
.33
.11
-.17

Extra-Organizational

Family-Spouse Relations
Financial Problems
Life Events

-.12
-.57
.21
.14

Individual Characteristics

Tolerance for Change
Age
Sex (Male)
Body Mass Index
Cigarettes Smoked per Day
Alcohol Consumption
Jogging

-.18
.13
.28
.31
.17
-.20
-.12

Intervening

Job Satisfaction
Job Stress
Emotional Exhaustion
Life Stress
Organizational Commitment
Cold/Flu Episode
Somatic Symptoms
Cholesterol Ratio

.46
-.11
.40
.24
-.19
.18
-.12
.22
.29
.26
.43
.30
.16
-.10

Outcomes

Performance
Absenteeism

-.17

TABLE 3   Full and Restricted Model R^2 Values for Each
Predictor Set

Criterion Variable	R^2 Full model	R^2 Restricted Model
Life Stress	.526	.446
Job Stress	.597	.425
Cholesterol Ratio	.515	.318
Job Satisfaction	.744	.606
Organizational Commitment	.460	.392
Performance	.240	.042
Emotional Exhaustion	.569	.402
Cold/Flu Episodes	.251	.090
Somatic Symptoms	.523	.352
Absenteeism	.449	.331

The major stress-related outcomes hypothesized to be ultimately affected by stress were absenteeism, performance, and the cholesterol ratio. Two of these that are of major concern to organizations are absenteeism and performance. The revised model suggests that absenteeism is affected by one's commitment to the organization and the extent that individuals experience illness (i.e., cold/flu episodes or somatic symptoms). The commitment-absenteeism link tends to be supported by the literature. For example, this result is consistent with the work of Cheloha and Farr (1980) who found, in reviewing studies on job involvement-absenteeism, a modest negative relationship between the two variables. Absenteeism can be due to many factors, illness being one of the most common reasons given for being absent from work. Steers and Porter (1982) also found a modest support for the commitment-absenteeism relationship when they reviewed the literature. Performance, on the other hand, was higher when the task was considered significant and it was lower with increases in emotional exhaustion.

Job stress and life stress were not related to the cholesterol ratio as hypothesized. This was not totally unexpected because so many other factors can affect the cholesterol ratio, and probably have greater effects than stress. For example, dietary fat, genetic predisposition, and exercise have been found to be related to the components of one's cholesterol ratio (Hegsted, 1982; Hendrix, 1985; Kromhout, 1983; Weintraub, Rosen, Otto, Eisenberg, & Breslow, 1989). These findings are also consistent with Hendrix et al. (1985) who found job and life stress not to be related to the cholesterol ratio. They are also consistent with Steffy and Jones (1988) who did not find job stressors and stressful life events to be related to a version of the cholesterol ratio.

Increased job stress was found to be related to decreased job satisfaction and increased emotional exhaustion, while life stress was positively related to somatic symptoms. These results are consistent with other research that has found job stress related to job dissatisfaction (Steffy & Jones, 1988), and stressful life events related to psychosomatic complaints (Frese, 1985; Steffy & Jones, 1988).

The revised model provides a framework that can be used by future researchers and practitioners. For practitioners, this research suggests interventions that can be used based on the specific health-related issue of concern. For example, if one's concern is for reducing risk of heart disease, then stress management efforts are not indicated. Instead, programs that would affect the traditional risk factors seem appropriate, for example, diet and nutrition, weight control, exercise, and smoking cessation programs. On the other hand, if the concern is for absenteeism, then dealing with stress would be desirable. Since job and life stress were related to absenteeism, both of these should be addressed through intervention programs. This research suggests that job stress reduction could be improved by programs dealing with improving supervision, reducing role conflict, reducing quantitative work overload, reducing factors that cause physical discomfort such as high noise levels, and reducing job boredom through job redesign efforts such as job enrichment. Reducing life stress would be helped by programs that focused on assisting families and spouses with their interpersonal problems and with assisting employees in finding solutions to their financial problems. Should the area of interest be job performance, attention can also be directed at providing stress management programs as well as job enrichment efforts to increase the significance and reduce the stressfulness of tasks being performed by employees.

As with all research, this study has limitations and strengths. The limitations suggest areas for future research to further refine and expand the model developed during this study. One major limitation of this research is that the study did not deal with stress coping behaviors. Although not explicitly indicated in the hypothesized model (Figure 1), coping behaviors would be incorporated under the Individual Characteristics category of the antecedent variables. Specifically, coping behaviors such as: using social support, possessing a "hardiness" type of personality, planning, taking direct action, and cognitive coping (e.g., justification) would be considered strategies that an individual would use to deal with stress. If coping variables were included, they might be depicted with a dotted line to job and life stress. The dotted line would indicate a variable that moderates the effect of stressors on the stress response. Another limitation of this research was that the data analyzed were primarily self-reported, cross-sectional/retrospective data. Also, the sample used in this study consisted of volunteers and therefore is subject to selection bias. The study did, however, include a wide range of age groups and racial groups, as well as white collar and managerial employees in the sample. Studies in the past tended to focus on blue-collar and clerical employees (e.g., Mowday, Porter, & Steers, 1982). Path analysis using regression, was performed in this research to develop a revised model to serve as a framework for future research. The use of a more complete LISREL-type analysis, with recursive paths to perform the path analysis, might have been more powerful. In particular, recursive paths for future research might include those of job and life stress leading back to smoking levels and job stress leading back to life stress.

Even with these limitations, we believe this model is sufficiently comprehensive to serve as a starting point for further refinement. To the best of our knowledge, the revised stress-based health promotion model represents the most comprehensive empirically-derived network that includes relationships among stresses (life and job) and their antecedents and consequences. Future models and research will benefit by including medically-oriented variables such as the cholesterol ratio, body mass index, somatic symptoms, cold/flu episodes, and stress hormones such as epinephrine and cortisol (Christensen & Jensen, 1994), as well as coping strategies (Lutgendorf, Antoni, Kumar, & Schneiderman, 1994), which are a part of this model. In addition, future research might include objective measures such as those incorporated in this model (i.e., cholesterol ratio, absenteeism, and performance). These variables lend credibility and external validity to the findings in addition to their previously discussed implications for practitioners.

The results of this research suggest that healthy organizations tend to contain healthy people. Organizations that create a hospitable work environment that has low physical stressors, low role conflict, enriched jobs, and good supervision will tend to have employees who have less stress, are more committed to the organization and who have lower absenteeism. Therefore, the specific causal paths established have implications for health and wellness programs, management development programs, as well as for organizational and job design efforts. In addition, the health of individual employees might be improved by screening for high cholesterol ratios followed by interventions to reduce them based on the results of this research. The results of this research suggest that by modification of the variables contained in the model, both individual health and organizational effectiveness (reduced absenteeism and increased performance) will be enhanced.

REFERENCES

Appels, A., Mulder, P., van't Hof, M., Jenkins, C.D., van Houtem, J., & Tan, F. (1987). A prospective study of the Jenkins Activity Survey as a risk indicator for coronary heart disease in the Netherlands. *Journal of Chronic Diseases, 40*(10), 959–965.

Barnett, R.C., Biener, L., & Baruch, G.K. (1987). *Gender and stress.* New York: Free Press.

Beehr, T.A., & Newman, J.E. (1978). Job stress, employee health, and organizational effectiveness: A facet analysis, model, and literature review. *Personnel Psychology, 31,* 665–699.

Bertera, P.H. (1990). The effects of workplace health promotion on absenteeism and employee cost in a large industrial population. *American Journal of Public Health, 80*(9), 1101–1105.

Bhagat, R.S. (1983). Effects of stressful life events on individual performance effectiveness and work adjustment processes within organizational settings: A research model. *Academy of Management Review, 8*(4), 660–671.

Brief, A.P., Schuler, R.S., & Van Sell, M. (1981). *Managing job stress.* Boston: Little, Brown.

Bromet, E.J., Dew, M.A., Parkinson, D.K., Cohen, S., & Schwartz, J.E. (1992). Effects of occupational stress on the physical and psychological health of women in a microelectronics plant. *Social Science & Medicine, 34*(12), 1377–1383.

Bryant, H.U., Story, J.A., & Yim, G.K.W. (1988). Assessment of endogenous opioid mediation in stress-induced hypercholesterolemia in the rat. *Psychosomatic Medicine, 50,* 576–585.

Burke, R.J., & Richardson, A.M. (1990). Sources of satisfaction and stress among Canadian physicians. *Psychological Reports, 67,* 1335–1344.

Caplan, R.D., Cobb, S., French, J.R.P., Jr., Van Harrison, R., & Pinneau, S.R., Jr. (1975). *Job demands and worker health: Main effects and occupational differences* (HEW NIOSH No. 75–160). Washington, DC: U.S. Government Printing Office.

Cheloha, R.S., & Farr, J.L. (1980). Absenteeism, job involvement and job satisfaction in an organizational setting. *Journal of Applied Psychology, 65,* 467–473.

Christensen, N.J., & Jensen, E.W. (1994). Effect of psychosocial stress and age on plasma norepinephrine levels: A review. *Psychosomatic Medicine, 56*(1), 77–83.

Cooper, C.L., & Marshall, J. (1976). Occupational sources of stress: A review of the literature relating to coronary heart disease and mental health. *Journal of Occupational Psychology, 49,* 11–28.

Corrigan, P.W., Holmes, P.E., Luchins, D., Buican, B., Basit, A., & Parks, J.J. (1994). Staff burnout in a psychiatric hospital: A cross-lagged panel design. *Journal of Organizational Behavior, 15,* 65–74.

DeBakey, M.E., Gotto, A.M., Jr., Scott, L.W., & Foreyt, J.P. (1984). *The living heart diet.* New York: Raven.

DeCotiis, T.A., & Summers, T.P. (1987). A path analysis of a model of the antecedents and consequences of organizational commitment. *Human Relations, 40*(7), 445–470.

Edwards, S.E., & Gettman, L.R. (1980, November). Determining the value of employee fitness programs. *Personnel Administrator, 61,* 41–44.

Epstein, F. (1965). The epidemiology of coronary heart disease: A review. *Journal of Chronic Diseases, 18,* 735–774.

Fox, M.L., Dwyer, D.J., & Ganster, D.C. (1993). Effects of stressful job demands and control on physiological and attitudinal outcomes in a hospital setting. *Academy of Management Journal, 36*(2), 298–318.

Frese, M. (1985). Stress at work and psychosomatic complaints: A causal interpretation. *Journal of Applied Psychology, 70,* 314–328.

Friedman, M., & Rosenman, R.H. (1959). Association of specific overt behavior pattern with blood and cardiovascular findings. *Journal of the American Medical Association, 169*(12), 96–105.

Friedman, M., Rosenman, R.H., & Carroll, V. (1958). Changes in serum cholesterol and blood clotting time in men subject to cyclical variation of occupational stress. *Circulation, 17,* 852–861.

Friend, K. (1982). Stress and performance: Effects of subjective work load and time urgency. *Personnel Psychology, 35,* 623–633.

Fripp, R.R, & Hodgson, J.L. (1987). Effect of resistive training on plasma lipid and lipoprotein levels in male adolescents. *Journal of Pediatrics, 111*(6), 926–931.

Gebhardt, D.L., & Crump, C.E. (1990). Employee fitness and wellness programs in the workplace. *American Psychologist, 45*(2), 262–272.

Greenglass, E.R. (1993). Social support and coping of employed women. In B.C. Long & S.E. Kahn (Eds.), *Women, work, and coping* (pp. 154–169). Montreal: McGill-Queens University Press.

Harris, M.M., & Heft, L.L. (1992). Alcohol and drug use in the workplace: Issues, controversies, and directions for future research. *Journal of Management, 18*(2), 239–266.

Hartman, S.R. (1983). *Psychological correlates of ratios of total cholesterol to HDL cholesterol: Type A behavior pattern, stress, and anger.* Unpublished master's thesis, Trinity University, San Antonio, TX.

Hegsted, D.M. (1982). What is a healthful diet? *Primary Care, 9,* 445–473.

Hendrix, W.H. (1984). Development of a contingency model organizational assessment survey for management consultants. *Journal of Experimental Education, 52,* 95–105.

Hendrix, W.H. (1985). Factors predictive of stress, organizational effectiveness, and coronary heart disease potential. *Aviation, Space, and Environmental Medicine, 56,* 654–659.

Hendrix, W.H., Ovalle, N.K., & Troxler, R.G. (1985). Behavioral and physiological consequences of stress and its antecedent factors. *Journal of Applied Psychology, 70*(1), 188–201.

Hendrix, W.H., & Spencer, B.S. (1989). Development and test of a multivariate model of absenteeism. *Psychological Reports, 64,* 923–938.

Hendrix, W.H., Steel, R.P., & Schultz, S.A. (1987). Job stress and life stress: Their causes and consequences. *Journal of Social Behavior and Personality, 2*(3), 291–302.

Holmes, T.H., & Rahe, R.H. (1967). The social readjustment rating scale. *Journal of Psychosomatic Research, 11*, 213–218.

House, J.S. (1974). Occupational stress and coronary heart disease: A review and theoretical integration. *Journal of Health and Social Behavior, 15*(1), 12–27.

Houston, B.K., Smith, M.A., & Cates, D.S. (1989). Hostility patterns and cardiovascular reactivity to stress. *Psychophysiology, 26*(3), 337–342.

Israel, B.A., House, J.S., Schurman, S.J., Heaney, C.A., & Mero, R.P. (1989). The relation of personal resources, participation, influence, interpersonal relationships and coping strategies to occupational stress, job strains, and health: A multivariate analysis. *Work and Stress, 3*(2), 163–194.

Ivancevich, J.M., & Matteson, M.T. (1980). *Stress and work: A managerial perspective.* Glenview, IL: Scott-Foresman.

Ivancevich, J.M., Matteson, M.T., Freedman, S.A., & Phillips, J.S. (1990). Worksite stress management interventions. *American Psychologist, 45*(2), 252–261.

Izraeli, D.N. (1993). Work/family conflict among women and men managers in dual-career couples in Israel. *Journal of Social Behavior and Personality, 8*(3), 371–388.

Jackson, S.E., Schwab, R.L., & Schuler, R.S. (1986). Toward an understanding of the burnout phenomenon. *Journal of Applied Psychology, 71*(4), 630–640.

Jenkins, C.D. (1976). Recent evidence supporting psychologic and social risk factors for coronary disease, Parts 1 and 2. *New England Journal of Medicine, 294*, 987–994, 1033–1038.

Johnson, A. (1978). Six differentials in coronary heart disease: The explanatory role of primary risk factors. *Journal of Health and Social Behavior, 18*, 46–54.

Kaltenbach, M., Lichten, P., & Friesinger, G.C. (Eds.). (1973). *Coronary heart disease: 2nd International Symposium at Frankfurt 1972.* Stuttgart, Germany: Georg Thieme.

Korabik, K., McDonald, L.M., & Rosin, H.M. (1993). Stress, coping, and social support among women managers. In B.C. Long & S.E. Kahn (Eds.), *Women, work, and coping* (pp. 133–154). Montreal: McGill-Queens University Press.

Kromhout, D. (1983). Body weight, diet, and serum cholesterol in 871 middle-aged men during 10 years of follow-up (the Zutpher study). *American Journal of Clinical Nutrition, 38*, 591–598.

Lauer, R., Lee, J., & Clarke, W.R. (1988). Factors affecting the relationship between childhood and adult cholesterol levels: The Muscatine Study. *Pediatrics, 82*(3), 309–318.

Lutgendorf, S.K., Antoni, M.H., Kumar, M., & Schneiderman, N. (1994). Changes in cognitive coping strategies predict EBV-antibody titre change following a stressor disclosure induction. *Journal of Psychosomatic Research, 38*(1), 63–78.

Malaspina, J.P., Bussiere, H., & LeCalve, G. (1981). The total cholesterol/HDL cholesterol ratio: A suitable atherogenesis index. *Atherosclerosis, 40*, 373–375.

Marshall, J.R., & Cooper, C.L. (1979). Work experience of middle and senior managers: The pressure and satisfaction. *International Management Review, 19*, 81–96.

Maslach, C. (1982). *Burnout: The cost of caring.* Englewood Cliffs, NJ: Prentice-Hall.

Matarazzo, J.D. (1984). Behavioral health: A 1990 challenge for the health services professions. In J.D. Matarazzo, S.W. Weiss, J.A. Herd, N.E. Miller, & S.M. Weiss (Eds.), *Behavioral health: A handbook of health enhancement and disease prevention* (pp. 3–40). New York: John-Wiley.

Miller, N.E., Bolton, C.H., & Hayes, T.M. (1988). Associations of alcohol consumption with plasma high density lipoprotein cholesterol and its major subfractons: The Caerphilly and Speedwell collaborative heart disease studies. *Journal of Epidemiology and Community Health, 42*(3), 220–225.

Motowidlo, S.J., Packard, J.S., & Manning, M.R., Jr. (1986). Occupational stress: Its causes and consequences for job performance. *Journal of Applied Psychology, 71*(4), 618–629.

Mowday, R.T., Porter, L.W., & Steers, R.M. (1982). *Employee organization linkages: The psychology of commitment, absenteeism, and turnover.* New York: Academic Press.

Myers, D.W. (1986). *Human resources management: Principles and practice.* Chicago, IL: Commerce Clearing House.

Nanas, S., Pan, W.H., & Stamler, J. (1987). The role of relative weight in the positive association between age and serum cholesterol in men and women. *Journal of Chronic Diseases, 40*(9), 887–892.

Niehouse, O.L. (1987). Life after burnout. *Business, 37*(3), 42–46.

Quick, J.C., & Quick, J.D. (1984). *Organizational stress and preventive management.* New York: McGraw-Hill.

Rahe, R.H., Rubin, T.T., & Arthur, R.J. (1974). The three investigations study: Serum uric acid, cholesterol, and cortisol variability during stresses of everyday life. *Psychosomatic Medicine, 36*(3), 258–268.

Schnall, P.L., Devereaux, R.B., Pickering, T.G., & Schwartz, J.E. (1992). The relationship between 'job strain' workplace diastolic blood pressure, and left ventricular mass index: A correction. *Journal of the American Medical Association, 267*(9), 1209.

Steers, R.M., & Porter, L.W. (1983). *Motivation and work behavior* (p. 445). New York: McGraw-Hill.

Steffy, B.D., & Jones, J.W. (1988). Workplace stress and indicators of coronary-disease risk. *Academy of Management Journal, 31*(3), 686–698.

Strube, M.J. (Ed.). (1990). Type A behavior [Special issue]. *Journal of Social Behavior and Personality, 5*(1).

Summers, T.P., DeNisi, A.S., & DeCotiis, T.A. (1994). A field study of some antecedents and consequences of felt job stress. In P.L. Perrewé & R. Crandall (Eds.), *Occupational stress: A handbook* (pp. 113–128). (Original work published 1989)

Uhl, G.S., Troxler, R.G., Hickman, J.T., & Clark, D. (1981). Relation between high density lipoprotein cholesterol and coronary artery disease in asymptomatic men. *American Journal of Cardiology, 48,* 903–910.

Verbrugge, L.M. (1993). Marriage matters: Young women's health. In B.C. Long & S.E. Kahn (Eds.), *Women, work, and coping* (pp. 170–192). Montreal: McGill-Queens University Press.

Weintraub, M.S., Rosen, Y., Otto, R., Eisenberg, S., & Breslow, J.L. (1989). Physical exercise conditioning in the absence of weight loss reduces fasting and postprandial triglyceride-rich lipoprotein levels. *Circulation, 79*(5), 1007–1014.

Zedeck, S., & Mosier, K.L. (1990). Work in the family and employing organization. *American Psychologist, 45*(2), 240–251.

Coping with Stressful Life Events:
An Empirical Analysis

Rabi S. Bhagat
Stephen M. Allie
David L. Ford, Jr.

During the past decade, much research has focused on the causes and consequences of organizational stress (Beehr & Newman, 1978; Cooper & Payne, 1978, 1980; Jamal, 1984; Parker & DeCotiis, 1983). Dependent variables in such research include both organizationally valued outcomes (such as job satisfaction, job performance, absenteeism, turnover, etc.) and personally valued outcomes (such as life satisfaction, feelings of depression and depersonalization, blood pressure, heart-rate, cholesterol level and heart disease, etc.). We know relatively little, however, about the effects of coping on the relationship between the experience of stress and the emergence of strains. This research was undertaken to explore the relative efficacies of two styles of coping in moderating relationships between the stressful events experienced at work and in life, and the emergence of strains.

A Conceptual Framework

Definitions of stress as employed in this investigation follow our earlier

Authors' Notes: The authors would like to acknowledge their gratitude to the teachers and the principals of the participating school districts of North Central Texas for their cooperation and assistance with this investigation. We would also like to express our appreciation to Irwin G. Sarason and George V. Coelho for their suggestions and encouragement.

conceptualizations (Bhagat, 1980, 1983; Bhagat & Allie, 1989; Bhagat, McQuaid, Lindholm, & Segovis, 1985). Organizational stress is defined as a problematic level of environmental demand that interacts with the individual to change (disrupt or enhance) his or her psychological or physiological condition such that the person (mind and/or body) is forced to deviate from normal functioning. Personal life is defined as an unresolved type of environmental demand that requires adaptive behaviors in the form of social readjustments (see, e.g., Bhagat, 1983; Dohrenwend & Dohrenwend, 1981; Holmes & Rahe, 1967; Johnson & Sarason, 1979; Rabkin & Streuning, 1976).

Life strains are defined as undesirable personal outcomes of the combined stressful experiences from the domains of work and non-work and consist of three kinds: cognitive, emotional/affective, and behavioral (Bhagat, 1983). Possible outcomes are numerous (Beehr & Newman, 1978) but those included in the present study appear to have the most theoretical relevance in the current organizational behavior literature. Satisfaction with one's work is an important indicator of one's mental health (Kornhauser, 1965; Locke, 1976). If one experiences considerably lower levels of satisfaction with one's work, then one is also experiencing an important life strain. Experiencing reduced performance effectiveness is also a significant source of strain that could result from either job or organizational stress (Jamal, 1984; Motowidlo, Packard, & Manning, 1986) or from stressful encounters in one's life (Bhagat, 1980, 1983; Bhagat et al., 1985). Emotional exhaustion, reduced feelings of personal accomplishment and intense feelings of depersonalization are all significant indicators of life strains (Corrigan et al., 1994; Cordes & Dougherty, 1993; Ford, Murphy, & Edwards, 1983; Jackson, Schwab, & Schuler, 1986; Maslach, 1982; Maslach & Jackson, 1981). Emotional exhaustion describes an affective state characterized by feelings of overextendedness and being consumed by one's work. Reduced personal accomplishment describes feelings of low levels of competence and achievement. Jackson et al. (1986) and Maslach and Jackson (1981) have shown that this strain is an important component of burnout. Feelings of depersonalization describe an unfeeling and impersonal response toward clients or recipients of one's work-related service or care. Incidence of illness can be an indicator of life strain (Holmes & Rahe, 1967; Johnson & Sarason, 1979). We have included it in the present analysis because our knowledge of the effects of organizational stress and life stress on such incidence of illness is almost non-existent in the literature (see Figure 1).

Coping is defined as efforts, both action-oriented and intra-psychic, which help the individual manage (i.e., to master, tolerate, reduce and minimize) environmental and internal demands, and conflicts among them, that result from stressful encounters with life and which tax or exceed a person's resources (Lazarus & Launier, 1978). Problem-focused coping focuses on those cognitive and problem-solving efforts, information seeking acts and behavioral strategies that are concerned with altering or managing the source of the stressful encounter or the problem itself (Folkman & Lazarus, 1980).

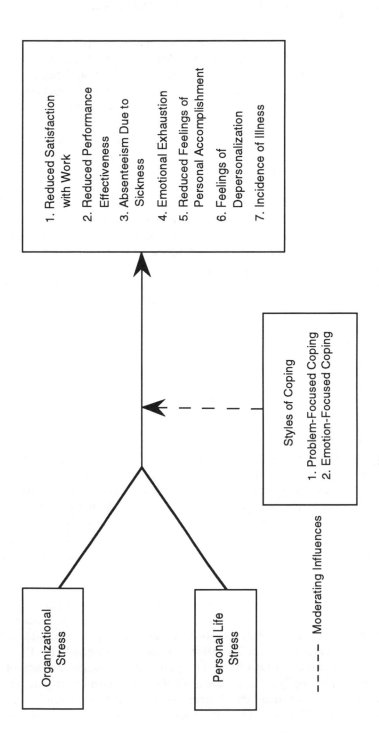

FIGURE 1   Organizational Stress, Personal Life Stress and the Moderating Role of Styles of Coping

95

Emotion-focused coping, on the other hand, is defined to include those cognitive, intra-psychic and behavioral efforts that are directed at reducing or managing emotional distress (Folkman & Lazarus, 1980). Emotion-focused coping is directed at regulation of stressful emotions. Strategies of intellectualization, social isolation, and suppression of reality are used in this mode of coping (Folkman & Lazarus, 1980; Lazarus & Launier, 1978).

In the present investigation, we focus on the effectiveness of both problem-focused and emotion-focused coping in moderating stress–outcome (i.e., strain) relationships. There is evidence in the literature of the growing importance of the moderating role of styles of coping in predicting important clinical and mental health-related outcomes (Folkman & Lazarus, 1980; Lazarus & DeLongis, 1983; Lee, Ashford, & Jamieson, 1993). However, relative predictive efficacies of each of these styles of coping in moderating stress–work outcome relationships have not been empirically examined. It is hypothesized that both problem-focused coping and emotion-focused coping moderate the relationship between (1) organizational stress–life strains and (2) personal life stress–life strains. Both problem-focused coping and emotion-focused coping are expected to account for a significant proportion of the variance in various measures of life strains. Individuals who are high in their predispositions to use either problem-focused or emotion-focused coping will experience lower levels of life strains.

METHOD

Participants

Three hundred four teachers from five suburban, urban and rural school districts in a large metropolitan area in North Central Texas were selected at random to participate in the study. During the initial planning phase of the study, principals in each of the schools were contacted for their permission to collect research data. Teachers were approached on an individual basis by the second author. They were given a composite questionnaire with measures of stressful life events, organizational stress, coping skills and life strains. The order of the various measures within this questionnaire was randomly determined. After one follow-up contact, 276 (91%) teachers returned usable questionnaires. The sample was predominantly female (82%), married (72%) and white (85%). The level of education was some coursework beyond the bachelor's degree required for certification. The mean age was 38.4 years with a standard deviation of 9.7. Teachers from all grade levels, i.e., elementary, middle and high, were represented, and the majority (72%) taught in the regular education program in their schools. The average tenure on the job was 11.7 years with a standard deviation of 7.4 years. The sample closely matched the population of teachers from which it was drawn on the above parameters. Numerous encouraging comments were made by some of these 276 teachers regarding the nature of this investigation and its relevance and 65% requested summary results of the data analyzed on an aggregate basis.

Measurement of Organizational Stress and Personal Life Stress

Organizational stress was measured by an adaptation of the "Teacher Stress Survey" (TSS) developed by Kyriacou and Sutcliff (1977, 1978). The TSS consists of 45 demanding organizational events that are often found in schools. Examples of events include dealing with excessive responsibility for pupils, shortages of equipment, maintaining discipline in the classroom, teaching students from a wide variety of socio-economic backgrounds, having to attend too many staff meetings, dealing with poor promotional and advancement related personnel policies, etc. The questions required respondents to indicate on a five-point scale the extent to which each of the events had been stressful to them in their teaching jobs during the last year. The scale anchor points were 0 (no stress) to 4 (extreme stress). Scores on all 45 items were summed to arrive at a total job stress scale. There were no significant differences in the mean levels of organizational stress reported among the schools represented in the study. This indicated that organizational stress as perceived by the teachers tended to be relatively uniform across all of the schools of the districts sampled.

Personal life stress was measured by using the Life Experiences Survey (LES; Sarason, Johnson, & Seigel, 1978) which consists of 47 stressful events that could conceivably occur during a person's life during any given year. The teachers responded by indicating all of the events that they had experienced in their lives during the exact calendar year before February, 1982. The Life Experiences Survey also required the respondents to indicate whether each life event experienced was positive or negative and the degree of positive or negative impact that the event exerted on their lives during the immediately preceding year. Ratings were made on 7-point scales ranging from -3 to +3, with a rating of -3 indicating an extremely negative impact, a rating of +3 an extremely positive impact and a rating of 0 signifying no impact in either direction. Separate scores for both negative and positive life stress were calculated. Examples of life events included in the survey were serious illness or injury of a close friend, death of a spouse, sexual difficulties, trouble with in-laws, major change in financial status for the worse, marriage, birth of a baby, reconciliation with one's mate, major change in financial status for the better, etc. Since this instrument is a checklist of life events, internal consistency and reliability in terms of homogeneity of the life events reported and calculated by Cronbach's formula does not have any psychometric significance.

Previous research by Bhagat (1980), Bhagat, McQuaid, Lindholm, and Segovis (1985), Sarason and Johnson (1979), and Vinokur and Selzer (1975) has shown that negative or undesirable stresses have much more detrimental effects on both health as well as organizationally valued outcomes than positive stresses. Therefore, for the present study, the authors chose to focus only on negative personal life stress and its relation to various indicators of life strains. The test-retest reliabilities of the LES reported by Sarason et al. (1978) are in the range of .56 to .88.

Measurement of Styles of Coping

In the present study, a 23-item adaptation of Pearlin and Schooler's (1978) instrument for the structure of coping was employed. Respondents rated the degree to which they utilized each of the 23 coping mechanisms on a five-point Likert scale (0 = not at all to 4 = quite often). According to Lazarus, Averill, and Opton (1974), much of the research on coping has emphasized psychological dispositions rather than situation-specific responses to various events that occur in life. Traditionally, coping ability has been assessed primarily in terms of possession of personality characteristics that help people defend against stressful events. Thus an individual with the "right" personality characteristics might be more effective than another individual who does not possess such a "right" personality predisposition in dealing with problems in life. In contrast to this personality characteristics-based approach, the approach of Lazarus and his associates (e.g., Folkman & Lazarus, 1980; Pearlin & Schooler, 1978) focuses on specific behavioral responses to various events. In the present investigation, a factor analysis of the 23-item adaptation of Pearlin and Schooler's (1978) instrument for the structure of coping yielded two factors which accounted for 65% of the variance. These two factors were labeled problem-focused coping and emotion-focused coping. Problem-focused coping consisted of six items that described an individual's commitment towards an action-oriented mode of resolving the difficulties that arise due to social readjustments brought about by life event changes. Items included such actions as trying to find a compromise, making a plan of action and following it, and taking some action to get rid of the problems, etc.

Emotion-focused coping consisted of seven items that dealt with strategies such as looking for the silver lining, trying to ignore difficulties and reminding oneself that things could be worse. Ilfeld (1980) found similar coping styles in a re-analysis of Pearlin and Schooler's (1978) data on the structure of coping.

Measurement of Life Strains

The first strain measured was dissatisfaction with work using the work itself scale of the Job Descriptive Index (Smith, Kendall, & Hulin, 1969). Dissatisfaction with work is an important indicator of life strain. The second strain was assessed in terms of reduced performance effectiveness conceptualized as a strain (Bhagat, 1983). Ratings were provided by the immediate supervisors (i.e., the principal) of the schools in which the teachers worked. A nineteen-item scale incorporating behavioral descriptors of teacher performance was designed specifically for this study. The principals responded on a five-point Likert-type scale, with scale anchors of 1 (strongly disagree) and 5 (strongly agree). Items included: "This teacher could be expected to provide a tension-free environment where students are orderly and attentive"; "This teacher could be expected to convey to students a sense of feeling of being needed and welcomed"; "This teacher could be expected to challenge the students to think, to question, to inquire, to examine and to use logic," etc. Performance effectiveness was assessed by summing all of the nineteen items ($\alpha = .94$). Absenteeism was measured by counting the number of sick days used by the teachers in the preceding year.

Emotional exhaustion, reduced feelings of personal accomplishment and depersonalization were assessed by using the Maslach and Jackson (1981) burnout scales. The scores on intensity and frequency dimension for each of the burnout items were multiplied and then added, leading to composite scores for the three subscales. The Cronbach alphas for emotional exhaustion, personal accomplishment, and depersonalization were .88, .80 and .61, respectively. Incidence of illness was measured by a slightly modified version of the Seriousness of Illness Scale (SIS) developed by Wyler, Masuda, and Holmes (1968) which is a checklist of 104 commonly recognized physical and mental symptoms, accidents and diseases. Examples include such ailments as sore throat, headache, depression, peptic ulcer, sinus infection, cancer, asthma, fainting, etc. This scale is widely used in clinical studies dealing with the effects of stress on the onset of illness (Dohrenwend & Dohrenwend, 1974; Kobasa & Puccetti, 1983) and has been found to be a reliable indicator of the degree of illness experienced (Kobasa & Puccetti, 1983; Wyler, Masuda, & Holmes, 1971). For the present study, several symptoms of obvious irrelevance to the sample (e.g., snakebites) were deleted. The respondents were asked to circle each illness experienced in the past twelve months. The severity weights associated with each illness circled were summed to yield a seriousness of illness score.

Confidant Method

In order to assess the validity of self-reported measures of stress, an additional data collection method was employed. In this method, each respondent was asked to provide the name of a confidant (a close friend, spouse or other relative) who might be in a position to judge the amount of stressful changes that had taken place in the work and personal lives of the respondent during the previous year. A brief questionnaire was then either mailed or delivered to the confidant who was requested to respond and then return it without discussing the nature of his or her assessment (i.e., confidant's report) with the focal subject. The confidant form thus provided another method of checking the validity of the self-reported data on organizational stress, personal life stress and coping skills.

Research on social support (Gottlieb, 1983) suggests that individuals seek informational help and emotional and structural support from their relatives and close associates (i.e., a confidant) during stressful life event changes. Therefore, the confidant often finds himself or herself in an unusually good situation to be aware of the extent of stress the focal subject experiences or has experienced. Because close friends and relatives could sometimes be in a position to provide a more objective assessment of a person's life than the focal person can, this method of measurement presents a potentially useful technique for assessing convergent validity of the self-appraised measures of stress in studies on human stress and cognition (Bhagat, 1980; Bhagat et al., 1985). Similar data have proven useful for components of life satisfaction (Crandall, 1976). The format of these questions was global, e.g., "What do you think is the total amount of personal life stress your friend has had during the past year?" "What do you think is the total amount of work-related stress your friend has had?" The questions were measured on 7-point scales anchored from almost none to extremely high amounts. Coping was measured by asking the confidant to report the degree to which they thought the focal subject "may never be able to cope" (coded 1) to "can easily cope" (coded 7).

Analytical Procedures

The research hypothesis was tested using moderated regression analyses (Cohen, 1978). The nature of the interactions were interpreted in accordance with the recommendations made by Peters and Champoux (1979). In this procedure, the STRESS x COPING interaction term was added to the regression equation which already contained the STRESS and COPING variables. The increment in variance due to the STRESS x COPING interaction term, above and beyond that due to the additive effects of STRESS and COPING, was then tested for significance. Separate analyses were conducted for problem-focused coping and emotion-focused coping.

RESULTS

Table 1 contains the means, standard deviations and reliabilities of all the variables.

An intercorrelation matrix of the independent, dependent, and moderator variables is presented in Table 2. An examination of Table 2 reveals numerous statistically significant correlations and several interesting trends. Negative personal life stress and organizational stress are correlated (.31, p < .001), suggesting that there might be a spillover effect, although the direction of the spillover cannot be assessed in this cross-sectional investigation. Organizational stress is negatively correlated with satisfaction with work (-.51, p < .001) and feelings of personal accomplishment (-.16, p < .01) and positively correlated with emotional exhaustion (.51, p < .001), feelings of depersonalization (.34, p < .001), and incidence of serious illness (.20, p < .001). Negative personal life stress is correlated positively with feelings of emotional exhaustion (.33, p < .001), depersonalization (.19, p < .001), and serious illness (.42, p < .001), and negatively with satisfaction with work (-.21, p < .001). The pattern of correlations reported in Table 2 clearly shows that both negative personal life stress and organizational stress are correlated significantly with most of the symptoms of life strains.

Job performance was significantly correlated with only two variables, prob-

TABLE 1   Means, Standard Deviations and Reliabilities of All Variables

Variables	Means[a]	Standard Deviation	No. of Items	Reliabilities[b]
Independent				
Organizational Stress	62.04	28.51	45	.95
Negative Personal Life Stress	6.56	4.69	50	n/a
Moderator				
Problem-Focused Coping	14.16	2.87	6	.65
Emotion-Focused Coping	13.50	3.34	7	.63
Dependent				
Satisfaction with Work	34.85	8.67	18	.81
Job Performance	75.50	10.36	19	.94
Absenteeism Due to Sickness	3.29	3.73	1	n/a
Emotional Exhaustion	93.97	62.30	9	.88
Reduced Personal Accomplishment	188.71	59.52	8	.80
Depersonalization	18.41	21.44	5	.61
Serious Illness	205.76	143.74	105	n/a

[a]$N = 276$ (Except for supervisory rating of job performance for which $N = 173$)
[b]Reliabilities are computed using the Cronbach's Alpha Method.

lem-focused coping and personal accomplishment. It appears that, for this sample, the better performers on the job feel they are accomplishing something and are more likely to engage in a problem-solving approach to stressful situations. Absenteeism correlated significantly with emotion-focused coping and depersonalization.

Serious illness seems to be influenced more by negative stressful events occurring in one's personal life as opposed to organizational influences, although both sources were significant. Neither problem-focused coping nor emotion-focused coping correlated with negative personal life stress, indicating that there are no systematic relationships among these variables. However, emotion-focused coping is related to organizational stress (.26, $p < .001$) and serious illness (.12, $p < .05$). Additionally, problem-focused coping is related to satisfaction with work (.18, $p < .001$), personal accomplishment (.36, $p < .001$) and job performance (.14, $p < .05$) and is unrelated with measures of serious illness.

Satisfaction with work is negatively correlated with emotional exhaustion (-.59, $p < .001$) and feelings of depersonalization (-.44, $p < .001$). The correlation of emotional exhaustion with feelings of personal accomplishment is -.15 ($p < .01$); with depersonalization, .61 ($p < .001$); and with serious illness, .34 ($p < .001$). All in all, the pattern of correlations among the measures of life strains is consistent with Maslach and Jackson's (1981) research with these strain symptoms. Additionally, the pattern of correlations among all the study variables was as one would expect, *a priori* in theoretical terms and in accordance with earlier trends involving

TABLE 2   Intercorrelation Matrix for All Variables

	1	2	3	4	5	6	7	8	9	10	11
Independent Variables											
1. Negative Personal Life Stress	1.00										
2. Organizational Stress	.31***	1.00									
Moderator Variables											
3. Problem-Focused Coping	.04	.05	.58***	1.00							
4. Emotion-Focused Coping	.02	.26***	.65***	.02	1.00						
Dependent Variables											
5. Satisfaction with Work	-.21***	-.51***	-.04	.18***	-.04	1.00					
6. Emotional Exhaustion	.33***	.51***	.23	-.03	.10	-.59***	1.00				
7. Personal Accomplishment	.04	-.16**	.15**	.36***	-.06	.32***	-.15**	1.00			
8. Depersonalization	.19***	.34***	.10*	.08	.16**	-.44***	.61***	-.16**	1.00		
9. Job Performance	.06	.01	.06	.14*	-.00	-.07	-.06	.24***	-.07	1.00	
10. Absenteeism	.09	.08	.13*	-.05	.16**	-.01	-.10	-.02	.14**	.04	1.00
11. Serious Illness	.42***	.20***	.14*	-.07	.12*	-.14*	.34***	-.12	.15**	-.11	.25***

$N = 276$ (except for supervisory rating of job performance for which $N = 137$)
*$p < .05$; **$p < .01$; ***$p < .001$

TABLE 3   Correlations Between Self-Report Measures with Confidant Report Measures

Confidant Reports	Self-Report	
	Organizational Stress	Negative Personal Life Stress
1. Life Event Changes Experienced by Respondent	.21***	.39***
2. Job Stress Experienced by Respondent	.38***	.17*
3. Personal Life Stress Experienced by Respondent	.17*	.31***
4. Coping Effectiveness of the Respondent	-.22***	-.19*

*$p < .05$; ***$p < .001$; $N = 234$

these variables. Correlations between self-report measures and confidant measures are reported in Table 3. These correlations provide some evidence of validity for the life event measures.

Moderating Effects of Problem-Focused Coping

The moderated regression analyses were concerned first with the role of problem-focused coping in moderating the five organizational stress–life strain relationships. Second, we examined the moderating role of this variable on negative personal life stress–life strain relationships. As has already been shown in Table 2, job performance was unrelated to either negative personal life stress or organizational stress. Furthermore, neither of the coping measures moderated the personal life stress–performance or organizational stress–performance relationships. Similarly, no significant main or moderating effects were found in the stress–absenteeism relationship. Therefore, performance and absenteeism as dependent variables are not included in Table 4. Table 4 presents the results of the moderated regression analyses for problem-focused coping. Problem-focused coping moderated the relationships between (1) organizational stress and emotional exhaustion, (2) organizational stress and feelings of depersonalization, and (3) organizational stress and serious illness. The gains in the amount of variance explained were 1%, 3%, and 5%, respectively. Problem-focused coping slightly moderates the adverse effects of organizational stress on serious illness, feelings of depersonalization and emotional exhaustion. In addition to these modest moderating effects, problem-focused coping also had main effects on satisfaction with work and feelings of accomplishment. Individuals who utilize this particular mode of coping also experience significantly higher levels of satisfaction with work and feelings of personal accomplishment in the domain of work.

TABLE 4   Moderated Regression Analyses Using Problem-Focused Coping as a Moderator of Organizational Stress, Personal Life Stress and Life Strain Relationships

Dependent Variable	Independent and Moderator	Multiple Correlation[a]	Beta Coefficients	F[b] Step
Satisfaction with Work	Organizational Stress	.51 (.26)	-.51	95.18***
	Problem-Focused Coping	.55 (.30)	.21	16.73***
	Organizational Stress x Problem-Focused Coping	.55 (.30)	.12	.25
Satisfaction with Work	Negative Personal Life Stress	.21 (0.04)	-.21	13.01***
	Problem-Focused Coping	.29 (.08)	.19	11.07**
	Negative Personal Life Stress x Problem-Focused Coping	.31 (.10)	.53	4.03*
Emotional Exhaustion	Organizational Stress	.51 (.26)	.51	97.97***
	Problem-Focused Coping	.52 (.27)	-0.05	.86
	Organizational Stress x Problem-Focused Coping	.53 (.28)	-.50	4.34*
Emotional Exhaustion	Negative Personal Life Stress	.34 (.12)	.34	34.61**
	Problem-Focused Coping	.34 (.12)	-.04	.47
	Negative Personal Life Stress x Problem-Focused Coping	.37 (.14)	-.71	7.44**
Feelings of Accomplishment	Organizational Stress	.16 (.03)	-.16	6.89**
	Problem-Focused Coping	.40 (.16)	.37	44.36***
	Organizational Stress x Problem-Focused Coping			(F level insufficient for further computation)

(continued)

TABLE 4 (continued)

Dependent Variable	Independent and Moderator	Multiple Correlation[a]	Beta Coefficients	F[b] Step
Feelings of Accomplishment	Negative Personal Life Stress	.04 (.002)	.04	.37
	Problem-Focused Coping	.36 (.13)	.36	40.84***
	Negative Personal Life Stress x Problem-Focused Coping	.36 (.13)	.04	.03
Feelings of Depersonalization	Organizational Stress	.34 (.12)	.34	35.15***
	Problem-Focused Coping	.35 (.12)	-.09	2.67
	Organizational Stress x Problem-Focused Coping	.39 (.15)	-.79	9.40**
Feelings of Depersonalization	Negative Personal Life Stress	.19 (.04)	.19	10.17**
	Problem-Focused Coping	.21 (.04)	-.90	2.07
	Negative Personal Life Stress x Problem-Focused Coping	.27 (.07)	-.79	8.73**
Serious Illness	Organizational Stress	.20 (.04)	.20	11.72***
	Problem-Focused Coping	.22 (.05)	-.08	1.74
	Organizational Stress x Problem-Focused Coping	.31 (.10)	-1.02	14.55***
Serious Illness	Negative Personal Life Stress	.42 (.18)	.42	59.31***
	Problem-Focused Coping	.43 (.18)	-.09	2.53
	Negative Personal Life Stress x Problem-Focused Coping	.45 (.20)	-.54	4.68*

[a]Numbers in parentheses indicate magnitude of squared multiple R for each step in the regression equation.
[b]F values for testing significance of incremental gains in multiple R.
*p < .05; **p < .01; ***p < .001; N = 276

Table 4 also indicates that problem-focused coping moderates four out of five relationships between negative personal life stress and life strains. These are between negative personal life stress and (1) satisfaction with work, (2) emotional exhaustion, (3) depersonalization, and (4) serious illness. The gains in the amount of variance explained were 2%, 2%, 3%, and 2%, respectively. Problem-focused coping only moderately reduces the adverse effects of negative personal life stress on life strains as well. In addition, it also has main effects on work satisfaction and feelings of accomplishment.

Moderating Effects of Emotion-Focused Coping

Compared to the moderating effects of problem-focused coping, moderating effects of emotion-focused coping on all of the ten regressions were considerably weaker. Emotion-focused coping moderated only the relationship between organizational stress and feelings of depersonalization (increase in $R^2 = 2\%$, $\beta = .57$, $F = 4.91$, $p < .05$) and had main effects on feelings of depersonalization in the moderated regression equation associated with negative personal life stress (increase in $R^2 = 2\%$, $\beta = .16$, $F = 7.54$, $p < .01$). It also had a main effect in the prediction of serious illness with negative personal life stress (increase in $R^2 = 1\%$, $\beta = .12$, $F = 5.09$, $p < .05$). The nature of the moderating effect shows that individuals who are more predisposed towards using emotion-focused coping skills in dealing with organizational stress tend to experience higher levels of depersonalization.

DISCUSSION

Our primary aim in this investigation was to examine the moderating role of problem-focused and emotion-focused coping strategies on a variety of stress–life strain relationships. The present results of the moderating role of personal styles of coping on both organizational stress–life strain and personal life stress–life strain relationships complement the findings of Folkman and Lazarus (1980). They reported a significant increase in the use of problem-solving techniques when subjects in their study confronted work-related stressful episodes. Emotion-focused coping was more prevalent while coping with stressful outcomes of physical disabilities and illness. The present results, however, indicate that problem-solving coping strategies moderated organizational stress–life strain and personal life stress–life strain relationships to a far greater extent than did emotion-focused coping strategies. Emotion-focused coping only moderated the relationship between organizational stress and feelings of depersonalization.

The results of this study as well as those of Folkman and Lazarus (1980) are particularly interesting in view of Pearlin and Schooler's (1978) finding that people use fewer problem-focused than emotion-focused strategies directed at changing the problems encountered in the context of work. They interpreted their finding by noting that the impersonal and chronic nature of problems at work are perhaps not easily ameliorated by attempting to change them. One is perhaps better off by coping with such problems by developing a renewed cognitive perspective on them and afterwards by changing one's emotional attachments to work-related stressful encounters. In other words, emotion-focused coping was a useful strategy for the subjects of the Pearlin and Schooler (1978) study for dealing with organizational stress.

We found the opposite to be true for the present sample. One might conjecture that the teachers in this sample are relatively "hardy," high in internal locus of control and, therefore, rely more on problem-focused modes of coping with their work-related stresses. In fact, the significant relationship between emotion-focused coping and depersonalization (see Table 2) suggests that reliance on emotion-focused modes of coping might make it even more difficult to deal effectively with the demands of organizational stress. As one employs more emotion-focused coping, feelings of depersonalization also increase. This could make it difficult to keep one's energies focused on one's work and organizational activities. Overall, it seems reasonable to conclude that problem-focused coping for the present sample of teachers serves to moderate both personal life stress–strain and organizational stress–strain relationships. However, since there is no *a priori* theoretical rationale as to why one or the other modes of coping would be the choice of coping for all people relative to all situations and with respect to all possible kinds of life strains, it is quite possible that emotion-focused coping could indeed be quite effective in some situations.

Brickman et al. (1982) suggested that the attribution of the responsibility for both problems and solutions is a crucial factor in understanding the role of coping and helping. Implicit throughout the explanation of attribution of locus of responsibility and hardiness is the notion that there are ways of thinking, feeling and acting toward one's particular stressful circumstances that enhance psychological and physiological resilience. Kobasa and Puccetti (1983) argued that individuals who view stressful events as opportunities for exercising personal and social resources and gaining mastery are more effective copers compared to those who view them as threats from which one should retreat. Looking for "the silver lining" and letting "time heal the problems" are examples of the kinds of emotion-focused coping strategies that this latter group of individuals might frequently employ. The results of the present study suggest that, despite their role in protecting the "self" from excessive hardships as Lazarus and Launier (1978) hypothesized, emotion-focused coping is not effective for the amelioration of the symptoms of life strains.

One interesting implication of the low impact of emotion-focused coping in moderating the various stress–life strain relationships could be due to its possible connection to personality, which, in the long-term, could either induce or reduce additional strains. It is possible that, if one uses predominantly emotion-focused coping in response to life events, one also risks inducing personality changes which could endanger further coping effectiveness in specific contexts. Kobasa

(1982) and Kobasa and Puccetti's (1983) research on hardiness seems to support this line of reasoning. Excessive reliance on emotional strategies such as optimistic self-talk, depending on families to allow oneself some self-pity, and distraction from one's major preoccupation, might be effective in the short-term. But it could induce feelings of learned helplessness which, over time, could lead to changes in one's personality functioning. At any rate, it is an interesting theme which needs attention in future studies in this area.

Another issue that merits some serious attention is the issue of variance accounted for in some of the life strains, particularly where illness is concerned. This is consistent with studies in the life events area which also reports a similar pattern of findings. Two explanations are possible. One could be that either the life events or the organizational stress measured in this study are not sufficiently high in magnitude or intensity to affect what one could call organismic debilitation or a kind of malfunctioning that would induce greater illness in the person.

Second, there may be a ceiling effect for the life events and coping styles in predicting strain. Error in recall of the experience of life events has already been reported (Tausig, 1982). Such memory-related effects might be present in our data as well, making them less predictive of strains. A more complete understanding of how stress affects health and illness should include, along with life events and organizational stress, such promising stress-resistance factors as one's diet, genetic and family medical history, immunological functioning, etc. It is perhaps, then, that we might be able to account for a greater percentage of variance in various symptoms of life strains and occurrence of illness episodes.

An important point that one might wish to address is the issue of gender differences in personal styles of coping. Folkman and Lazarus (1980) found that men used a problem-focused mode longer than women before accepting the fact that nothing can be done. Furthermore, they reasoned that, even when it appears that nothing can be done, men might be more predisposed to think of various strategies to solve the problem than women. In a recent study, Defares, Brandjes, Nass, and Van der Ploeg (1984) report that men resort to a more active and cognitive based coping style compared to women. In contrast, women compared to men resort, to a greater extent, to social support related mechanisms in dealing with problems at work. However, in the context of our predominantly female sample of school teachers, problem-focused coping seems to be effective in reducing life strains, indicating that there may not be any significant gender differences in terms of coping style.

Statistical comparisons by tests of the means of each of the items of the coping scale by sex show that, with the exception of items 1, 8, 9 and 21, there were no significant differences. Items 8 and 9 reflect problem-focused strategies while items 1 and 21 reflect emotion-focused strategies. Thus, there are no overall differences between the sexes in terms of the relative use of either of these coping styles.

It is possible, as Folkman and Lazarus (1980) noted, that whatever differences might be present are perhaps more attributable to sources of stress (i.e., work, personal life, health related, retirement, etc.) rather than gender differences in styles of coping within similar contexts (i.e., work or personal life or health). But they could, indeed, differ in terms of the *contexts* in which their stressful experiences might occur. For example, in the Folkman and Lazarus (1980) investigation, women reported more health-related episodes than men and health-related events were associated with gender emotion-focused coping. However, they also found males and females did not differ in their respective uses of emotion-focused techniques in the *specific* context of coping with health-related stressful episodes.

We need additional investigations on this important theme before we know for sure how males and females differ, either by contexts or by styles of coping, or if they do differ at all. The results of this are generalizable to academic contexts and to other human services agencies. These patterns may not emerge in predominantly manufacturing or high tech organizations.

Lack of significant relationships between measures of stress, performance and absenteeism merits some discussion. Performance of the teachers was measured by supervisory ratings from their principals. Possible scores could range from 19 to 95 with a sample range of 48–95, mean of 75.50 and a standard deviation of 10.36. Thus, this sample of teachers was rated as being relatively high in terms of rated performance and this restriction of range has possibly affected some of the correlations.

It is possible that the levels of stress measured in this investigation might not be high or intense enough to cause debilitation to such a point where an individual teacher is either unable to function effectively, to perform his/her teaching related duties, or becomes too sick to even report to work. Bhagat (1983) has argued that, in some organizational settings, an employee just has to maintain his/her performance regardless of his/her level of experienced stress in order to conform to the expectations or culture of that setting. It is conceivable that, in this setting, the teachers experienced an implicit norm to perform at an adequate level regardless of their level of experienced stress, either in the context of their organizational life or personal life. Future research in this area should follow the suggestions made by Latack and Havlovic (1992) and Endler and Parker (1994). Their recommendations are focused on the assessment of multidimensional aspects of coping and we believe that this new trend will provide additional insights into the dynamics of coping with stressful life events.

The effects of stress on performance and absenteeism are perhaps best investigated in a longitudinal research design which allows for a fuller understanding of how various stressful experiences slowly and over repeated encounters might affect both performance and absenteeism. In a similar vein, the moderating effects of both problem-focused and emotion-focused coping are also perhaps best understood in a longitudinal design. The processes relating to how one might cope and whether one is better off using either a problem-focused mode or an emotion-focused mode in the context of one's work and personal life might unfold better over time. If learned helplessness is an important element in this process, its influence on emotion-focused coping and performance could be better examined and interpreted in future longitudinal designs.

Implications for Future Research

These results have several implications for research in the area of occupational stress and coping. First, personal orientations in terms of idiosyncrasies of styles of coping should be studied in greater depth as possible moderators of various stress–strain relationships. Second, it is important to develop a more precise theoretical argument as to when and under what circumstances problem-focused coping styles might be relatively more effective than emotion-focused coping styles in the work arena and vice-versa. Third, a typology of boundary conditions, as well as personality differences which evoke the use of problem-focused vs. emotion-focused coping strategies, needs to be developed and examined in future studies. Finally, the investigation of other forms of life-strains (e.g., frustration, etc.) or different measures of the present life strains might prove useful in future studies to help determine the convergent validity of the present methods and measures.

Kahn (1981) and Katz and Kahn (1978), in their attempts at integrating the diverse body of empirical literature on stress, underscored the importance of work, non-work and personality factors in the prediction of work- and health-related outcomes. The results of the present investigation indicate that all three of these factors are important in predicting employee reactions to stressful life events.

REFERENCES

Beehr, T.A., & Newman, J.E. (1978). Job stress, employee health, and organizational effectiveness: A facet analysis, model, and literature review. *Personnel Psychology, 31,* 665–699.

Bhagat, R.S. (1980). *Effects of personal life stress upon individual performance effectiveness and work adjustment processes within organizational settings.* James McKeen Cattell Invited Address delivered to the Division of Industrial and Organizational Psychological Association (J.R. Hackman, Chair), Montreal.

Bhagat, R.S. (1983). Effects of stressful life events on individual performance effectiveness and work adjustment processes within organizational settings: A research model. *Academy of Management Review, 8*(4), 660–671.

Bhagat, R.S., & Allie, S.M. (1989). Organizational stress, personal life stress and symptoms of life strains: An examination of the moderating role of sense of competence. *Journal of Vocational Behavior, 35,* 231–253.

Bhagat, R.S., McQuaid, S.J., Lindholm, H., & Segovis, J.C. (1985). Total life stress: A multimethod validation of the construct and its effects on organizationally valued outcomes and withdrawal behaviors. *Journal of Applied Psychology, 70*(1), 202–214.

Brickman, P., Rabinowitz, V.C., Karuza, J.K., Jr., Coates, D., Cohn, E., & Kidder, L. (1982). Models of helping and coping. *American Psychologist, 37,* 368–384.

Burke, R.J., & Weir, T. (1980). Coping with the stress of managerial occupations. In C.L. Cooper & R. Payne (Eds.), *Current concerns in occupational stress* (pp. 299–335). Chichester: Wiley.

Cohen, J. (1978). Partial products are interactions: Partial powers are curve components. *Psychological Bulletin, 85*(4), 858–866.

Cooper, C.L., & Payne, R. (Eds.). (1978). *Stress at work.* New York: Wiley.

Cooper, C.L., & Payne, R. (Eds.). (1980). *Current concerns in occupational stress.* New York: Wiley.

Cordes, C.L., & Dougherty, T.W. (1993). A review and an integration of research on job burnout. *Academy of Management Review, 18*(4), 621–656.

Corrigan, P.W., Holmes, E.P., Luchins, D., Buican, B., Basit, A., & Parks, J.J. (1994). Staff burnout in a psychiatric hospital: A cross-lagged panel design. *Journal of Organizational Behavior, 15,* 65–74.

Crandall, R. (1976). Validation of self-report measures using ratings by others. *Sociological Methods and Research, 4,* 380–400.

Defares, P.B., Brandjes, M., Nass, C.H., & Van der Ploeg, J.D. (1984). Coping styles and vulnerability of women at work in residential settings. *Ergonomics, 27,* 527–545.

Dohrenwend, B.S., & Dohrenwend, B.P. (Eds.). (1981). *Stressful life events and their contexts.* New York: Prodist.

Endler, S.N., & Parker, J.D.A. (1994). Assessment of multidimensional coping: Task, emotion and avoidance strategies. *Psychological Assessment, 6*(1), 50–60.

Folkman, S., & Lazarus, R.S. (1980). An analysis of coping in a middle-aged community sample. *Journal of Health and Social Behavior, 21,* 219–239.

Ford, D.L., Murphy, C.J., & Edwards, K.L. (1983). Exploratory development and validation of a perceptual job burnout inventory: Comparison of corporate sector and human services professionals. *Psychological Reports, 52,* 995–1006.

Gottlieb, B.H. (1983). Social support as a focus for integrative research in psychology. *American Psychologist, 38,* 278–287.

Holmes, T.H., & Rahe, R.H. (1967). The social readjustment rating scale. *Journal of Psychosomatic Research, 11,* 213–218.

Ilfeld, F.W. (1980). Coping styles of Chicago adults. *Journal of Human Stress, 6,* 2–10.

Jackson, S.E., Schwab, R.L., & Schuler, R.S. (1986). Toward an understanding of the burnout phenomenon. *Journal of Applied Psychology, 71*(4), 630–640.

Jamal, M. (1984). Job stress and job performance controversy: An empirical assessment. *Organizational Behavior and Human Performance, 33*(1), 1–21.

Johnson, J.H., & Sarason, I.G. (1979). Recent developments in research on life stress. In V. Hamilton & D.M. Warburton (Eds.), *Human stress and cognition: An information processing approach.* New York: Wiley.

Kahn, R.L. (1981). *Work and health.* New York: Wiley.

Katz, D.M., & Kahn, R.L. (1978). *The social psychology of organizations* (2nd ed.). New York: Wiley.

Kobasa, S.C. (1982). The hardy personality: Toward a social psychology of stress and health. In J. Suls & G.S. Sanders (Eds.), *The social psychology of health and illness.* Hillsdale, NJ: Erlbaum.

Kobasa, S.C., & Puccetti, M.C. (1983). Personality and social resources in stress resistance. *Journal of Personality and Social Psychology, 45,* 839–850.

Kornhauser, A. (1965). *Mental health of the industrial worker.* New York: Wiley.

Kyriacou, C., & Sutcliffe, J. (1977). Teacher stress: A review. *Educational Review, 29,* 299–306.

Kyriacou, C., & Sutcliffe, J. (1978). Teacher stress: Prevalence, sources and symptoms. *British Journal of Educational Psychology, 2,* 159–167.

Latack, J.C., & Havlovic, S.J. (1992). Coping with job stress: A conceptual evaluation framework for coping measures. *Journal of Organizational Behavior, 13,* 479–508.

Lazarus, R.S., Averill, J.R., & Opton, E.M., Jr. (1974). The psychology of coping: Issues of research and assessment. In G.V. Coelho, D.A. Hamburg, & J.F. Adena (Eds.), *Coping and adaptation* (pp. 249–315). New York: Basic Books.

Lazarus, R.S., & DeLongis, A. (1983). Psychological stress and coping in aging. *American Psychologist, 38*(3), 245–254.

Lazarus, R.S., & Launier, R. (1978). Stress-related transactions between person and environment. In L.A. Pervin & M. Lewis (Eds.), *Perspectives in interactional psychology* (pp. 287–327). New York: Plenum.

Lee, C., Ashford, S.J., & Jamieson, L.F. (1993). The effects of Type A behavior dimensions and optimism on coping strategy, health and performance. *Journal of Organizational Behavior, 14,* 143–157.

Locke, E.A. (1976). The nature and consequences of job satisfaction. In M. Dunnette (Ed.), *Handbook of industrial and organizational psychology* (pp. 1297–1350). Chicago: Rand McNally.

Maslach, C. (1982). Understanding burnout: Definitional issues in analyzing a complex phenomenon. In W.S. Paine (Ed.), *Job stress and burnout: Research, theory, and intervention perspectives* (pp. 29–41). Beverly Hills, CA: Sage.

Maslach, C., & Jackson, S.E. (1981). The measurement of experienced burnout. *Journal of Occupational Behavior, 2,* 99–113.

McLean, A.A. (1979). *Work stress.* Reading, MA: Addison-Wesley.

Motowidlo, S.J., Packard, J.S., & Manning, M.R., Jr. (1986). Occupational stress: Its causes and consequences for job performance. *Journal of Applied Psychology, 71*(4), 618–629.

Near, J.P. (1984). Predictive and explanatory models of work and non-work. In M.D. Lee & R.N. Kanungo (Eds.), *Management of work and personal life.* New York: Praeger Scientific.

Newman, J.E., & Beehr, T.A. (1979). Personality and organizational strategies for handling job stress: A review of research and opinion. *Personnel Psychology, 32,* 1–44.

Parker, D.F., & DeCotiis, T.A. (1983). Organizational determinants of job stress. *Organizational Behavior and Human Performance, 32,* 160–177.

Pearlin, L.I., & Schooler, C. (1978). The structure of coping. *Journal of Health and Social Behavior, 19,* 2–21.

Peters, W.S., & Champoux, J.E. (1979). The role and analysis of moderator variables in organizational research. In R.T. Mowday & R.M. Steers (Eds.), *Research in organizations: Issues and controversies.* Santa Monica: Goodyear.

Rabkin, J.G., & Streuning, E.L. (1976). Life events, stress and illness. *Science, 194,* 1013–1020.

Sarason, I.G., & Johnson, J.H. (1979). Life stress, organizational stress and job satisfaction. *Psychological Reports, 44,* 75–79.

Sarason, I.G., Johnson, J.H., & Seigel, J.M. (1978). Development of the life experiences survey. *Journal of Consulting and Clinical Psychology, 46,* 932–946.

Schuler, R.S. (1980). Definition and conceptualization of stress in organizations. *Organizational Behavior and Human Performance, 25,* 184–215.

Smith, P.C., Kendall, L.M., & Hulin, C.L. (1969). *The measurement of satisfaction in work and retirement.* Chicago: Rand-McNally.

Tausig, M. (1982). Measuring life events. *Journal of Health and Social Behavior, 23*(1), 52–64.

Vinokur, A., & Selzer, M. (1975). Desirable versus undesirable life events: Their relationship to stress and mental distress. *Journal of Personality and Social Psychology, 66,* 297–333.

Wyler, A.R., Masuda, M., & Holmes, T.H. (1968). Seriousness of illness rating scale. *Journal of Psychosomatic Research, 11,* 363–375.

Wyler, A.R., Masuda, M., & Holmes, T.H. (1971). Magnitude of life events and seriousness of illness. *Psychosomatic Medicine, 32,* 115–122.

A Field Study of Some Antecedents
and Consequences of Felt Job Stress

Timothy P. Summers
Thomas A. DeCotiis
Angelo S. DeNisi

In addition to its financial costs, job stress has been linked to other serious consequences for both the individual (e.g., high blood pressure and heart disease) and the organization (e.g., decrements in performance, and increased turnover and absenteeism) (Bedeian & Armenakis, 1981; Beehr & Newman, 1978; Fox, Dwyer, & Ganster, 1993; Gupta & Beehr, 1979; Hendrix, Ovalle, & Troxler, 1985; Hendrix, Steel, Leap, & Summers, 1994; Parasuraman & Alutto, 1984; Schnall, Devereaux, Pickering, & Schwartz, 1992). Hence, the current interest in job stress on the part of both researchers and practitioners is not surprising. What is very surprising, however, is that this interest has resulted in little agreement about the content, causes, and consequences of job stress.

For example, researchers have defined the construct of job stress in broad terms (Gibson, Ivancevich & Donnelly, 1985; McDonald & Korabik, 1991), in narrow terms (O'Hanlon, 1978), and occasionally not at all (Barling & Rosenbaum, 1986). They have treated it as a stimulus variable (Lazarus & Launier, 1978; Steffy & Laker, 1991), a response (Quick & Quick, 1984), the environment (Caplan, Cobb, French, Harrison & Pinneau, 1975), and a person–

Authors' Notes: The authors would like to acknowledge the helpful suggestions of William H. Hendrix.

environment interaction (Gibson et al., 1985). Sometimes they have used several definitions simultaneously (cf., Beehr & Newman, 1978; or Katz & Kahn, 1978).

This variety of treatments may speak to the universal nature and importance of job stress. However, the absence of any consensus on the concept and the process of job stress makes it difficult to either understand or control job stress and its consequences. The purpose of the present paper is to propose a new conceptualization and a new measure of job stress, and to report on an initial test of it.

The Concept of Job Stress

The concept of stress we will use in the present study is a simple one: Stress is the manifestly uncomfortable *feeling* that an individual experiences when he or she is forced to deviate from normal or desired patterns of functioning. Job stress, then, is simply the set of such feelings that derives from forces found in the workplace. This conceptualization is similar to Quick and Quick's (1984) definition of strain, but the emphasis here is on the individual's feelings resulting from a deviation in patterns of functioning, not the deviation per se. Our definition suggests that there is no stress unless the person is aware of it (Gaylin, 1979; Parker & DeCotiis, 1983). Thus, the definition is limited to the simpler case of felt stress which allows for the possibility that a given situation can be differentially stressful to different people. This individual differences approach was incorporated in the social readjustment scale of Holmes and Rahe (1967), but they measured 43 "stressful" life events (or stressors), not felt stress per se.

Our definition of job stress contrasts with previous ones, particularly in its emphasis on *the individual experience and feeling* of it. In contrast, Lazarus and Launier (1978) refer to the *causes* of stress by defining it as any *event* that takes or exceeds normal adaptive responses. At the other end (on a continuum) of stress definitions (beginning at the left with the causes of stress and ending on the right with its consequences) is the definition of Quick and Quick, "The *stress response* is the generalized, patterned, unconscious mobilization of the body's natural energy resources when confronted with a stressor," (1984, p. 3). The operational definition of Caplan et al. (1975) is, in contrast, far to the left of the continuum because job stress is operationalized as work load, job complexity, role conflict, and role ambiguity, along with several other variables regarded as stressors in more "centrist" definitions. The proposed definition is intended to come closer to the center of the continuum by permitting more direct measurement of perceived felt job stress.

Additional support for this definition of job stress and the proposed model can be found in the results of a study by Hendrix, et al. (1994), who defined job stress along the lines suggested here—i.e., as felt job stress. Those researchers employed a less comprehensive, three-item measure of job stress, but they did include a similar, three-item measure of life stress. Importantly, their model of antecedents and consequences of stress contained many of the same variables and was strongly supported by their data.

Our definition has implications for the measurement of stress. It implies individual differences in reactions to potentially stressful situations, and individual differences in the way stress is experienced. Seyle (1976) identified two categories of individual modifiers of the response to stress, internal conditioning factors, and external conditioning factors. Internal factors include past experiences, personality, age and sex, while external factors include diet, drugs, climate, and social setting. Thus, self reports of stress would have to allow for self-definitions as well. That is, any self report measure of stress must allow the individual to indicate his or her unique reaction to a situation, and must allow for potentially unique manifestations of stress.

Potential Causes of Job Stress

As noted earlier, the concept of job stress has to do with its content, while the process of becoming stressed has to do with its causes. In the latter situation we are interested in environmental events or characteristics (stressors) that alter the individual's state of well-being. When an individual is exposed to those stressors for prolonged periods of time, he or she reacts in a way that may have negative consequences for the individual as well as for the organization. These reactions can be physiological (Christensen & Jensen, 1994; Fox, et al., 1993; Hendrix, et al., 1985; Howard, Cunningham, & Rechnitzer, 1986; Moss, 1981), psychological (Cordes & Dougherty, 1993; Parker & DeCotiis, 1983), behavioral (Cordes & Dougherty, 1993; Jamal, 1984), and/or social (Epstein, 1976).

Figure 1 presents a general research model of job stress within which our concept of stress is embedded. The model is based on two typical models of the antecedents and consequences of job stress. Those models hypothesize a set of organizational stressors, a set of personal characteristics, an individual stress response (a new measure of which is provided in this paper), and a set of individual consequences (Ivancevich & Matteson, 1980; Quick & Quick, 1984). Each of the variables hypothesized to be organizational or role stressors or individual consequences has been included in at least one of the referent models. The model consists of four categories of variables, all of which have been implicated as factors which lead to job stress: personal characteristics, structural organizational characteristics, procedural organizational characteristics, and role characteristics (Hendrix et al., 1994). Figure 2 contains the hypothesized paths among specific variables within our model.

Personal characteristics include sex, tenure in present job, tenure in the company, and number of dependents. Included in organizational (structural) characteristics are formalization and centralization in structure. Organizational characteristics (procedural) refer to the amount and quality of communications, the quality of training, the equity of the reward system, the nature of decision making, the quality of the performance appraisal and feedback system, and hours worked per week. Role characteristics include job level, leadership received, role conflict, and role ambiguity.

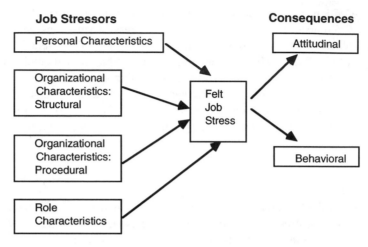

FIGURE 1   General Research Model: The Causes and Consequences
of Job Stress

The role related factors, such as role conflict and role ambiguity, have been linked to stress by Caplan, et al. (1975). These role demands were identified in the models of Hendrix et al. (1994), Ivancevich and Matteson (1980) and Quick and Quick (1984). Organizational characteristics linked to job stress by Hendrix et al. (1985) included organizational control, supervision, communications, and participation. The organizational (structure) stressors were modeled by Ivancevich and Matteson (1980), while the organizational (procedural) stressors were modeled by either Ivancevich and Matteson (1980) or by Quick and Quick (1984). Personal and family factors have been implicated in several studies and models (Burke, Weir, & DuWors, 1979; Ivancevich & Matteson, 1980; Marshall & Cooper, 1979; Matteson & Ivancevich, 1979).

Figures 1 and 2 also include two categories of consequences of job stress: attitudinal and behavioral. The linkages of the consequences in Figure 2 are based on Mowday, Porter, and Steers (1982). Attitudinal consequences include intrinsic and extrinsic satisfaction, organizational commitment, motivation, and intention to leave, while the behavioral consequence modeled is voluntary turnover. Stress has been implicated as a factor which leads to job dissatisfaction, absenteeism, intent to leave, and voluntary turnover (Hendrix, et al., 1985; Hendrix, et al., 1994; Ivancevich, Matteson, & Preston, 1982). In terms of the attitudinal consequences, dissatisfaction and low motivation were referred to by Quick and Quick (1984) as indirect costs of job stress through a loss of vitality. In addition to dissatisfaction and low motivation, Ivancevich and Matteson (1980) included in their model the negative effect of job stress on organizational commitment hypothesized in Figure 2.

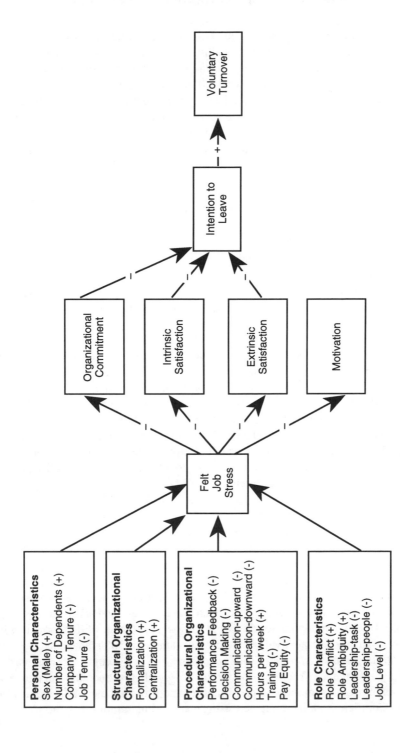

FIGURE 2   Hypothesized Path Model

117

The purpose of the present study is to develop, through exploratory path analysis, a model of stress using a unique measure of felt stress that allows for individualized reactions to, or manifestations of, potential stressors on the job. The specific hypotheses of this research are illustrated in Figure 2 in the path analytic framework. Paths are indicated by the directions of the arrows, and the plus or minus signs indicate the directions of the hypothesized relationships between factors. Absences of arrows indicate hypotheses of no relationships. The curved, two-headed arrows usually provided between exogenous variables were omitted, but the intercorrelations were examined during the analysis.

METHOD

Subjects

The present study was conducted in a nationwide restaurant chain headquartered in the southwest. Surveys were mailed to 1,043 managerial employees from five levels of management. The questionnaires were completed anonymously and returned to the researchers. The company's accounting department provided data on voluntary turnover. Questionnaires were returned by 365 managers for a 35 percent response rate. The 200 managers who provided their names were tracked for eight months to determine voluntary turnover. Nineteen, 9.5 percent, left voluntarily.

Some characteristics of the final sample of 365 were that the mean hours worked per week was 63 (SD = 10.4), the mean time in the present job was 17 months (SD = 20), the mean time with the company was 46 months (SD = 66), and the mean number of dependents was 1.5 (SD = 1.4). Of the 365 managers, only 53 (14.5 percent) were women and 25 (6.8 percent) were racial minorities. Managers at all levels were paid bonuses which comprised between 40 percent and 60 percent of their total pay. The bonuses, and to a lesser extent, base salaries, were based on the restaurants' profitability.

Measures

Personal characteristics. Personal characteristics of respondents included sex, number of dependents, tenure in the organization, and tenure in the present job. Each variable was measured with a single item.

Organizational characteristics (structural). Organizational structure was assessed with a five item measure of *formalization* and a five item measure of *centralization.* These measures had been used earlier by Parker and DeCotiis (1983). Coefficient alphas for the two scales were .76 and .66 for formalization and centralization respectively in the present sample of 365 managers.

Organizational characteristics (processes). A ten item scale was used to assess *performance feedback*, the extent to which respondents' superiors set performance goals, monitored performance, and provided clear and useful feedback. Coefficient alpha computed using the present sample was .90. *Decision making* was measured with Parker and DeCotiis' (1983) four item scale, and indicated the extent to which respondents felt they could participate in decision making, and their perceptions that decisions were consistent and well thought out. Coefficient alpha using this sample was .65. The adequacy and openness of both *upward and downward communications* were assessed using a three item scale and a two item scale respectively, both developed by Parker and DeCotiis (1983). Coefficient alphas, using the present sample, were found to be .74 and .79 for upward and downward communications, respectively. Respondents were asked, using a single item, to report the average number of *hours* they worked *per week*. The adequacy and quality of the *training programs* were assessed using an eight item scale developed for this study. Coefficient alpha was .73.

Pay equity. The equitability of the pay system was measured with 9 items, which were developed by Summers and DeNisi (1990). The measure required respondents to indicate, on a one–seven scale, whether their pay was "much too low" (1) or "much too high" (7) compared to nine possible referent others. Coefficient alpha for the present sample was .87.

Role characteristics. Four separate measures of potential stressors associated with the respondents' role in the organization were also included. *Role conflict* was measured with five items and *role ambiguity* with four items of the scales developed by Rizzo, House, and Lirtzman (1970). Coefficient alphas in the present study were .70 for conflict and .76 for ambiguity. Four-item measures of *leaders' task and people orientations*, used by Parker and DeCotiis (1983) were used to assess perceptions of the type of leadership provided. Coefficient alphas for the present sample were .74 for task orientation and .84 for people orientation. Respondents' level in the organization was assessed using a single item indicating job title. There were six job titles corresponding to the six levels of management represented in the present sample. Job level scores therefore ranged from one to six, with higher scores indicating higher levels in the organization.

Job stress. The present measure assumed that individuals do not all experience stress in the same way, and so allowed for individualized measures of stress. Therefore, respondents were given a list of 52 potential symptoms of job stress. The list was generated through a series of open-ended questionnaires and interviews of managers in several organizations, and included such symptoms as "feel sleepy," "feel a tightness in my chest," "lose my appetite," "lack energy," and "smoke more than usual." The list of job stress items was examined by an expert scholar of job stress (with 9 articles and two book chapters on the topic) who found the list to be a content-valid one. For each potential symptom, respondents indicated how often, if ever, they had experienced the symptom (0 to 6 scale), and how uncomfortable the symptom was if they had experienced it (0 to 6 scale). Individual stress scores were computed by summing the products of the two ratings for each item. Thus, a person who never experienced any of the symptoms received a score of 0, while a person who frequently experienced all the symptoms, and experienced them with great discomfort, received a score of 1,872. The actual range of scores in the present sample was 50 to 1,064 with a mean of 266 (SD = 177). This approach allowed stress to be operationalized differently for each person depending on the kinds of symptoms he or she experienced as a result of job stress.

In an initial assessment of the reliability and dimensionality of this measure of felt job stress, three procedures were undertaken: factor analysis, assessment of the internal consistency of the predominant factor, and the computation of a test-retest coefficient. Because the 52 individual stress items were not thought to be orthogonal, factor analysis using oblique rotation was employed. Forty-eight of 52 items loaded on a single factor whose eigenvalue was 14.9. With an eigenvalue for the second factor of only 2.3, and no apparent (conceptual) clustering of items for any of the 12 other factors, (which carried eigenvalues between 1.0 and 2.1), these data suggest the stress measure is primarily unidimensional with one major factor accounting for most of the variance. In addition, the coefficient alpha reliability of .95, computed using this sample of 365 managers' responses to all 52 items, suggests a high degree of interrelatedness among items.

A sample of 30 undergraduate students (in a human resources management course at a southeastern U.S. university) was used to compute a test-retest coefficient. The instructions for the job stress measure were reworded to reflect "school" stress. The two administrations of the stress measure were separated by one week. The test-retest coefficient was .71, indicating a reasonable degree of stability over time, especially considering expected changes in levels of felt stress.

Consequences. Six potential consequences of job stress were also included. *Extrinsic satisfaction* was comprised by 18 items, measuring satisfaction with: pay, opportunities for advancement, supervision, co-workers, and benefits. Both the extrinsic and intrinsic satisfaction measures were based on the Minnesota Satisfaction Questionnaire (Weiss, Dawis, England & Lofquist, 1966). *Intrinsic satisfaction* was measured with four items addressing the actual work performed. For this sample, coefficient alpha for extrinsic satisfaction was .91, for intrinsic satisfaction, .80. *Organizational commitment* was assessed using a six item scale developed by DeCotiis and Summers (1987). Coefficient alpha in the present sample was .72. *Motivation* was measured with three items dealing with the respondent's willingness to put in additional hours and expend additional effort for the organization. Coefficient alpha for this sample was computed to be .78. A three item measure was used to assess respondents' *intention to leave* the organization. Coefficient alpha for the present sample was .90. V*oluntary turnover* after eights months was reported to the researchers by the organization's accounting department.

Analysis

An exploratory path analysis was conducted to determine which of the hypothesized paths (see Figure 2) were statistically significant so that a revised model could be developed for testing in future research. While path analyses are useful for model testing, the results should be interpreted with caution when the model contains

correlated factors, since *extreme* multicollinearity can lead to path coefficient instability (Asher, 1976). Table 1 presents the means, standard deviations, and intercorrelations for all study variables. Based on an examination of the intercorrelations
among the exogenous variables, i.e., variables 1–18 in Table 1, instability was not
assumed to be a problem in the present study. Generally, the intercorrelations were
not high; they ranged, in absolute value, from 0 to .83, with a mean absolute value of
.28. Only seventeen of the 153 correlations were above .60 (absolute value). In
addition, the use of a large sample size probably reduced the problem of
multicollinearity by decreasing the standard error (Billings & Wroten, 1978).

RESULTS

Path Analysis

The first stage of the path analysis involved regression of each variable in the
model (see Figure 2) against all preceding variables in order to produce a revised
model. The standardized beta weights, which indicate the relative strength of
variables in the model, served as path coefficients. Variables that did not have
statistically significant ($p < .05$) path coefficients were removed from the model.

The second step in the path analysis was to rerun the regression equations
using the retained variables. The result was a revised path model which is shown
in Table 2 and Figure 3. Table 2 is included in order to simplify presentation of the
paths from the exogenous variables.

The revised model, presented in Table 2 (and in Figure 3), indicates that many
of the hypothesized paths from the exogenous factors were not supported. Most
notably, the paths from the organizational structure characteristics of formalization and centralization were not supported. Among the organizational processes,
the hypothesized links from decision making, communications, hours, and pay
equity to job stress were also not supported. Perhaps most surprising was the lack
of support for the hypothesized path from role ambiguity to job stress. The paths
from exogenous variables to job stress that received support were: the number of
dependents, performance feedback, training, role conflict, and supervisors' people
leadership skills. Of the 26 statistically significant direct paths from the exogenous variables, only three were from the combination of personal characteristics
and structural organizational characteristics.

The revised path model was much more supportive of the links between
endogenous factors, which are reported in Figure 3. These data supported the
hypothesized paths from job stress to intrinsic satisfaction, extrinsic satisfaction,
and motivation. Two of the three hypothesized paths to intention to leave were
supported, i.e., from organizational commitment and intrinsic satisfaction. Finally, the path from intention to leave to voluntary turnover was also significant,
with sex unexpectedly emerging as a significant predictor of voluntary turnover
with females *more* likely to leave.

TABLE 1 Means, Standard Deviations, and Intercorrelations Among All Variables

	M	SD	1	2	3	4	5	6	7	8	9	10	11	12	13	14	15	16	17	18	19	20	21	22	23	24	25
1 Sex	NA	NA	—																								
2 Dependents	1.5	1.4	27	—																							
3 Job Level	2.5	1.1	07	24	—																						
4 Co. Tenure	46.2	66.2	-15	21	60	—																					
5 Job Tenure	16.8	20.4	11	17	27	49	—																				
6 Hours	63.0	10.4	01	05	23	25	05	—																			
7 Formalization	4.9	0.9	04	-06	-17	-07	-12	-23	—																		
8 Centralization	3.8	1.0	-10	05	-05	06	09	13	-29	—																	
9 Perf. Feedback	4.7	1.1	03	-06	-12	-04	-13	-28	83	-40	—																
10 Decision Making	4.7	1.0	00	-06	-11	-05	-16	-16	67	-55	70	—															
11 Role Conflict	3.7	1.0	00	11	01	-03	03	12	-37	46	-41	-57	—														
12 Role Ambiguity	2.8	1.0	-13	01	01	01	07	13	-73	47	-71	-61	52	—													
13 Leadership (People)	4.9	1.2	08	08	-08	-04	-17	-23	66	-43	79	62	-40	-61	—												
14 Leadership (Task)	5.3	0.9	08	-09	-13	-08	-18	-23	80	-40	78	67	-40	-66	77	—											
15 Comm. (Upward)	4.9	1.2	00	-08	-05	00	-16	-17	59	-58	68	07	-49	-58	73	64	—										
16 Comm. (Downward)	5.1	1.2	03	-06	02	04	-09	-14	53	44	57	57	-44	-54	59	55	63	—									
17 Training	5.2	0.8	-05	-15	-18	-13	-20	-21	57	-37	54	57	-47	-52	47	60	56	49	—								
18 Pay Equity	3.5	0.7	07	06	09	04	04	-10	13	-10	17	25	-17	-23	18	14	23	15	19	—							
19 Ex. Satisfaction	4.9	1.0	-02	-06	-06	-09	-17	-24	60	-47	70	70	-51	-61	58	60	55	53	63	49	—						
20 Int. Satisfaction	5.2	1.0	00	-04	-05	-07	-15	-18	55	-50	64	62	-46	-57	58	56	68	38	56	32	78	—					
21 Commitment	5.7	0.7	06	04	05	04	-09	-06	43	-31	39	48	-28	-41	30	48	67	33	50	15	54	53	—				
22 Motivation	5.8	0.9	02	-03	07	02	-08	-15	33	-29	34	42	-34	-39	28	35	40	33	43	19	52	58	61	—			
23 Intention to Leave	2.1	1.3	-02	01	-01	-03	12	13	-44	31	-46	-51	37	45	-40	-45	-47	-40	-42	-28	-61	-62	-59	-63	—		
24 Vol. Turnover	NA	NA	-19	-11	-15	-14	02	-08	-01	00	-08	-09	08	12	-09	-05	-04	-08	04	-15	-14	-10	-09	-11	-26	—	
25 Stress	266.4	177.2	-06	-07	-02	-02	-01	15	-24	28	-31	-28	32	28	28	-21	-30	-30	-10	-35	-37	-37	-17	-31	-28	-02	—

Note: Decimals are omitted. Although sample sizes vary due to missing data, all correlations are based on at least 357 cases, except for the voluntary turnover correlations where the correlations are based on 200 cases. Sex was coded 0 = females, 1 = males; voluntary turnover was coded 0 = stay, 1 = leave .24, correlations ≥ .09, p < .05; correlations ≥ .12, p < .01; correlations ≥ .16, p < .001 for rows 1-23 and 25. For row 24, correlations ≥ .11, p < .05; correlations ≥ .16, p < .01.

TABLE 2   Paths From Exogenous Variables to Stress and its Consequences

Factor and Variable	Path Coefficients					
	Job Stress	Organizational Commitment	Intrinsic Satisfaction	Extrinsic Satisfaction	Motivation	Voluntary Turnover
Personal Characteristics						
Sex (male)						
Number of Dependents						-.19
Company Tenure	-.12	.10		-.06		
Procedural						
Organizational Characteristics						
Performance Feedback	-.25	.27	.27	.35		
Decision Making			.31	.15	.16	
Communication-upward		.31	.17	.16		
Training	-.14		.16	.19	.26	
Pay Equity				.32		
Role Characteristics						
Role Conflict	.22			-.07		
Leadership-Task		.33				
Leadership-People	-.16	-.24		-.09		
Job Level					.15	

Note: Only statistically significant paths are included (p < .05).

123

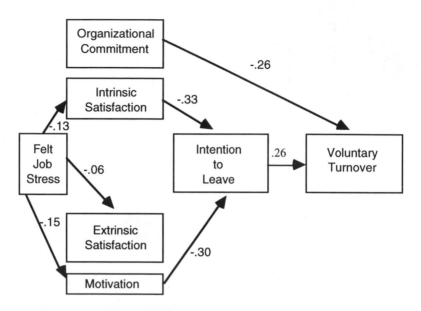

FIGURE 3   Revised Path Model – Relationships Among
Endogenous Variables

Confirming Hendrix et al. (1985) the lack of a significant (direct) relationship between stress and intention to leave (or turnover) suggests an indirect effect of stress on intention to leave and voluntary turnover, through satisfaction.

The final stage of the path analysis was to test the model for goodness of fit to determine whether the retained paths adequately fit the data. The large-sample chi-square test suggested by Kim and Kohout (1975) was employed. This L test involves a comparison of the unexplained variance in the revised model (i.e., the retained paths) to the unexplained variance in the full model (i.e., all possible prior predictors). If L is not significant the full and revised models explain similar amounts of variance, and therefore the revised model adequately fits the data. In this case, L = 0, ns, indicating a good fit of the model to the data.

Table 3 contains the percentage of variance explained by the full model and the revised model for each endogenous variable (which was regressed on those variables prior to the regressed variable).

DISCUSSION

A path analytic model was proposed that hypothesized that personal, organizational, and role characteristics relate to job stress and its attitudinal and behavioral consequences.

TABLE 3   Variance Explained by Full and Revised Models

Dependent Variable	Full Model	Revised Model
Job Stress	.22	.18
Organizational Commitment	.40	.36
Intrinsic Satisfaction	.60	.58
Extrinsic Satisfaction	.76	.75
Motivation	.31	.29
Intention to Leave	.57	.53
Voluntary Turnover	.21	.10

Note: n = 357–365 for all but voluntary turnover, where n = 200.

While this exploratory path analysis cannot be considered as confirmation of the revised model, it does offer useful suggestions for both theory development and research.

It is especially interesting to note that procedural-organizational characteristics and role characteristics appeared to be much more important as stressors than were personal characteristics or structural-organizational characteristics. This suggests that policies and procedures introduced by organizations are an important source of job stress. For example, these results suggest that the quality of training and performance feedback are particularly important to felt job stress. Although not a major focus of the present study, these results are also noteworthy in that the measure of performance feedback was found to be strongly related to a number of variables of interest, e.g., role conflict, role ambiguity, extrinsic satisfaction, and intention to leave the organization. This is important because, although there have been frequent calls for performance feedback and regular goal setting as part of a system of performance management, there has been little empirical research to support these calls.

A major purpose of the present study was to test the usefulness of a new, individualized measure of job stress. It seems reasonable that individuals experience stress in somewhat unique ways, and the present measure allowed for the expression of that uniqueness. The patterns of relationships obtained suggest that our measure of felt job stress is potentially useful. In fact, although the relationships obtained in the present study were not any stronger than those obtained in other studies of stress, it is important to note that most of those studies did not actually measure stress. Instead, it has been common to measure stressors, correlate them with consequences of stress, and infer stress (cf. Caplan, et al., 1975). Thus, we believe the present study presents a more accurate picture of what does cause stress on the job and what the consequences of that stress are. One final note on the magnitude of the relationships obtained—the present study, as is typical of job stress research, only considered potential stressors on the *job*. Yet stress, as experienced on the job, is also partly a function of factors off the job. By not considering these "life events" we will always be limited in how well we can predict stress in any setting.

Additional support for this definition of job stress and the proposed model can be found in the results of a study by Hendrix, et al. (1994), who defined job stress along the lines suggested here—i.e., as felt job stress. Those researchers employed a less comprehensive, three-item measure of job stress, but they did include a similar, three-item measure of life stress. Importantly, their model of antecedents and consequences of stress contained many of the same variables and was strongly supported by their data.

In addition to support for its role within the hypothesized path model, the proposed measure of job stress appeared to be unidimensional, with a high degree of internal consistency. Another strength of the measure is its resistance to contamination. While a number of the items will always be inapplicable to a given respondent (not stress for him/her), those items will simply add zero to the total stress score. Thus, it would be counterproductive to delete items from the measure. On the other hand, deficiency is possible in that all symptoms may not be represented for a given subject. A lengthier list of symptoms might add content validity.

The lack of any sex differences in job stress, or stress-related outcomes is consistent with the literature review by Korabik, McDonald, and Rosin (1993). They concluded that, despite being subjected to more stressors than their male counterparts, women managers do not appear to experience more stress, nor do women experience more negative outcomes of stress. Greenglass (1993) proposed that sex differences in coping mechanisms and resources account for the lack of sex differences in stress. Future job stress models, particularly those hypothesizing gender effects, would benefit by the inclusion of coping variables.

The present study suffered from limitations. Because the majority of variables were measured using a questionnaire, a certain amount of common-method bias was, no doubt, present. However, the length of the questionnaire, its widely varied response format, and the large number of measures included, were thought to have kept this problem to a practical minimum. We believe it to be unlikely that the obtained relationships are due primarily to response bias. Nonetheless, some caution is advised in interpreting the results. The 35 percent response rate was also a disappointment, although it was similar to that obtained in much survey research involving the use of mail questionnaires (cf. Lee & Mowday, 1987). Nevertheless, such a response rate, from a single organization, suggests caution in generalizing from these results. Caution is also suggested by the use of a sample of restaurant managers, who were mostly white males and were not typical of the general population. In fact, the lengthy work week suggests a higher-than-average stress level for this sample. Further normative data would be of value.

Research needs to examine this new measure of stress more closely. It represents an intuitively appealing approach to measuring stress because of its ability to account for individual differences and a variety of stress responses. Furthermore, future stress research needs to rely upon a wider range of organizational (and role) characteristics. In the present study, for example, performance feedback, a variable not usually considered in job stress research, appeared to be both an important and interesting factor. Research and models of stress would also benefit by inclusion of physiological, as well as attitudinal and behavioral consequences. The relationships between stress and such physiological variables as cholesterol levels (Hendrix et al., 1985; Hendrix et al., 1994) and heart disease are of obvious importance. Finally, job stress research must turn away from one-shot correlational studies, and begin to examine the longer term effects of job stress in longitudinal designs. Better measures of felt stress, coupled with a broader range of organizational variables would shed further light on the causes of job stress so that steps can be taken to control stress and eliminate its consequences.

REFERENCES

Asher, H.B. (1976). *Causal modeling.* Sage University Paper Series on Quantitative Applications in the Social Sciences. Beverly Hills, CA: Sage.

Barling, J., & Rosenbaum, A. (1986). Work stressors and wife abuse. *Journal of Applied Psychology, 7,* 346–348.

Beehr, T.A., & Newman, J.E. (1978). Job stress, employee health, and organizational effectiveness: A facet analysis, model, and literature review. *Personnel Psychology, 31,* 665–699.

Bedeian, A.G., & Armenakis, R.M. (1981). A path-analytic study of the consequences of role conflict and ambiguity. *Academy of Management Journal, 24,* 417–424.

Billings, R.S., & Wroten, S.P. (1978). Use of path analysis in industrial/organizational psychology: Criticisms and suggestions. *Journal of Applied Psychology, 63,* 677–688.

Burke, R.J., Weir, T., & DuWors, R.E., Jr. (1979). Type A behavior of administrators and wives' reports of marital satisfaction and well being. *Journal of Applied Psychology, 64,* 57–65.

Caplan, R.D., Cobb, S., French, J.R.P., Jr., Van Harrison, R., & Pinneau, S.R., Jr. (1975). *Job demands and worker health: Main effects and occupational differences* (HEW NIOSH No. 75–160). Washington, DC: U.S. Government Printing Office.

Christensen, N.J., & Jensen, E.W. (1994). Effect of psychosocial stress and age on plasma norepinephrine levels: A review. *Psychosomatic Medicine, 56*(1), 77–83.

Cordes, C.L., & Dougherty, T.W. (1991). A review and an integration of research on job burnout. *Academy of Management Review, 18*(4), 621–656.

DeCotiis, T.A., & Summers, T.P. (1987). A path analysis of a model of the antecedents and consequences of organizational commitment. *Human Relations, 40*(7), 445–470.

Epstein, S. (1976) in, I.G. & Spieberger, C.D. (Eds.). *Stress and anxiety.* New York: Wiley and Sons.

Fox, M.L., Dwyer, D.J., & Ganster, D.C. (1993). Effects of stressful job demands and control on physiological and attitudinal outcomes in a hospital setting. *Academy of Management Journal, 36*(2), 298–318.

Gaylin, W. (1979). *Feelings: Our vital signs.* New York: Harper & Row.

Gibson, J.L., Ivancevich, J.M., & Donnelly, J.H. (1985). *Organizations: Behavior, structure, and processes.* Plano, TX: Business Publications, Inc.

Goodman, P.S. (1974). An examination of referents used in the evaluation of pay. *Organizational Behavior and Human Performance, 12,* 170–195.

Greenglass, E.R. (1993). Social support and coping of employed women. In B.C. Long & S.E. Kahn (Eds.), *Women, work, and coping* (pp. 154–169). Montreal: McGill-Queens University Press.

Gupta, N., & Beehr, T.A. (1979). Job stress and employee behaviors. *Organizational Behavior and Human Performance, 23,* 373–387.

Hendrix, W.H., Ovalle, N.K., & Troxler, R.G. (1985). Behavioral and physiological consequences of stress and its antecedent factors. *Journal of Applied Psychology, 70*(1), 188–201.

Hendrix, W.H., Steel, R.P., Leap, T.L., & Summers, T.P. (1994). Antecedents and organizational effectiveness outcomes of employee stress and health. In P.L. Perrewé & R. Crandall (Eds.), *Occupational stress: A handbook* (pp. 73–92). Washington, DC: Taylor & Francis. (Original work published 1991)

Holmes, T.H., & Rahe, R.H. (1967). The social readjustment rating scale. *Journal of Psychosomatic Research, 11,* 213–218.

Howard, J.H., Cunningham, D.A., & Rechnitzer, P.A. (1986). Role ambiguity, Ttype A behavior, and job satisfaction: Moderating effects on cardiovascular and biochemical responses associated with coronary risk. *Journal of Applied Psychology, 71*, 95–101.

Ivancevich, J.M., & Matteson, M.T. (1980). *Stress and work: A managerial perspective.* Glenview, IL: Scott, Foresman.

Ivancevich, J.M., Matteson, M.T., & Preston, C. (1982). Occupational stress, Type A behavior, and physical well-being. *Academy of Management Journal, 25*, 373–391.

Jamal, M. (1984). Job stress and job performance controversy: An empirical assessment. *Organizational Behavior and Human Performance, 33*(1), 1–21.

Katz, D.M., & Kahn, R.L. (1978). *The social psychology of organizations* (2nd ed.). New York: Wiley.

Kim, J., & Kohout, F.K. (1975). Special topics in general linear models. In Nie, N.H., Hull, C.H., Jenkins, J.G., Steinbrenner, K., & Bent, D.H. (Eds.), *Statistical package for the social sciences.* New York: McGraw-Hill.

Klein, S.M. (1973). Pay factors as predictors to satisfaction: A comparison of reinforcement, equity, and expectancy theories. *Academy of Management Journal, 16*, 598–610.

Korabik, K., McDonald, L.M., & Rosin, H.M. (1993). Stress, coping, and social support among women managers. In B.C. Long & S.E. Kahn (Eds.), *Women, work, and coping* (pp. 133–154). Montreal: McGill-Queens University Press.

Lazarus, R.S., & Launier, R. (1978). Stress-related transactions between person and environment. In L.A. Pervin & M. Lewis (Eds.), *Perspectives in international psychology* (pp. 287–327). New York: Plenum.

Lee, T.W., & Mowday, R.T. (1987). Voluntarily leaving an organization: An empirical investigation of Steers and Mowday's model of turnover. *Academy of Management Journal, 30*, 721–743.

Marshall, J.R., & Cooper, C.L. (1979). Work experience of middle and senior managers: The pressure and satisfaction. *International Management Review, 19*, 81–96.

Matteson, M.T., & Ivancevich, J.M. (1979). Organizational stressors and heart disease: A research model. *Academy of Management Review, 4*, 347–357.

McDonald, L.M., & Korabik, K. (1991). Sources of stress and ways of coping among male and female managers. In P.L. Perrewé (Ed.), Handbook on job stress [Special issue]. *Journal of Social Behavior and Personality, 6*(7), 185–198.

Mowday, R.T., Porter, L.W., & Steers, R.M. (1982). *Employee organization linkages: The psychology of commitment, absenteeism, and turnover.* New York: Academic Press.

O'Hanlon, J.F. (1978). *Performance and physiological reactions to monotony in simulated industrial inspection.* Paper presented at XIXth International Congress of Applied Psychology, Munich (August).

Parasuraman, S., & Alutto, J.A. (1984). Sources and outcomes of stress in organizational settings: Toward the development of a structural model. *Academy of Management Journal, 27*, 330–350.

Parker, D.F. & DeCotiis, T.A. (1983). Organizational determinants of job stress. *Organizational Behavior and Human Performance, 32*, 160–177.

Quick, J.C., & Quick, J.D. (1984). *Organizational stress and preventive management.* New York: McGraw-Hill.

Rizzo, J.R., House, R.J., & Lirtzman, S.I. (1970). Role conflict and ambiguity in complex organizations. *Administrative Science Quarterly, 15*, 150–163.

Schnall, P.L., Devereaux, R.B., Pickering, T.G., & Schwartz, J.E. (1992). The relationship between 'job strain' workplace diastolic blood pressure, and left ventricular mass index: A correction. *Journal of the American Medical Association, 267*(9), 1209.

Seyle, H. (1976). *The stress of life* (rev. ed.). New York: McGraw-Hill.

Steffy, B.D., & Laker, D.R. (1991). Workplace and personal stresses antecedent to employees' alcohol use. In P.L. Perrewé (Ed.), Handbook on job stress [Special issue]. *Journal of Social Behavior and Personality, 6*(7), 115–126.

Summers, T.P., & DeNisi, A.S. (1990). In search of Adams' other: Reexamination of referents used in the evaluation of pay. *Human Relations, 43*(6), 497–511.

Weiss, D.J., Dawis, R.V., England, G.W., & Lofquist, C.H. (1966). Manual for the Minnesota Satisfaction Questionnaire. *Minnesota Studies in Vocational Rehabilitation, 22.*

Relationship of Work and Family Stressors to Psychological Distress: The Independent Moderating Influence of Social Support, Mastery, Active Coping, and Self-Focused Attention

Michael R. Frone
Marcia Russell
M. Lynne Cooper

Since the seminal research of Kahn, Wolfe, Quinn, Snoek, and Rosenthal (1964), organizational researchers have come to view job-related stress as an important area of study. Prior research has documented relationships between a variety of job stressors and the psychological and physical health of workers (e.g., Beehr & Newman, 1978; Ganster & Schaubroeck, 1991; House, 1974; Hurrell & Murphy, 1992; Jex & Beehr, 1991). However, if the goal of occupational stress research is to foster the development of comprehensive models of stress, prior research has suffered from at least two important limitations. First, research on job stress has generally failed to view workers from the broader context of other domains of life such as family. Thus, we know little about the independent influence of work and nonwork (e.g., family) stressors on an individual's well-being (e.g., Burke, 1986). Second, the typical job stress study examines several work-related stressors in conjunction with a *single* psychosocial moderator variable (e.g., social support, self-focused attention). However, the various psychosocial moderators are likely to be significantly intercorrelated. Therefore, we also lack a

Authors' Notes: This research was supported by the National Institute of Alcohol Abuse and Alcoholism, Grant #AA05702, awarded to the second author. We would like to thank the editors and the two anonymous reviewers for their helpful comments on an earlier version of this article.

129

clear understanding of which variables moderate stressor–distress relationships apart from their relationship to other potential moderator variables. The purpose of this study was to address both of these limitations using data from a randomly drawn community survey.

Work and Family Stressors

Although several models of work stress have represented the putative influence of nonwork stressors on distress (e.g., Bhagat, 1983; Matteson & Ivancevich, 1987), the most detailed model in this regard has been proposed by Greenhaus and Parasuraman (1986). These authors posit that three categories of stressors can be distinguished: (1) stressors stemming from one's job (e.g., work overload, role ambiguity), (2) stressors related to conditions within nonwork domains of life such as the family (e.g., marital inequity, emotional distance, child problems), and (3) stressors involving the interface of work and nonwork domains (e.g., work-family conflict). According to this model, all three types of stressors may make a unique contribution to an individual's well-being.

In keeping with Greenhaus and Parasuraman's (1986) contention, research has recently begun to examine the unique predictive powers of job stressors, family stressors, and work-family conflict with regard to psychological distress (e.g., Barnett & Marshall, 1992; Bedeian, Burke, & Moffett, 1988; Bromet, Dew, & Parkinson, 1990; Burke, 1988; Frone, Russell, & Cooper, 1992; Higgins, Duxbury, & Irving, 1992; Klitzman, House, Israel, & Mero, 1990; Kopelman, Greenhaus, & Connolly, 1983; Parasuraman, Greenhaus, & Granrose, 1992; Phelan et al., 1991). Although results are mixed, this growing body of research suggests that each stressor has important implications for psychological well-being. Nonetheless, prior research suffers from two major limitations. First, few studies have controlled adequately for the potential confounding influence of sociodemographic characteristics and indicators of psychosocial resources or vulnerabilities. For example, having young children in the house or having limited access to supportive relationships (social support) may be associated with higher levels of both work-family conflict and psychological distress. Second, most studies have typically employed convenience samples drawn from a single organization or occupation. Thus, until stronger controls and more heterogeneous and representative samples are utilized, the results of prior studies are vulnerable to third variable explanations and their generalizability may be limited. These limitations suggest that additional research taking into account a broad range of potential confounding variables and using more broadly representative samples would be useful.

Psychosocial Moderators

Most models of job stress agree that various psychosocial resources and vulnerabilities may exert important stress-buffering or stress-exacerbating influences, respectively (e.g., Kahn et al., 1964; Greenhaus & Parasuraman, 1986; Hurrell & Murphy, 1992). However, of the 10 studies cited earlier that examined the independent relationships of work stressors, family stressors, and work-family conflict to psychological distress, only two (Parasuraman et al., 1992; Phelan et al., 1991) have examined the moderating influence of psychosocial resource or vulnerability factors. This finding supports our contention that little attention has been paid to the independent stress-moderating function of multiple psychosocial resource/vulnerability variables.

Our selection of psychosocial moderators was guided principally by the typology presented by Pearlin and Schooler (1978). These authors suggest that a fundamental distinction can be made between social resources, psychological resources, and coping responses. We therefore examine a set of psychosocial resources (social support, mastery, and an active coping style) that represent each of these three general categories of stress moderators. In addition, we examine self-focused attention because recent job stress research has found it to be an important stress-related vulnerability factor.

Social support. Of all the potential stress moderators, social support has probably received the most research attention. Social support has been broadly defined as "the resources provided by other persons" (Cohen & Syme, 1985, p. 4). In terms of its stress-moderating function, the most prevalent hypothesis is that social support buffers the relationship between stressors and stress outcomes. That is, a high level of social support is expected to attenuate the magnitude of stressor–distress relationships. A recent comprehensive review of the social support literature (Cohen & Wills, 1985) lends qualified support to the stress-buffering influence of social support. In addition, Thoits (1986) recently concluded that "considerable research now indicates that social support reduces, or buffers, the adverse psychological impacts of exposure to stressful life events and ongoing life strains" (p. 416). In contrast, other reviewers (e.g., Alloway & Bebbington, 1987; Barrera, 1988; Beehr & McGrath, 1992; Callaghan & Morrissey, 1993) are more cautious, pointing out that considerable variability exists among studies testing this hypothesis. For example, numerous research studies have either failed to support the buffering hypothesis or have found support for counter-buffering. Clearly, the debate concerning the stress-buffering properties of social support is far from over.

Most research examining the buffering role of social support has used cumulative counts of stressful life events that combine events from several domains of life. And, of the few studies examining domain-specific stressors, most have focused exclusively on job-related stressors. Thus, with the exception of two studies (Parasuraman et al., 1992; Phelan et al., 1991), relatively little is known about the stress-buffering role of social support when paired with measures of family stressors or work-family conflict. Parasuraman et al. (1992) examined whether social support buffered the relationships of work stressors, family stressors, and work-family conflict to psychological distress. The results of their study, however, provided little support for the stress-buffering role of social support. Likewise, Phelan et al. (1991) failed to find a buffering influence of social support on the relationships of work and family stressors to depression.

Mastery. Mastery, or the belief that one has control over the forces that affect one's life, can best be considered a psychological resource (Menaghan, 1983; Pearlin & Schooler, 1978). Psychological resources are generalized attitudes and skills that are considered advantageous across many situations, and include attitudes about self (e.g., esteem), attitudes about the world (e.g., belief in mastery), intellectual skills, and interpersonal skills (Menaghan, 1983). Because psychological resources are generally regarded as adaptive (Menaghan, 1983; Pearlin & Schooler, 1978), one would expect mastery to buffer the relationship between stressors and distress. However, little research has examined the stress-buffering role of mastery in regard to work and family stressors.

A study by Pearlin and Schooler (1978) purportedly examined the stress-reducing properties of mastery (in conjunction with two other psychological resources) on the relationships between work, parental, and marital stressors and their respective domain-specific indicators of psychological distress. In testing the stress-reducing properties of mastery on these relationships, Pearlin and Schooler documented that the relationships between stressors and distress became smaller as mastery was stepped into the regression equations. However, these authors failed to explicitly test interactions between the stressors and mastery. Thus, they failed to provide a test of the potential stress-buffering influence of mastery, a point made also by Marshall (1979) in a critique of this study. In contrast, Phelan et al. (1991) tested interactions between mastery and both work and family stressors. Their results, however, failed to support the buffering hypothesis because none of the interactions were significantly related to depression.

Active coping. The last category of stress moderators proposed by Pearlin and Schooler (1978) was coping responses. In contrast to psychological resources which represent personality dispositions, coping responses refer to what individuals do when they experience problems (Pearlin & Schooler, 1978). Menaghan (1983) has made a further distinction between coping responses that are relatively stable across problems or life domains (i.e., coping styles) and coping responses that are specific to certain problems within a given domain of life (i.e., coping behaviors). In the present study, we use a measure of active coping style. Although coping styles can be either adaptive or maladaptive, a style reflecting an active, problem-solving approach is generally regarded as adaptive (e.g., Menaghan, 1983; Pearlin & Schooler, 1978). Thus, one would expect an active coping style to buffer the relationship between stressors and distress. As with psychological resources, though, little research has examined the stress-buffering role of an active coping style in regard to work and family stressors. For example, the study by Pearlin and Schooler (1978) also purportedly examined the stress-reducing properties of several domain-specific coping styles. However, similar to their examination of psychological resources, they failed to provide an explicit test for the stress-moderating role of coping styles.

Self-focused attention. Self-focused attention refers to a dispositional tendency to focus attention internally on such covert aspects of the self as thoughts, feelings, attitudes, and internal bodily sensations (Carver & Scheier, 1981; Scheier, Carver, & Matthews, 1983). In both the clinical and social psychological literatures there has been a long history of theoretical arguments hypothesizing that self-focused attention may play an important etiological role in the onset of a variety of psychopathologies (cf. Ingram, 1990). In support of this contention, research has found that, relative to low self-focused individuals, highly self-focused individuals attend more closely to and are more accurate in detecting changes in their emotional experience and bodily sensations (Carver and Scheier, 1981; Scheier et al., 1983). Given these findings, one role self-focused attention may play in the etiology of psychological distress is as a moderator of relationships between negative environmental events and psychological distress.

A recent review of the self-focused attention literature (Ingram, 1990) suggests that self-focused attention moderates the relationship between negative environmental events and psychological distress. Specifically, Ingram (1990) found considerable empirical evidence indicating that high levels of self-focused attention make an individual more vulnerable to the negative effects of laboratory-induced stressors. Moreover, once negative affect is triggered, highly self-focused individuals are vulnerable to an exacerbation of the affect. These two findings collectively suggest that self-focused attention may represent an important vulnerability factor, exacerbating the relationship between stressors and psychological distress. In support of this conclusion, recent job stress research has found that self-focused attention's role as a vulnerability factor in stressor–psychological distress relationships extends beyond the laboratory to at least one class of "real-life" stressors (Frone & McFarlin, 1989; Hollenbeck, 1989; Innes & Kitto, 1989). No research, however, has examined the moderating role of self-focused attention with respect to family stressors or work-family conflict.

Hypotheses

Based on our earlier discussion, the following two general hypotheses will be examined:

Hypothesis 1. Work stressors, family stressors, and work-family conflict will each be independently and positively related to psychological distress.

Given the limitations of past research, however, we examine this hypothesis both before and after statistically controlling for several sociodemographic characteristics and several indicators of psychosocial resources/vulnerabilities. Examining Hypothesis 1 at varying levels of statistical control will help clarify the extent to which the hypothesized stressor–psychological distress relationships may be vulnerable to third variable explanations.

Hypothesis 2. Social support, mastery, active coping, and self-focused attention will independently moderate the relationship of work stressors, family stressors, and work-family conflict to psychological distress.

With regard to the direction of the hypothesized moderator effects, we expect that social support, mastery, and active coping will buffer the stressor–psychological distress relationships. That is, stressors and psychological distress will be more strongly related at low levels of these three variables. In contrast, self-focused attention is expected to exacerbate the stressor–psychological distress relationships such that stressors and psychological distress will be more strongly related at high levels of this variable.

METHOD

Sample

Respondents in this study were drawn from the longitudinal follow-up of a random sample survey of 1,933 household residents in Erie County, New York. Designated respondents were identified in a three-stage probability sample, designed to yield equal representation of two racial groups (Blacks and non-Blacks) and three education levels (less than high school graduate, high school graduate, at least some college). Of the 1,616 respondents (83.6% of the initial sample) who were interviewed during the second wave of data collection, 596 met the following criteria for selection into this study: (1) were employed at least 20 hours per week, (2) were married/living as married or had children living at home, and (3) provided valid data on all measures described below. We use data only from the second wave of data collection because not all the major variables used in this study (e.g., work-family conflict, self-focused attention) were assessed in the first wave. Table 1 provides a detailed description of the demographic characteristics of this sample.

Procedures

Data for this study were collected by a team of 20 professionally trained interviewers during the spring and summer of 1989 as part of a larger study of stress processes. Interviews were conducted in respondents' homes using a highly structured interview schedule that contained both interviewer- and self-administered sections. The complete interview required approximately 90 minutes to administer. Respondents were compensated $25 for their time.

Measures

A total of 20 variables were employed to assess four broad categories of variables: (1) sociodemographic characteristics (11 indicators), (2) psychosocial moderators (four measures), (3) chronic work and family stressors (three measures), and (4) psychological outcomes (two measures). Each of these measures is described below. Descriptive statistics for the sociodemographic characteristics (i.e., percentages, means, and standard deviations) are summarized in Table 1, whereas descriptive statistics for the major study variables (i.e., number of items, means, standard deviations, ranges, and reliability estimates) are summarized in

TABLE 1 Sociodemographic Profile of the Sample

Variable	Percent	Mean	SD
Sex			
Male	44		
Female	56		
Race			
Black	55		
White	43		
Other	2		
Job Type			
Blue-Collar	49		
White-Collar	51		
Marital Status			
Married	73		
Not Married	27		
No. of Children at Home		1.50	1.28
0	22		
1	34		
2	27		
3	11		
4+	6		
Age of Youngest Child			
Less than 6 Years	22		
6-12 Years	22		
13-18 Years	19		
Over 18 Years	15		
No Children	22		
Age		40.50	10.30
Education		13.28	2.32
Family Income		$36,397.00	$24,185.00
Job Tenure		8.72	8.61
Number of Work Hours/Week		44.39	11.47

Table 2. In addition, a zero-order correlation matrix among the major study variables is presented in Table 3. Unless otherwise noted, all variables are scored so that a high score represents higher levels of the construct. A complete zero-order correlation matrix, which includes the 11 sociodemographic covariates, and copies of each of the measures are available from the first author.

Sociodemographic Characteristics

The following sociodemographic characteristics were used as covariates: sex (1 = male, 2 = female), race (1 = white; 2 = nonwhite), age (in years), education (in years), family income (in dollars), job type (1 = white-collar, 2 = blue-collar), job tenure (in years), number of hours worked per week (in hours), marital status (1 =

TABLE 2   Psychometric Properties of the Stressor, Psychosocial Moderator, and Psychological Distress Measures

Variable	No. of Items	Mean	SD	Alpha	Range Potential/Actual
Psychosocial Resources/ Vulnerabilities					
Social Support	24	3.52	.35	.89	1-4 / 2.33-4.00
Mastery	7	3.12	.46	.75	1-4 / 1.86-4.00
"John Henryism" Active Coping	8	3.43	.40	.81	1-4 / 2.25-4.00
Self-Focused Attention	9	2.61	.54	.72	1-4 / 1.00-4.00
Stressors					
Job Stressor Index	20	2.05	.36	.72	1-4 / 1.21-3.25
Family Stressor Index	11	1.60	.42	.74	1-4 / 1.00-3.40
Work-Family Conflict	4	1.80	.70	.67	1-5 / 1.00-4.50
Psychological Distress					
Depression	20	1.58	.39	.87	1-4 / 1.00-3.40
Somatic Symptoms	12	1.46	.42	.78	1-4 / 1.00-3.58

married/living as married, 2 = not married), number of children living at home (0 to 4 or more), and age of youngest child living at home. Following a coding scheme outlined by Bedeian et al. (1988), we coded age of youngest child into the following five categories that represented increasing levels of parental demands: (1) no children, (2) youngest child over 18 years of age (3) youngest child 13-18 years of age, (4) youngest child 6-12 years of age, and (5) youngest child less than 6 years of age.

Psychosocial Moderators

Social support. Perceived social support was measured using an abbreviated version of the Interpersonal Support Evaluation List (ISEL; Cohen & Hoberman, 1983; Cohen, Mermelstein, Kamarck, & Hoberman, 1985), which included 24 of the originally published 40 items. The ISEL contains items that assess the perceived availability of (1) tangible support (i.e., the availability of tangible assistance or material aid), (2) appraisal support (i.e., the availability of a confidant and trusted advisor), (3) belonging support (i.e., the availability of someone with whom to socialize or relax), and (4) self-esteem support (i.e., the availability of a positive comparison when comparing one's self to others). A 4-point response format was used that required respondents to indicate how true or false each statement was in describing their relationships with other people. Detailed psychometric and validity information for the ISEL is provided by Cohen and Hoberman (1983) and Cohen et al. (1985).

Mastery. Mastery was assessed with seven items developed by Pearlin and Schooler (1978). This measure assessed the extent to which individuals see themselves as controlling the forces that affect their lives. Each item was answered

TABLE 3   Intercorrelations Among Major Study Variables

Variables	1	2	3	4	5	6	7	8	9
Psychosocial Resources/									
Vulnerabilities									
1. Social Support	—								
2. Mastery	.35**	—							
3. Active Coping	.37**	.22**	—						
4. Self-Focused Attention	-.01	.01	.10*	—					
Stressors									
5. Job Stressor Index	-.19**	-.16**	-.20**	.03	—				
6. Family Stressor Index	-.33**	-.27**	-.20**	.12*	.19**	—			
7. Work-Family Conflict	-.19**	-.12*	-.11*	.12*	.33**	.30**	—		
Psychological Distress									
8. Depression	-.38**	-.52**	-.20**	.18**	.28**	.40**	.37**	—	
9. Somatic Symptoms	-.16**	-.30**	-.23**	.07	.19**	.28**	.30**	.50**	—

*$p \leq .01$; **$p \leq .001$

on a 4-point agree/disagree scale. Pearlin, Menaghan, Lieberman, and Mullan (1981) present evidence indicating that the scale items represent a single underlying factor, and that the scale scores are relatively stable across time (test-retest correlation over four years = .44).

"John Henryism" (JH) active coping style. Active coping style was assessed with eight items developed by James, Strogatz, Wing, and Ramsey (1987). The JH active coping style is a stress-coping style characterized by the belief that one can control one's environment coupled with direct and active efforts to do so. A 4-point response format was used that required respondents to indicate how true or false each statement was in describing their typical approach to problems in their life. James et al. (1987) reported a coefficient alpha of .72 for this scale.

Self-focused attention. Self-focused attention was assessed with the nine-item measure of private self-consciousness taken from Scheier and Carver's (1985) revised Self-Consciousness scale. This measure assesses an individual's dispositional tendency to focus attention on covert aspects of the self such as thoughts, feelings, attitudes, and internal bodily sensations (e.g., Scheier & Carver, 1985; Scheier et al., 1983). Using a 4-point scale, respondents were asked to indicate how well each statement described them. The revised version of the original Private Self-Consciousness scale (Fenigstein, Scheier, & Buss, 1975) was developed so that it was more easily understood by general (i.e., noncollege student) population samples. Psychometric data presented by Scheier and Carver (1985) indicate that their revised measure of private self-consciousness is reliable and that scores from the revised measure correlate highly with scores from the original instrument. Reliability and validity information for the original instrument is summarized in Carver and Scheier (1981) and Scheier et al. (1983).

Chronic Work and Family Stressors

Job stressor index. Three dimensions of work stressors were assessed: work pressure, lack of job control, and role ambiguity. Work pressure (8 items) assessed the frequency with which individuals perceive high job-related demands resulting from heavy workloads and responsibilities. Lack of job control (6 items) assessed the frequency with which individuals perceive constraints on their ability to function autonomously and influence important job parameters. Role ambiguity (6 items) assessed the frequency of being confused or unclear about day-to-day tasks and expectations, and job-related goals. The 20 items comprising these three scales were taken from several previously published measures of work stressors (Beehr, 1976; House, McMichael, Wells, Kaplan, & Landerman, 1979; Insel & Moos, 1974; Pearlin & Schooler, 1978; Rizzo, House, & Lirtzman, 1970; Sims, Szilagyi, & Keller, 1976). Each item used a 4-point frequency-based response scale. Because we were interested in comparing job stressors to stressors stemming from other life domains, rather than in comparisons among the various job stressor dimensions themselves, all analyses use an overall job stressor index created by averaging the 20 individual items.

Family stressor index. The family stressor index was composed of six parenting stressor items and five marriage-related stressor items. Of the parenting stressor items, two were developed by Kessler (1985) and four were developed specifically for this study. Parenting stressor items assessed lack of parent-child emotional bonding, parental workload, and extent of child(ren)'s misbehavior. Marital stressor items (Kessler, 1985) assessed lack of emotional closeness, quality of communication, and degree of tension in the relationship. These 11 items were combined to create an overall family stressor index. This index was calculated as the average of the five marital stressor items among respondents who were married only, or as the average of the six parental stressor items among respondents who were single parents, or as the average of the 11 marital and parental stressor items among respondents who were both married and parents. Each of the parental and marital stressor items used a 4-point frequency-based response scale. Although this index represented a different combination of stressor items for the three different subgroups (married only, parents only, married and parents), preliminary analyses suggested that the results reported below for the family stressor index were generalizable to the separate marital and parental stressor scales.

Work-family conflict. Four items were developed to assess work-family conflict: two items assessed the degree to which a respondent's job interfered with his or her home-life, and two items assessed the degree to which a respondent's home-life interfered with his or her job. Each item provided a number of specific referents so that this fairly broad construct could be adequately captured with a relatively small set of items. For example, representative items are "How often does your job or career interfere with your responsibilities at home such as yard work, cooking, cleaning, repairs, shopping, paying the bills, or child care?" and "How often does your home-life keep you from spending the amount of time you would like to spend on job- or career-related activities?" Each item used a 5-point frequency-based response scale.

Psychological Distress

Depression. Depression was assessed with the 20-item Center for Epidemiologic Studies Depression Scale (CES-D; Radloff, 1977). A 4-point scale was used to determine how frequently each of the 20 symptoms were experienced during the past month. The items in this scale assess several dimensions of depressive symptomatology such as depressed mood, feelings of guilt and worthlessness, feelings of helplessness and hopelessness, psychomotor retardation, loss of appetite, and sleep disturbance. The CES-D was selected because it was developed specifically for general population samples. Reliability and validity information concerning the CES-D is presented by Ensel (1986).

Somatic symptoms. Somatic symptoms were assessed with the 12-item Somatization subscale from the SCL-90-R (Derogatis, 1977; Derogatis & Cleary, 1977). Somatization reflects psychological distress arising from perception of bodily dysfunction. The items represent symptoms that have strong autonomic mediation (e.g., cardiovascular, gastrointestinal, and respiratory symptoms). A 5-point scale was used to indicate the amount of distress each symptom caused the respondent during the past month. The SCL-90-R has been shown to be sensitive to low levels of symptomatology in normal population samples (Derogatis, 1977; Derogatis & Cleary, 1977).

RESULTS

Relationships of Job Stressors, Family Stressors, and Work-Family Conflict to Psychological Distress

Hypothesis 1, which predicted that job stressors, family stressors, and work-family conflict would have independent positive relationships to psychological distress, was tested using a series of hierarchical multiple regression analyses (Cohen & Cohen, 1983).[1] A series of hierarchical regression analyses was used because we were also interested in examining the extent to which the relationships between stressors and psychological distress could be accounted for by sociodemographic and/or psychosocial resource/vulnerability variables. Results of these analyses are presented in Tables 4 and 5. By moving from left to right in Tables 4 and 5, one can observe the contributions of the three stressors to

---

[1] *Prior conceptual discussions of work-family processes have suggested that gender may influence the magnitude of relationships between work and family stressors and psychological distress (e.g., Eckenrode & Gore, 1990; Kline & Cowan, 1988). Therefore, we conducted a set of exploratory moderated regression analyses (e.g., Cohen & Cohen, 1983; Jaccard, Turrisi, & Wan, 1990) to examine this issue. For each measure of psychological distress, the main*

(continued on next page)

TABLE 4   Multiple Regression Analyses Predicting Depression from Job Stressors, Family Stressors, and Work-Family Conflict

Predictors	Zero-Order Correlation	Equation 1 ΔR^2	Equation 1 Beta	Equation 2 ΔR^2	Equation 2 Beta	Equation 3 ΔR^2	Equation 3 Beta
Block of Sociodemographic Characteristics				.135***		.135***	
Sex							.17***
Race							.02
Age							-.15**
Education							-.20***
Family Income							-.08
Marital Status							-.08
No. of Children							.04
Age of Youngest Child							.00
Job Tenure							.01
Job Type							.02
No. of Work Hours/Week							.07
Block of Resources/Vulnerabilities						.289***	
Social Support							-.20***
Mastery							-.41***
Active Coping							-.04
Self-Focused Attention							.17***
Block of Stressors		.248***		.182***		.064***	
Job Stressor Index	.28***		.14***		.15***		.08*
Family Stressor Index	.40***		.30***		.24***		.12***
Work-Family Conflict	.37***		.23***		.23***		.19***

Note: Because the beta coefficients for the sociodemographic characteristics are identical in Equation 2 and Equation 3, they are only reported for Equation 3.
*p < .05; **p < .01; ***p < .001

psychological distress under increasing levels of statistical control: no control (zero-order correlations), minimal control (Equation 1: three stressors controlling for each other), moderate control (Equation 2: three stressors controlling for each other and the sociodemographic variables), and high control (Equation 3: three stressors controlling for each other and both the sociodemographic and psychosocial resource/vulnerability variables).

(Footnote 1 continued)
    effects for all sociodemographic, psychosocial resource/vulnerability, and stressor variables entered the regression equations on Step 1, followed by a block of three stressor-by-gender interactions on Step 2. The results of these analyses revealed that none of the six stressor-by-gender interactions were statistically significant. These results support a growing body of research (e.g., Bedeian et al., 1988; Coverman, 1989; Greenhaus et al., 1989; Harris, Heller, & Braddock, 1988; Rice, Frone, & McFarlin, 1992) suggesting that there is little empirical support for statistically significant gender differences in the magnitude of relationships between work and family stressors and psychological well-being.

TABLE 5   Multiple Regression Analyses Predicting Somatic Symptoms from Job Stressors, Family Stressors, and Work-Family Conflict

Predictors	Zero-Order Correlation	Equation 1		Equation 2		Equation 3	
		ΔR^2	Beta	ΔR^2	Beta	ΔR^2	Beta
Block of Sociodemographic Characteristics				.063***		.063***	
Sex							.19***
Race							-.05
Age							-.09
Education							-.12*
Family Income							-.05
Marital Status							-.04
No. of Children							.05
Age of Youngest Child							.06
Job Tenure							.03
Job Type							.03
No. of Work Hours/Week							.04
Block of Resources/Vulnerabilities						.107***	
Social Support							-.01
Mastery							-.23***
Active Coping							-.19***
Self-Focused Attention							.09*
Block of Stressors		.134***		.113***		.064***	
Job Stressor Index	.19***		.08*		.09*		.05
Family Stressor Index	.28***		.20***		.16***		.10*
Work-Family Conflict	.30***		.21***		.23***		.22***

Note: Because the beta coefficients for the sociodemographic characteristics are identical in Equation 2 and Equation 3, they are only reported for Equation 3.
*$p < .05$; **$p < .01$; ***$p < .001$

An examination of the zero-order correlations and beta weights indicate that each stressor makes a unique positive contribution to the prediction of psychological distress. However, one can also see that the magnitude of the individual coefficients is substantially attenuated as one moves (from left to right) across the table. Compared to the zero-order correlations, the beta weights for the stressors from Equation 1 suggest that a fairly substantial part of each stressor's relationship to psychological distress is due to its association (i.e., shared variance) with the remaining two stressors. In contrast, a comparison of the increments in R^2 and beta weights associated with the block of stressors from Equation 2 to those from Equation 1 indicates that the 11 sociodemographic characteristics had a relatively small attenuating effect. However, a comparison of the increments in R^2 and beta weights associated with the block of stressors from Equation 3 to those from Equation 2 indicates that the four psychosocial resources/vulnerabilities had a relatively large attenuating effect.

To summarize, the magnitude of the zero-order correlations between the stressors and psychological distress tend to be inflated due to: (1) significant covariation among the stressors, and (2) significant covariation between the psychosocial resources/vulnerabilities and both the stressors and psychological distress. Nonetheless, each stressor made a significant contribution to psychological distress after controlling for the remaining two stressors and the sociodemographic and psychosocial resource/vulnerability variables. Thus, Hypothesis 1 was confirmed.

Stress-Moderating Influence of Psychosocial Resources/Vulnerabilities

The independent moderating influence of social support, mastery, active coping, and self-focused attention (Hypothesis 2) on each of the stressor–distress relationships was tested using a series of hierarchical moderated regression analyses (Cohen & Cohen, 1983; Jaccard, Turrisi, & Wan, 1990). For each outcome, the main effects for all sociodemographic, psychosocial resource/vulnerability, and stressor variables (a total of 18 variables) were entered on the first step, followed by a block of stressor by resource/vulnerability interactions on the second step. However, because we were interested in testing 12 interactions per outcome variable (three stressors by four resources/vulnerabilities), we estimated Step 2 separately for each of the three stressors and for both of the outcomes. Thus, a total of six regression equations was estimated. The results of these analyses are presented in Table 6.

In order to control for Type 1 error without unduly sacrificing statistical power, we conducted follow-up analyses of significant interactions only when the block of interactions attained significance at the conventional $p < .05$ level. Significant interactions were probed following procedures outlined by Cohen and Cohen (1983) and Jaccard et al. (1990). Specifically, separate regression lines were generated from the overall regression equation to represent the stressor–distress relationship at relatively high (+1 SD) and relatively low (-1 SD) levels of the moderator variable.

An examination of Table 6 reveals that four of the six blocks of interactions were associated with a significant increment in R^2. Examining the individual beta weights within each of these blocks indicated that self-focused attention moderated the relationship between each stressor and depression. In addition, it moderated the relationship between the family stressor index and somatic symptoms. In contrast, social support, mastery, and active coping failed to moderate the stressor–distress relationships.

To examine the form of the significant stressor-by-self-focused attention interactions, we plotted (as described earlier) the stressor–distress relationships at high and low levels of self-focused attention. The four interactions are presented in Figure 1. An examination of this figure reveals that, as hypothesized, high levels of self-focused attention exacerbated the stressor–distress relationships.

To summarize, the most general form of Hypothesis 2 was not supported because all four stress moderators failed to have independent moderating effects.

TABLE 6   Moderated Regression Results Summarizing Stressor by Resource/
Vulnerability Interactions

	Depression		Somatic Symptoms	
	ΔR^2	Beta	ΔR^2	Beta
Step 1	.489***		.234***	
All Main Effects				
Equation 1, Step 2	.008*		.008	
JSI X Social Support		-.06		-.04
JSI X Mastery		.00		.04
JSI X Active Coping		.03		.08
JSI X Self-Focused Attention		.06*		.03
Equation 2, Step 2	.008*		.015*	
FSI X Social Support		.02		.01
FSI X Mastery		.02		-.05
FSI X Active Coping		-.02		-.01
FSI X Self-Focused Attention		.08**		.11**
Equation 3, Step 2	.013**		.008	
WFC X Social Support		-.06		.06
WFC X Mastery		-.04		-.08
WFC X Active Coping		.02		-.03
WFC X Self-Focused Attention		.09**		.04

Note: JSI = Job Stressor Index, FSI = Family Stressor Index, WFC = Work-Family Conflict. On Step 1, all main effects representing the 11 sociodemographic characteristics, 4 psychosocial resource/ vulnerability variables, and 3 stressors were entered.
*$p < .05$, **$p < .01$, ***$p < .001$

However, self-focused attention had a significant stress-exacerbating influence on each of the three stressors, thereby providing partial support for Hypothesis 2.

DISCUSSION

The present paper addressed two general issues concerning work and family stress. The first issue was whether job stressors, family stressors, and work-family conflict make independent contributions to the prediction of psychological distress. Our results indicated that each of these stressors was positively related to psychological distress, even after controlling for the remaining two stressors and for the sociodemographic and psychosocial resource/vulnerability variables. However, our data suggest an important caveat concerning the stressor–distress relationships. We found that one may risk overestimating the magnitude of the relationship between stressors from a particular life domain (e.g., work) and psychological distress if one fails to control for the concurrent influence of stressors emanating from other life domains (e.g., family, work-family interface) and for individual differences in psychosocial resources/vulnerabilities. In contrast, our findings suggest that there is little risk of overestimating the magnitude of relationships between work and family stressors and psychological distress if one fails to control for sociodemographic variables.[2]

Collectively, the present findings in conjunction with those from prior research (e.g., Barnett & Marshall, 1992; Bedeian et al., 1988; Bromet et al., 1990; Burke, 1988; Frone et al., 1992; Higgins et al., 1992; Klitzman et al., 1990; Kopelman et al., 1983; Parasuraman et al., 1992; Phelan et al., 1991) lend strong support to the model proposed by Greenhaus and Parasuraman (1986). As suggested by their model, a complete understanding of the psychological well-being of workers will only come from job stress research that adopts a broad ecological perspective that acknowledges the sources of stress emanating from other domains of life. Furthermore, a full appreciation of the importance of the workplace with regard to psychological well-being will only be developed through job stress research that examines the impact of work-related stressors independent of stressors in other life domains. Based on these arguments, future job stress research should be expanded to include stressors from nonwork domains other than the family (e.g., insufficient leisure time, quality of one's neighborhood, or insufficient finances).

The second issue addressed in this study was whether several psychosocial resources/vulnerabilities might independently moderate the relationships between work and family stressors and psychological distress. Our results suggested that social support, mastery, and an active coping style did not buffer the stressor–distress relationships, though the correlations and regression results indicate that these psychosocial resources are each negatively related to psychological distress. In contrast, there was consistent support for the stress-exacerbating function of high levels of self-focused attention. The present results replicated previous research showing that relationships between job stressors and psychological distress were strongest among individuals who were highly self-focused (Frone & McFarlin, 1989; Hollenbeck, 1989; Innes & Kitto, 1989). In addition, this study extended the generalizability of these prior findings by showing that self-focused attention exacerbated the relationship of family stressors and work-family conflict

---

[2] *One reviewer was concerned that there may be an arbitrary quality to our results concerning the stressor–distress relationships because of the order in which variables were forced into the regression analyses (e.g., see Table 3, Equation 3). That is, the regression analyses presented in Table 3 imply a specific causal order among the three sets of variables. Given this implied causal order, explained variance in the outcomes that is shared by the sociodemographic and psychosocial resources is attributed to the sociodemographic variables. Similarly, explained variance in the outcomes that is shared by the stressors and both the sociodemographic and psychosocial resources is attributed to the latter two sets of variables. To the extent that this implied causal order is not correct, some shared variance may actually belong to the "second" set of variables. Although a comprehensive response to this essentially epistomological concern is beyond the scope of this paper, two points may help place this concern in perspective. First, the causal order implied by our regression analyses is compatible with prior models of job stress (e.g., Beehr & McGrath, 1992; Beehr & Newman, 1979; Cooper, Russell, & Frone, 1990; Frone, Russell, & Cooper, 1990; Matteson & Ivancevich, 1987). Second, more stable variables are typically modeled as causally antecedent to less stable variables (Davis, 1985).*

to psychological distress. These findings suggest that self-focused attention may be a particularly pervasive stress exacerbator.

Future research on the moderating function of social support, mastery, and active coping style should begin to examine boundary conditions that may influence whether or not these three variables act as stress buffers. A growing number of stress researchers are postulating a moderator/stressor specificity hypothesis (e.g., Cohen & Hoberman, 1983; Cooper et al., 1990; Frone, 1990; Swindle, Heller, & Lakey, 1988). The specificity hypothesis posits that many resources/vulnerabilities may not have generic power to moderate any stressor–strain relationship. Rather, a conceptual fit may be necessary between a stressor and a moderator before any person–environment interaction is observed. Therefore, future research should begin to explore the moderating role of domain-specific measures of psychosocial resources/vulnerabilities (e.g., work-related mastery vs. family-related mastery) rather than assuming cross-situational consistency as implied by general measures of resources/vulnerabilities. It should be noted, however, that, although Parasuraman et al. (1992) and Phelan et al. (1991) used domain-specific measures of social support (i.e., supervisor/co-worker and spouse/friend support), neither study found evidence of stress-buffering. This lack of evidence for the importance of domain-specific social support may suggest that, in addition to domain-specificity, future research may also need to take into account whether individuals value social support or possess the capabilities to effectively utilize it (Cummins, 1989, 1990).

In contrast to the hypothesized buffering influence of the three psychosocial resources, the stress-exacerbating function of self-focused attention was consistent across each of the three domain-specific stressors. Because self-focused attention appears to be a pervasive vulnerability factor, future research should begin to focus on the processes that underlie its stress-exacerbating influence. Ingram (1990) has suggested two potential explanations for why self-focused attention may exacerbate stressor–psychological distress relationships. First, stressors may create some initial level of negative affect in all individuals that is subsequently intensified among highly self-focused individuals. Second, self-focused attention may serve to trigger negative cognitive schemas (e.g., negative cognitive appraisals) in response to stressors that then precipitate heightened negative affect. In addition, work by Carver and Scheier (Carver & Scheier, 1981; Scheier et al. 1983) suggests that, if a stressor is perceived to be chronic or intractable, high levels of self-focused attention might be related to a reduced likelihood of problem-focused coping in response to that stressor (see Frone & McFarlin, 1989 for a more detailed discussion of this issue). It should be noted that these three explanations for the stress-exacerbating influence of self-focused attention are not mutually exclusive; in fact, each of these processes may be operating simultaneously.

In summary, this study had two major strengths relative to past research in this area. First, our hypotheses were tested within the context of a strong multivariate framework that decreases the likelihood that our findings are attributable to third-

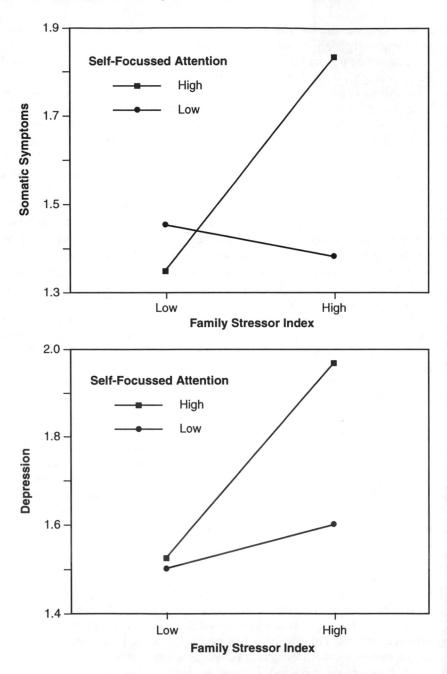

FIGURE 1   Relationships Between Stressors and Distress at High and Low
Levels of Self-Focused Attention

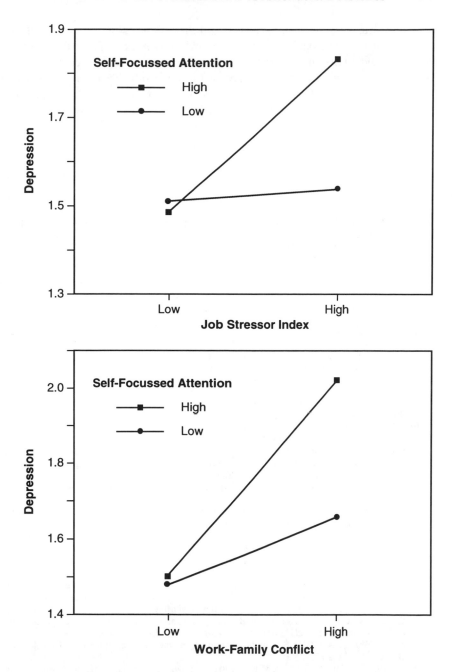

FIGURE 1   (continued)

variable influences. Second, the generalizability of our findings was strengthened by using data from a randomly selected community sample. These two strengths notwithstanding, it is important to note that, because our data are cross-sectional, direction of causality cannot be addressed in this study. Therefore, conclusions regarding direction of causality will have to await future research employing longitudinal and quasi-experimental designs.

REFERENCES

Alloway, R., & Bebbington, P. (1987). The buffer theory of social support: A review of the literature. *Psychological Medicine, 17,* 19–108.

Barnett, R.C., & Marshall, N.L. (1992). Worker and mother roles, spillover effects, and psychological distress. *Women and Health, 18,* 9–40.

Barrera, M., Jr. (1988). Models of social support and life stress: Beyond the buffering hypothesis. In L.H. Cohen (Ed.), *Life events and psychological functioning: Theoretical and methodological issues.* Newbury Park, CA: Sage.

Bedeian, A.G., Burke, B.G., & Moffett, R.G. (1988). Outcomes of work–family conflict among married male and female professionals. *Journal of Management, 14,* 475–491.

Beehr, T.A. (1976). Perceived situational moderators of the relationship between subjective role ambiguity and role strain. *Journal of Applied Psychology, 61,* 35–40.

Beehr, T.A., & McGrath, J.E. (1992). Social support, occupational stress and anxiety. *Anxiety, Stress, and Coping, 5,* 7–19.

Beehr, T.A., & Newman, J.E. (1978). Job stress, employee health, and organizational effectiveness: A facet analysis, model, and literature review. *Personnel Psychology, 31,* 665–699.

Bhagat, R.S. (1983). Effects of stressful life events on individual performance effectiveness and work adjustment processes within organizational settings: A research model. *Academy of Management Review, 8*(4), 660–671.

Bromet, E.J., Dew, M.A., & Parkinson, D.K. (1990). Spillover between work and family: A study of blue-collar working wives. In J. Eckenrode & S. Gore (Eds.), *Stress between work and family.* New York: Plenum.

Burke, R.J. (1986). The present and future status of stress research. *Journal of Organizational Behavior Management, 8,* 249–267.

Burke, R.J. (1988). Some antecedents and consequences of work–family conflict. In E.B. Goldsmith (Ed.), Work and family: Theory, research, and applications [Special issue]. *Journal of Social Behavior and Personality, 3*(4), 287–302.

Callaghan, P., & Morrissey, J. (1993). Social support and health: A review. *Journal of Advanced Nursing, 18,* 203–210.

Carver, C.S., & Scheier, M.F. (1981). *Attention and self-regulation: A control theory approach to human behavior.* New York: Springer-Verlag.

Cohen, J., & Cohen, P. (1983). *Applied multiple regression/correlation analysis for the behavioral sciences* (2nd ed.). Hillsdale, NJ: Erlbaum.

Cohen, S., & Hoberman, H.M. (1983). Positive events and social support as buffers of life change stress. *Journal of Applied Social Psychology, 13,* 99–125.

Cohen, S., Mermelstein, R., Kamarck, T., & Hoberman, H.M. (1985). Measuring the functional components of social support. In I.G. Sarason & B.R. Sarason (Eds.), *Social support: Theory, research and applications.* The Hague, Holland: Martines Niijhoff.

Cohen, S., & Syme, S.L. (1985). Issues in the study and application of social support. In S. Cohen & S.L. Syme (Eds.), *Social support and health.* New York: Academic Press.

Cohen, S., & Wills, T.A. (1985). Stress, social support, and the buffering hypothesis. *Psychological Bulletin, 98,* 310–357.

Cooper, M.L., Russell, M., & Frone, M.R. (1990). Work stress and alcohol effects: A test of stress-induced drinking. *Journal of Health and Social Behavior, 31,* 260–276.

Coverman, S. (1989). Role overload, role conflict, and stress: Addressing consequences of multiple role demands. *Social Forces, 67,* 965–982.

Cummins, R. (1989). Locus of control and social support: Clarifiers of the relationship between job stress and job satisfaction. *Journal of Applied Social Psychology, 19,* 772–788.

Cummins, R. (1990). Job stress and the buffering effect of supervisory support. *Group and Organizational Studies, 15,* 92–104.

Derogatis, L.R. (1977). *SCL–90 administration, scoring and procedures: Manual–I for the revised version and other instruments of the psychopathology rating scale series.* Baltimore, MD: Johns Hopkins University School of Medicine.

Derogatis, L.R., & Cleary, P.A. (1977). Confirmation of the dimensional structure of the SCL–90: A study in construct validation. *Journal of Clinical Psychology, 33,* 981–989.

Eckenrode, J., & Gore, S. (1990). Stress and coping at the boundary of work and family. In J. Eckenrode & S. Gore (Eds.), *Stress between work and family.* New York: Plenum.

Ensel, W.M. (1986). Measuring depression: The CES–D scale. In N. Lin, A. Dean, & W.M. Ensel (Eds.), *Social support, life events, and depression* (Vol. 1). New York: Academic Press.

Fenigstein, A., Scheier, M.F., & Buss, A.H. (1975). Public and private self-consciousness: Assessment and theory. *Journal of Consulting and Clinical Psychology, 43,* 522–527.

Frone, M.R. (1990). Intolerance of ambiguity as a moderator of the occupational role stress–strain relationship: A meta-analysis. *Journal of Organizational Behavior, 11,* 309–320.

Frone, M.R., & McFarlin, D.B. (1989). Chronic occupational stressors, self-focused attention, and well-being: Testing a cybernetic model of stress. *Journal of Applied Psychology, 74,* 876–883.

Frone, M.R., Russell, M., & Cooper, M.L. (1990, August). *Occupational stressors, psychosocial resources, and psychological distress: A comparison of black and white workers.* Paper presented at the annual meeting of the Academy of Management, San Francisco.

Frone, M.R., Russell, M., & Cooper, M.L. (1992). Antecedents and outcomes of work–family conflict: Testing a model of the work–family interface. *Journal of Applied Psychology, 77,* 65–78.

Ganster, D.C., & Schaubroeck, J. (1991). Work stress and employee health. *Journal of Management, 17,* 235–271.

Greenhaus, J.H., & Parasuraman, S. (1986). A work–nonwork interactive perspective of stress and its consequences. *Journal of Organizational Behavior Management, 8,* 37–60.

Harris, M.M., Heller, T., & Braddock, D. (1988). Sex differences in psychological well-being during a facility closure. *Journal of Management, 14,* 391–402.

Higgins, C.A., Duxbury, L.E., & Irving, R.H. (1992). Work-family conflict in the dual-career family. *Organizational Behavior and Human Decision Processes, 51,* 51–75.

Hollenbeck, J.R. (1989). Control theory and the perception of work environments: The effects of focus of attention on affective and behavioral reactions to work. *Organizational Behavior and Human Decision Processes, 43,* 406–430.

House, J.S. (1974). Occupational stress and coronary heart disease: A review and theoretical integration. *Journal of Health and Social Behavior, 15*(1), 12–27.

House, J.S., McMichael, A.J., Wells, J.A., Kaplan, H.B., & Landerman, L.R. (1979). Occupational stress and health among factory workers. *Journal of Health and Social Behavior, 20,* 139–160.

Hurrell, J.J., & Murphy, L.R. (1992). Psychological job stress. In W.M. Rom (Ed.), *Environmental and occupational medicine* (2nd ed.). Boston, MA: Little, Brown.

Ingram, R.E. (1990). Self-focused attention in clinical disorders: Review and conceptual model. *Psychological Bulletin, 107,* 156–176.

Innes, J.M., & Kitto, S. (1989). Neuroticism, self-consciousness and coping strategies, and occupational stress in high school teachers. *Personality and Individual Differences, 10,* 303–312.

Insel, P.M., & Moos, R.H. (1974). *Work environment scale, form R.* Palo Alto, CA: Consulting Psychologist Press.

Jaccard, J., Turrisi, R., & Wan, C.K. (1990). *Interaction effects in multiple regression.* Newbury Park, CA: Sage.

James, S.A., Strogatz, D.S., Wing, S.B., & Ramsey, D.L. (1987). Socioeconomic status, John Henryism, and hypertension in blacks and whites. *American Journal of Epidemiology, 126,* 664–673.

Jex, S.M., & Beehr, T.A. (1991). Emerging theoretical and methodological issues in the study of work-related stress. In K. Rowland & G. Ferris (Eds.), *Research in personnel and human resources management* (Vol. 9, pp. 311–365). Greenwich, CT: JAI.

Kahn, R.L., Wolfe, D.M., Quinn, R.P., Snoek, J.D., & Rosenthal, R.A. (1964). *Organizational stress: Studies in role conflict and ambiguity.* New York: Wiley.

Kessler, R.C. (1985). *1985 Detroit area survey.* Unpublished questionnaire. Ann Arbor, MI: Institute for Social Research, University of Michigan.

Kline, M., & Cowan, P.A. (1988). Re-thinking the connections among "work" and "family" and well-being: A model for investigating employment and family work contexts. In E.B. Goldsmith (Ed.), Work and family: Theory, research, and applications [Special issue]. *Journal of Social Behavior and Personality, 3*(4), 61–90.

Klitzman, S., House, J.S., Israel, B.A., & Mero, R.P. (1990). Work stress, nonwork stress, and health. *Journal of Behavioral Medicine, 13,* 221–243.

Kopelman, R., Greenhaus, J.H., & Connolly, T. (1983). A model of work, family, and interrole conflict: A construct validation study. *Organizational Behavior and Human Performance, 32,* 198–215.

Marshall, J.R. (1979). Stress, strain, and coping. *Journal of Health and Social Behavior, 20,* 200–201.

Menaghan, E.G. (1983). Individual coping efforts: Moderators of the relationship between life stress and mental health outcomes. In H.B. Kaplan (Ed.), *Psychosocial stress: Trends in theory and research.* New York: Academic Press.

Parasuraman, S., Greenhaus, J.H., & Granrose, C.S. (1992). Role stressors, social support, and well-being among two-career couples. *Journal of Organizational Behavior, 13,* 339–356.

Pearlin, L.I., Menaghan, E.G., Lieberman, M.A., & Mullan, J.T. (1981). The stress process. *Journal of Health and Social Behavior, 22,* 337–356.

Pearlin, L.I., & Schooler, C. (1978). The structure of coping. *Journal of Health and Social Behavior, 19,* 2–21.

Phelan, J., Schwartz, J.E., Bromet, E.J., Dew, M.A., Parkinson, D.K., Schulberg, H.C., Dunn, L.O., Blane, H., & Curtis, C. (1991). Work stress, family stress, and depression in professional and managerial employees. *Psychological Medicine, 21,* 999–1012.

Radloff, L.S. (1977). The CES–D scale: A self-report depression scale for research in the general population. *Applied Psychological Measurement, 1,* 385–501.

Rice, R.W., Frone, M.R., & McFarlin, D.B. (1992). Work–nonwork conflict and the perceived quality of life. *Journal of Organizational Behavior, 13,* 155–168.

Rizzo, J.R., House, R.J., & Lirtzman, S.I. (1970). Role conflict and ambiguity in complex organizations. *Administrative Science Quarterly, 15,* 150–163.

Scheier, M.F., & Carver, C.S. (1985). The Self-Consciousness scale: A revised version for use with general populations. *Journal of Applied Social Psychology, 15,* 687–699.

Scheier, M.F., Carver, C.S., & Matthews, K.A. (1983). Attentional factors in the perception of bodily states. In J.T. Cacioppo & R.E. Petty (Eds.), *Social psychophysiology: A sourcebook.* New York: Guilford.

Sims, H.P., Szilagyi, A.D., & Keller, R.T. (1976). The measurement of job characteristics. *Academy of Management Journal, 19,* 547–559.

Swindle, R.W., Jr., Heller, K., & Lakey, B. (1988). A conceptual reorientation to the study of personality and stressful life events. In L.H. Cohen (Ed.), *Life events and psychological functioning: Theoretical and methodological issues.* Newbury Park, CA: Sage.

Thoits, P.A. (1986). Social support as coping assistance. *Journal of Consulting and Clinical Psychology, 54,* 416–423.

PART THREE

The Roles of Coping and Dispositional Influences in Occupational Stress Research

The Impact of Persistence on the Stressor–Strain and Strain–Intentions to Leave Relationships: A Field Examination

Wayne A. Hochwarter
Pamela L. Perrewé
Russell L. Kent

Much research has assessed the relationships among environmental, psychological, and physiological stressors, and behavioral outcomes (Jex, Spector, Gudanowski, & Newman, 1994; Perrewé, 1991; Wolfgang, 1994). Overall, however, the findings of these studies have been less than conclusive (see Cooper & Marshall, 1976; Matteson & Ivancevich, 1979; Van Sell, Brief, & Schuler, 1981, for reviews of this literature). One reason for these inconclusive findings may be that past research has focused on situational determinants while disregarding the impact of individual difference variables on these relations.

It has long been suggested that stress and stress-related outcomes can best be understood by examining the interaction between the individual and the situation (Girodo, 1991; Lazarus & Laurier, 1978; Pervin, 1989). However, little research has taken this approach. The purpose of this study is to reexamine the stressor-strain-outcome relationship using the individual difference variable of persistence as a moderator variable. Specifically, persistence is viewed as moderating both the relationship between job stressors (i.e., quantitative job demands) and job strains

Authors' Notes: The authors thank William P. Anthony for his assistance during the data collection phase of this study.

(burnout, dissatisfaction, and tension) as well as the subsequent relationship between job strain and intentions to turnover.

In this study, stressors refer to the objective and perceived environmental demands while strains refer to the psychological responses made by individuals to environmental demands. Although previous research assessing the impact of moderating variables in the relation of stressors and strains has been limited, recent support exists. For example, Perrewé and Ganster (1989) found that task control moderated the relationship between job demands and perceived anxiety. In a study assessing the relations between work load, tension, and coping, Kirmeyer and Dougherty (1989) concluded that perceptions of supervisors' support was a significant moderating variable. The authors argued that perceptions of supervisors' support can serve to ease the trauma that occurs due to variations in employees' workload and increased tension under specific conditions (Kirmeyer & Dougherty, 1989).

Less research has been aimed at identifying how and under what conditions individual personality variables moderate the reaction of stressors and strains. Only a few exceptions exist, notably Type A behavior patterns (Howard, Cunningham, & Rechnitzer, 1986; Ivancevich, Matteson, & Preston, 1982), hardiness (Kobasa, 1979), activity level and locus of control (Perrewé, 1986). Identifying the differential patterns of strain reactions and behaviors when presented with comparable stressful stimuli remains an interesting problem.

To address these problems, this study will examine the moderating effects of one individual difference variable, persistence, on the stressor-strain and strain-outcome relationships. Specifically, it is hypothesized that the impact of job stressors on reported job strain will vary with the individual persistence level. In addition, individual persistence is viewed as a moderator in the relationship between strains and intent to turnover. Finally, strain variables will be examined as to their intervening effect between perceived job demands and intentions to turnover. Following is a discussion of the research pertaining to each of the variables used in conducting the field study.

Persistence

A variety of definitions of persistence have been proposed in many different bodies of literature (e.g., MacArthur, 1955; Mukherjee, 1974). However, most agree with MacArthur's denotation that persistence is the continued steadfast pursuit of an objective in spite of opposition. Feathers (1962) provided a through discussion of the various facets of persistence that have been presented in early literature. For example, he notes that persistence can be viewed as: (1) a trait or disposition, (2) a behavioral response associated with specific situational variables, and (3) a motivational outcome resulting from an interaction of personality traits and situational characteristics.

In this study, persistence is viewed as a stable characteristic that predisposes individuals to perceive and react to stimuli in a certain consistent manner. In accepting this definition, we take a stance similar to the one proposed by Staw and Ross (1985) in their investigation of dispositional variables and job attitudes. Specifically, this approach goes beyond the traditional "trait" approach that has been so often criticized in the literature (Davis-Blake & Pfeffer, 1989). The approach recognizes that an individuals' predisposition may be the result of a socialized or learned response to a broad class of situations. With respect to the present study, an individual may learn that persistence often leads to favorable outcomes across a broad range of situations. This individual may tend to persist longer than someone who has not experienced the same results.

The notion that individuals are predisposed to behave in a certain way has been the topic of an extensive debate in recent years (Chatman, 1989; Monson, Hesley, & Chernick, 1982; Pervin, 1989; Staw & Ross, 1985). Although the debate continues, one general conclusion appears to be receiving a great deal of acceptance. Specifically, it has been suggested that individual difference variables (dispositions, traits, etc.) interact with situational variables to determine how an individual perceives and reacts to environmental stimuli. This is the position taken in this study. The individual difference variable, persistence, is hypothesized to interact with perceptions of a situational factor (job demands) to determine strains (burnout, job dissatisfaction, and tension). In addition, persistence is also conceptualized as a moderating variable that will have an overall effect on the relationship between these strains and the behavioral outcome of turnover.

Persistence in Organizational Settings

Persistence is generally associated with positive outcomes. In fact, several studies have shown that managers who are perceived as persistent are evaluated more favorably than others, are seen as possessing more leadership qualities, and are viewed as being more responsible for favorable results (Staw & Ross, 1980). In addition, persistence has also been shown to be a major factor in individual success (Zufelt, 1988), managerial success (Kanter, 1988), effective leader behavior (Kent & Martinko, 1989), sales success (Sager, Futrell, & Varadarajan, 1989), and even creativity (Jacobson, 1989).

It should be noted, however, that both researchers and practitioners acknowledge that persistence may at times lead to unfavorable outcomes such as continuing to commit resources to a project when success is impossible or highly unlikely (Staw, 1976). These findings suggest that persistence, as an individual difference variable, may interact with situational factors to determine whether the resulting behavior is beneficial or non-beneficial.

Although much of the existing literature has focused on persistence as a behavioral outcome (i.e., length of time working on a task), the above discussion suggests that persistence has also received considerable attention as a predictor of a variety of organizational-induced outcomes. In addition to those variables noted above, persistence has also been linked theoretically to a number of other variables of interest to organizational researchers such as depression (Abrahamson, Seligman, & Teasdale, 1978), stress (Martinko & Gardner, 1982), and turnover (Seligman & Schulman, 1986).

Job Demands

One of the most common stressors in the workplace and also the recipient of much research is perceived job demands (Hendrix, Steel, Leap, & Summers, 1994). Increased job demands leading to job overload can take one of two forms: qualitative overload, which occurs when the work exceeds the skills and abilities of the worker, and quantitative overload, which comes about as a result of work expectations greater than what an individual can accomplish in a given period of time. This study focuses specifically on the effects of quantitative job demands on perceptions of strain and the subsequent effect on intent to turnover.

Research assessing the impact of increased job demands has taken two distinct paths in the literature. A substantial amount of research has examined the effects of increased job demands on health and quality of life issues. Specifically, a direct positive correlation between increased quantitative job requirements and smoking (French & Caplan, 1972), higher cholesterol levels (Friedman, Rosenman, & Carroll, 1958), and cardiac problems (Caplan & Jones, 1975) has been shown in the literature.

The second predominant stream of research in this area has looked at the cumulative effects of increased job demands on organizational outcomes, such as greater job dissatisfaction (Beehr, Walsh, & Taber, 1976), lower performance (Sales, 1970), and feelings of anger, anxiety, and tension (Perrewé & Ganster, 1989). Finally, Jackson, Turner, and Brief (1987) found that perceptions of job demands led to emotional exhaustion, which is one of the major components of job burnout. Thus, perceptions of increased job demands have been associated with numerous negative strain responses. Overall, the relation of job demands and strain has been fairly well-established in the literature. The question, however, is whether certain individuals (i.e., those with high versus low levels of persistence) experience more strain given a similar level of perceived job demands.

Job Demands, Persistence, and Strain Variables

Although the idea of using persistence as a moderator of these relationships has face validity, little research has taken this approach. The work in the area of learned helplessness (Abrahamson et al., 1978) provides some insights. In this body of literature, helplessness is viewed as a lack of persistence. Also, Cherniss (1980) suggests that burnout among health professionals is the result of unpredictable work environments which cause workers to become helpless, or non-persistent. As these "helpless" individuals continue to face the increased job demands of their environment, they experience burnout and job dissatisfaction. Given this awareness, an interactive effect of job demands and persistence on strain variables (e.g., burnout and dissatisfaction) is suggested. Martinko and Gardner (1982) also suggest a direct relationship between the work environment and persistence levels on affective responses. Specifically, less persistent individuals are seen as more susceptible to the negative effects of stressors such as increased job demands when compared to more persistent individuals.

Turnover Intentions

Employee turnover has been one of the most researched topics in the organizational sciences over the last twenty years (Blau & Boal, 1989). The costs that must be absorbed by organizations in dealing with this phenomenon are a major reason for the research (Wanous, 1980). Past research has taken two distinct avenues: research on turnover within specific industries/occupations and research to identify organization-based antecedents of turnover. With respect to occupation-based turnover, research has examined a number of job categories including salespersons (Ivancevich & Donnelly, 1974; Sager et al., 1989), retail managers (Gable, Hollon, & Dangello, 1984), bank tellers (Vecchio, 1985), nurses (Saleh, Lee, & Prien, 1965) and manual laborers (Wild, 1970). This stream of research has sought to identify specific job features or individual traits that led to higher levels of turnover within the specific occupation or industry.

The second predominant stream of turnover research has been undertaken to identify precursors that explain employee turnover (e.g., Waters, Roach, & Waters, 1976). For example, much research has shown that employee dissatisfaction frequently leads to increased intentions to turnover. Dissatisfaction with such job facets as pay (Futrell & Parasuraman, 1984), supervision (Hom & Hulin, 1981), co-workers (Mobley, Horner, & Hollingsworth, 1978), recent promotions or promotional opportunities (Newman, 1974), as well as the work itself (Hom, Katerberg, & Hulin, 1979; Mobley et al., 1978) have all been associated with increased turnover intentions. In addition, research has shown that employees who reported feelings of burnout or experience stress were more likely to turnover (Jackson, 1983; Parasuraman, 1982).

The relation between employees' strain and intentions to turnover has been fairly well-established. However, the question remains unclear whether certain individuals have greater intentions to leave the organization given the same level of experienced strain. This study will seek to address this question by examining the intervening role of strain in the job demand—intent to turnover relationship as well as the moderating role of persistence.

METHOD

Subjects

The subjects used in this field study were 238 managers and non-managerial personnel employed at six different firms. These firms included: an electrical utility (n = 73), a supermarket/convenience store chain (n = 24), a real estate company (n = 18), a bank/credit union (n = 42), a state tax collection office (n = 52), and a manufacturer of consumer products (n = 28). Professionals (24.2%) comprised the greatest number of survey respondents, followed by first-line supervisors (22.2%), middle managers (20.6%), top managers (19.1%), non-managers (8.1%), and technicians (5.8%). The sample included 106 men (44.5%) and 117 women (49.1%). (Fifteen respondents (6.3%) did not provide this information.) The ages of respondents ranged from 19 to 65, with an average of 40.7 (standard deviation of 10.7).

Measures

All scales used a 5-point Likert scale format with responses ranging from *strongly agree* to *strongly disagree*. Quantitative job-demands (stressor) was measured using a 4-item scale developed by Caplan, Cobb, French, Van Harrison, and Pinneau (1975). Caplan and Jones (1975) reported an alpha coefficient of .77 for this scale.

Job-induced tension was measured using a 7-item sub-scale of an anxiety-stress instrument developed by House and Rizzo (1972). Suitable reliability measurements have been shown for this scale ranging from .73 (Rizzo, House, & Lirtzman, 1970) to .83 (House & Rizzo, 1972). Job dissatisfaction and intentions to turnover were each measured using a 3-item scale developed by Cammann, Fishman, Jenkins, and Klesh (1979). The authors report an initial alpha coefficient of .77 for this scale. A 7-item scale was developed by the researchers to measure individual perceptions of burnout (e.g., "By the time I'm ready to leave, I feel all used up"). Our reliability coefficient for this scale is .78. Items contained in this scale can be found in the Appendix. Mukherjee's (1974) 20-item scale was used to measure individuals' persistence. Previously reported reliabilities for this scale range from .79 to .85 (Dubey, 1986; Kent, 1990). Originally, this scale was designed with a "yes-no" format. However, the format was changed to a Likert scale to increase variability of responses.

Procedures

Questionnaires were administered at the conclusion of strategic planning workshops at each work site. Because these workshops were primarily attended by managers, the sample contains a much higher proportion of managers than non-managers. Employees had the choice of either completing the questionnaire immediately following the workshop or at home. To assure confidentiality, self-addressed envelopes were provided each respondent. A total of 150 respondents returned completed questionnaires following the workshop while 88 returned their surveys through the mail. The overall response rate was 79 percent. The purpose of the survey was to solicit worker responses pertaining to a variety of work related questions. However, none of the variables measured were discussed during the workshops.

TABLE 1   Means, Standard Deviations, Reliabilities and Correlations for
Measured Variables

Variables	M	SD	1	2	3	4	5	6
1. Persistence	3.78	.40	(.84)					
2. Job Demands	3.45	.83	.27	(.81)				
3. Burnout	2.63	.65	-.18	.22	(.78)			
4. Job Tension	2.92	.73	-.09	.33	.48	(.74)		
5. Dissatisfaction	1.79	.68	-.23	.07	.50	.28	(.86)	
6. Turnover Intentions	2.06	1.02	-.19	.17	.51	.33	.69	(.91)

p < .01, r > .16; p < .001, r > .21
Coefficient alphas in parentheses

RESULTS

Table 1 displays descriptive statistics, correlations among the study variables and reliability indices for each measure. Consistent with past research, perceptions of high job demands were correlated with reports of burnout ($r = .22$, $p < .001$) and tension ($r = .33$, $p < .001$). Interestingly, job demands were not associated with satisfaction ($r = .07$, ns). Thus, it does not appear that perceptions of increased job demands were associated with the amount of satisfaction derived from the job. Job dissatisfaction ($r = .69$, $p < .001$), burnout ($r = .50$, $p < .001$), and job-induced tension ($r = .33$, $p < .001$) were all highly correlated with intentions to turnover. These findings support past research that has reported comparable findings. Finally, persistence was positively correlated with perceived job demands ($r = .27$, $p < .001$) and negatively with self-reports of burnout ($r = -.18$, $p < .01$), job dissatisfaction ($r = -.23$, $p < .001$), and turnover intentions ($r = -.19$, $p < .01$).

Preliminary Analysis

As mentioned above, respondents in the sample came from seven distinct organizations. These organizations were included in the sample because the researchers were interested in determining the moderating effects of persistence across organizations. Thus, before moderating and intervening effects were determined, it was necessary to ascertain whether there were organizational, occupational level, or gender differences within the sample. Analysis of variance tests (ANOVAs) were conducted for each variable to determine if differences existed within the sample. With respect to the persistence measure, results indicated no differences for organizational effect ($F = .26$), occupational level ($F = .41$), or gender effects ($F = 1.02$). Although analyses for the other variables in the study revealed some mean differences, these differences were not significant.

Intervening Effects

Using regression techniques, perceptions of quantitative job demands had a positive significant main effect on both job tension ($\beta = .30$, $p < .001$, $R^2 = .11$) and burnout ($\beta = .11$, $p < .01$, $R^2 = .03$). Job demands did not have a significant effect

TABLE 2   Interactive Effects for Persistence

Variables	df	β	R^2	ΔR^2	F
Burnout					
A. Demands			.84	.05	
Persistence	(2,210)	.17	.10		12.78*
B. Demand x Persistence	(3,209)	-.16	.11	.01	1.88
Constant	1.23				
Dissatisfaction					
A. Demands			.25	.01	
Persistence	(2,210)	-.32	.07		7.97*
B. Demand x Persistence	(3,209)	-.04	.07	.00	.08
Constant	3.42				
Tension					
A. Demands			.93	.13	
Persistence	(2,210)	.21	.16		20.51**
B. Demands x Persistence	(3,209)	-.15	.17	.01	1.50
Constant	.92				
Turnover					
A. Burnout			1.05	.23	
Persistence	(2,210)	-.02	.24		33.82**
B. Burnout x Persistence	(3,209)	-.09	.24	.00	.22
Constant	.28				
Turnover					
A. Dissatisfaction			.48	.46	
Persistence	(2,210)	-.33	.46		90.97**
B. Dissat. x Persistence	(3,209)	.14	.47	.00	.74
Constant	1.49				
Turnover					
A. Tension			2.17	.10	
Persistence	(2,210)	.92	.13		15.41**
B. Tension x Persistence	(3,209)	-.46	.15	.02	5.22*
Constant	-2.72				

$*p < .01; **p < .001$

on job dissatisfaction. In regard to the proposed intervening variables, both job dissatisfaction ($\beta = .96$, $p < .001$) and job tension ($\beta = .16$, $p > .029$) had a main effect on intentions to turnover. In addition, job dissatisfaction and job tension explained approximately 50% of the variance in turnover intentions. Finally, job burnout did not significantly affect intentions to turnover. Thus, two of the three hypothesized relationships between the strain variables and intentions to turnover were substantiated in the data. Job demands did not have a significant effect on turnover intentions after the strain variables were partialed out of the regression equation. In summary, the results indicate that only job tension served as an intervening variable between job demands and intentions to turnover.

Moderating Effects

Table 2 contains a summary of the hierarchical regression used in this study. Following the guidelines presented by Cohen and Cohen (1983), variables were entered in the regression equation in hierarchical sequence to partial main effects from interactive effects. For example, in determining the effect of persistence on the job demands-job burnout relationship, job demands were entered in the equation first followed by persistence. Finally, the interactive effect of job demands and persistence was entered. Similar steps were completed for the other analyses as well.

As can be seen in Table 2, persistence moderated the relationship between job tension and intentions to turnover ($F = 5.22$, $p < .01$), by increasing R^2 from .13 to .15. This relationship is presented graphically in Figure 1. Under conditions of high tension, intentions to turnover increased as individual persistence levels declined. In other words, less persistent individuals were more likely to report increased turnover intentions under conditions of high tension. Conversely, highly persistent respondents' intentions to turnover did not change significantly under varying levels of tension.

In addition, results from this study indicted that persistence had no moderating effect with quantitative job demands on any of the measures of strain. This suggests that reactions to increased job demands seen as increases in burnout ($F = 1.88$), job dissatisfaction ($F = .08$), and tension ($F = 1.50$) were not affected by individual persistence levels.

Finally, persistence did not moderate the relations between job burnout ($F = .22$) and dissatisfaction ($F = .74$) on intentions to turnover. These findings indicate that individual persistence levels did not impact subsequent intentions to turnover when each of these work related strains increased.

DISCUSSION

This field study examined the relationship between perceived job demands, persistence, experienced strain and intentions to turnover. Critical to this study is the role that persistence plays as a moderating variable in the stressor-strain and strain-turnover relationships. Results from this study indicated that persistence moderated the tension-intentions to turnover relationship. This finding suggests that employees with low levels of persistence were significantly more likely to intend to quit when experiencing job tension. Conversely, it was shown that highly persistent individuals' intentions to turnover were unaffected by the level of job tension. Persistent individuals may simply view job tension as another obstacle to overcome, thus, they would be less likely to quit. Since research in the area of persistence is still at the preliminary stages of development, extensive explanations for this finding would be speculative.

Consistent with past research, perceptions of job demands were associated with job burnout and tension. However, perceptions of job demands were not associated

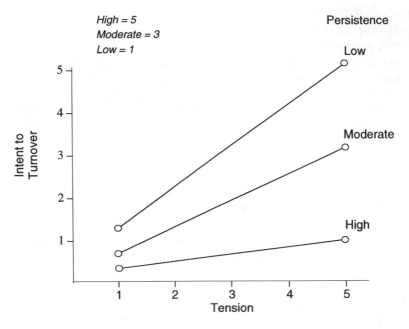

FIGURE 1   The Interactive Effects of Persistence and Tension on Intention to
Turnover

with job dissatisfaction. This finding is inconsistent with past research which has
shown a strong relationship between these two variables (Beehr et al., 1976). In
addition, self-reports of job strain (i.e., burnout, tension, and dissatisfaction) were
significantly correlated with intentions to turnover. Surprisingly, however, job burnout was not related to intentions to turnover when the other strain variables (i.e.,
dissatisfaction and tension) were entered into the regression equation first. Thus, the
results of this field study suggest that burnout did not explain a significant amount of
unique variance other than that explained by job tension and dissatisfaction. Moreover, the findings suggest that job tension was the only strain variable that served as
an intervening variable between job demands and intentions to turnover.

No moderating effect for persistence was found in the job demands-strain
variables relationship. Although persistent individuals reported high quantitative
job demands, they also reported less job burnout and job dissatisfaction and
expressed fewer intentions to turnover. Some light may be shed on this issue by
analyzing the correlation matrix. Persistence is significantly correlated with burnout, job tension, and job dissatisfaction, indicating that persistence may directly
affect these variables instead of moderating the effect of other variables.

Another possible explanation for the limited findings related to persistence may
relate to the questionnaire measure used. Little information on the scale is available

due to its limited use in published research. In addition, the loadings on a general persistence factor reported in the original factor analytic study were adequate but not overwhelming (.30 to .51).

Another issue impacting the findings of this study relates to the sample used. Although it was advantageous to use a sample that consisted primarily of managers for generalizability purposes, this may also be a limitation. As suggested earlier, most researchers have concluded that being persistent is beneficial (e.g., Kanter, 1988; Zufelt, 1988). For example, a direct link between persistence and career mobility has been established in the literature (Stier & Grusky, 1990). Thus, it is likely that a sample of managers would report significantly higher levels of persistence than would non-managers. In addition, it is plausible to assume that the amount of variability for this measure would be much less for a sample that was managerial-based. Findings reported in Table 1 substantiate this conclusion. Not only did the persistence measure display the highest mean value, it also exhibited the lowest amount of variability when compared to the other variables in the study.

One last potential limitation of this study should be noted. Since this study used questionnaire measures exclusively, the common methods variance may be a problem.

In summary, additional research is needed to examine the relationship between persistence and stress-related variables. For example, definitions of Type A behavior have included persistence as a component (cf. Strube, 1990). Given the health-related risk factors associated with hostility and aggression as dimensions of Type A behavior (Greenglass & Burke, 1991), an examination of other components may be fruitful as well. In addition, research is needed to examine other potential individual difference variables in the job stressors-strain relation. The notion that individuals are predisposed to behave in certain ways (Schaubroeck & Ganster, 1991; Staw & Ross, 1985) may open new avenues for researchers interested in job stress. Identifying employees who are more susceptible to job stressors or who react differently to certain stressors and strains would be valuable information for managers and those involved with managing stress. Further afield, persistence may be an important variable in other areas such as divorce.

REFERENCES

Abrahamson, L., Seligman, M., & Teasdale, J. (1978). Learned helplessness in humans: Critique and reformulation. *Journal of Abnormal Psychology, 87,* 49–74.

Beehr, T.A., Walsh, J., & Taber, T. (1976). Relationship of stress to individually and organizationally valued states: Higher order needs as moderators. *Journal of Applied Psychology, 61,* 41–47.

Blau, G., & Boal, K. (1989). Using job involvement and organizational commitment interactively to predict turnover. *Journal of Management, 15,* 115–127.

Cammann, C., Fishman, M., Jenkins, D., & Klesh, J. (1979). *Michigan Organizational Assessment Questionnaire.* Unpublished manuscript, University of Michigan, Ann Arbor, MI.

Caplan, R.D., Cobb, S., French, J.R.P., Jr., Van Harrison, R., & Pinneau, S.R., Jr. (1975). *Job demands and worker health: Main effects and occupational differences* (HEW NIOSH No. 75–160). Washington, DC: U.S. Government Printing Office.

Caplan, R.D., & Jones, K.W. (1975). Effects of workload, role ambiguity, and type A personality on anxiety, depression, and heart rate. *Journal of Applied Psychology, 60,* 713–719.

Chatman, J. (1989). Improving interactional organizational research: A model of person-organization fit. *Academy of Management Review, 14,* 333–349.

Cherniss, C. (1980). *Staff burnout: Job stress in the human services.* Beverly Hills, CA: Sage.

Cohen, J., & Cohen, P. (1983). *Applied multiple regression/correlation analysis for the behavioral sciences* (2nd ed.). Hillsdale, NJ: Erlbaum.

Cooper, C.L., & Marshall, J. (1976). Occupational sources of stress: A review of the literature relating to coronary heart disease and mental health. *Journal of Occupational Psychology, 49,* 11–28.

Davis-Blake, A., & Pfeffer, J. (1989). Just a mirage: The search for dispositional effects in organizational research. *Academy of Management Review, 14,* 385–400.

Dubey, R. (1986). Persistence-disposition among different types of Indian leaders. *Indian Psychological Review, 30,* 10–16.

Feathers, N. (1962). The study of persistence. *Psychological Bulletin, 59,* 94–115.

French, J.R.P., Jr., & Caplan, R.D. (1972). Organizational stress and individual strain. In A.J. Marrow (Ed.), *The failure of success* (pp. 30–66). New York: Amacom.

Friedman, M., Rosenman, R.H., & Carroll, V. (1958). Changes in serum cholesterol and blood clotting time in men subject to cyclical variation of occupational stress. *Circulation, 17,* 852–861.

Futrell, C., & Parasuraman, A. (1984). The relationship of satisfaction and performance to sales force turnover. *Journal of Marketing, 48,* 33–40.

Gable, M., Hollon, C., & Dangello, F. (1984). Predicting voluntary managerial trainee turnover in a large retailing organization from information on employment application blank. *Journal of Retailing, 60,* 43–63.

Girodo, M. (1991). Personality, job stress, and mental health in undercover agents: A structural equation analysis. In P.L. Perrewé (Ed.), Handbook on job stress [Special issue]. *Journal of Social Behavior and Personality, 6*(7), 375–390.

Greenglass, E.R., & Burke, R.J. (1991). The relationship between stress and coping among Type As. In P.L. Perrewé (Ed.), Handbook on job stress [Special issue]. *Journal of Social Behavior and Personality, 6*(7), 361–374.

Hendrix, W.H., Steel, R.P., Leap, T.L., & Summers, T.P. (1994). Antecedents and organizational effectiveness outcomes of employee stress and health. In P.L. Perrewé & R. Crandall (Eds.), *Occupational stress: A handbook* (pp. 73–92). Washington, DC: Taylor & Francis (Original work published 1991)

Hom, P., & Hulin, C. (1981). A competitive test of the prediction of reenlistment by several models. *Journal of Applied Psychology, 66,* 23–29.

Hom, P., Katerberg, R., & Hulin, C. (1979). Comparative examination of three approaches to the prediction of turnover. *Journal of Applied Psychology, 64,* 280–290.

House, R., & Rizzo, J. (1972). Role conflict and ambiguity as critical variables in a model of organizational behavior. *Organizational Behavior and Human Performance, 7,* 467–505.

Howard, J.H., Cunningham, D.A., & Rechnitzer, P.A. (1986). Role ambiguity, Type A behavior, and job satisfaction: Moderating effects on cardiovascular and biochemical responses associated with coronary risk. *Journal of Applied Psychology, 71,* 95–101.

Ivancevich, J.M., & Donnelly, J. (1974). A study of role clarity and need for clarity for three occupational groups. *Academy of Management Journal, 17,* 28–36.

Ivancevich, J.M., Matteson, M.T., & Preston, C. (1982). Occupational stress, Type A behavior, and physical well-being. *Academy of Management Journal, 25,* 373–391.

Jackson, S. (1983). Participation in decision making as a strategy for reducing job-related strain. *Journal of Applied Psychology, 68,* 3–19.

Jackson, S., Turner, J., & Brief, A. (1987). Correlates of burnout among public service lawyers. *Journal of Occupational Behavior, 8,* 3–19.

Jacobson, G. (1989). Carlson's timeless lessons on innovation. *Management Review, 78,* 19–36.

Jex, S.M., Spector, P.E., Gudanowski, D., & Newman, R. (1994). Relations between exercise and employee responses to work stressors: A summary of two studies. In P.L. Perrewé & R. Crandall (Eds.), *Occupational stress: A handbook* (pp. 283–301). Washington, DC: Taylor & Francis. (Original work published 1991)

Kanter, R.M. (1988). How to be an entrepreneur without leaving your company. *Working Women, 11*, 44–50.

Kent, R. (1990). Individual differences in the study of persistence. *Proceedings of the Annual Meeting of the Institute for Management Sciences-Southeastern Chapter* (pp. 358–360). Myrtle Beach, SC.

Kent, R., & Martinko, M. (1989). *The relationship between persistence and effective leader behavior.* Paper presented at the National Decision Sciences Meeting, New Orleans, LA.

Kirmeyer, S.L., & Dougherty, T.W. (1988). Work load, tension, and coping: Moderating effects of supervisor support. *Personnel Psychology, 41*, 125–139.

Kobasa, S.C. (1979). Stressful life events, personality and health: An inquiry into hardiness. *Journal of Personality and Social Psychology, 4*, 1–11.

Lazarus, R.S., & Launier, R. (1978). Stress-related transactions between person and environment. In L.A. Pervin & M. Lewis (Eds.), *Perspectives in interactional psychology* (pp. 287–327). New York: Plenum.

MacArthur, R. (1955). An experiential investigation of persistence in secondary school boys. *Canadian Journal of Psychology, 9*, 43–54.

Martinko, M., & Gardner, W. (1982). Learned helplessness: An alternative explanation for performance deficits. *Academy of Management Review, 7*, 195–204.

Matteson, M.T., & Ivancevich, J.M. (1979). Organizational stressors and heart disease: A research model. *Academy of Management Review, 4*, 347–357.

Mobley, W., Horner, S., & Hollingsworth, A. (1978). An evaluation of the precursors of hospital employee turnover. *Journal of Applied Psychology, 63*, 408–414.

Monson, T., Hesley, J., & Chernick, L. (1982). Specifying when personality traits can predict behavior: An alternative to abandoning the attempt to predict single act criteria. *Journal of Personality and Social Psychology, 43*, 385–399.

Mukherjee, B. (1974). A questionnaire measure of persistence disposition. *Indian Journal of Psychology, 49*, 263–278.

Newman, J. (1974). Predicting absenteeism and turnover: Field comparison of Fishbein's modeling traditional job attitude measures. *Journal of Applied Psychology, 59*, 612–615.

Parasuraman, S. (1982). Predicting turnover intention and turnover behavior: A multivariate analysis. *Journal of Vocational Behavior, 21*, 111–121.

Perrewé, P.L. (1986). Locus of control and activity level as moderators in the quantitative job demands—satisfaction/psychological anxiety relationship: An experimental analysis. *Journal of Applied Social Psychology, 16*, 620–632.

Perrewé, P.L. (Ed.). (1991). Handbook on job stress [Special issue]. *Journal of Social Behavior and Personality, 6*(7).

Perrewé, P.L., & Ganster, D. (1989). The impact of job demands and behavioral control on experienced job stress. *Journal of Organizational Behavior, 60*, 213–229.

Pervin, L. (1989). Persons, situations, interactions: The history of a controversy and a discussion of the theoretical models. *Academy of Management Review, 14*, 350–360.

Rizzo, J.R., House, R.J., & Lirtzman, S.I. (1970). Role conflict and ambiguity in complex organizations. *Administrative Science Quarterly, 15*, 150–163.

Sager, J., Futrell, C., & Varadarajan, R. (1989). Exploring salesperson turnover: A causal model. *Journal of Business Research, 18*, 303–326.

Saleh, S., Lee, R., & Prien, E. (1965). Why nurses leave jobs: An analysis of female turnover. *Personnel Administrator, 28*, 25–28.

Sales, S. (1970). Some effects of role overload and role underload. *Organizational Behavior and Human Performance, 5*, 592–608.

Schaubroeck, J., & Ganster, D. (1991). The Role of negative affectivity in work-related stress. In P.L. Perrewé (Ed.), Handbook on job stress [Special issue]. *Journal of Social Behavior and Personality, 6*(7), 319–330.

Seligman, M., & Schulman, P. (1986). Explanatory styles as a predictor of productivity and quitting among life insurance sales agents. *Journal of Personality and Social Psychology, 50*, 832–838.

Staw, B.M. (1976). Knee-deep in the big muddy: A study of escalating commitment to a chosen course of action. *Organizational Behavior and Human Performance, 16*, 27–44.

Staw, B.M., & Ross, J. (1980). Commitment in an experimenting society: An experiment on the attribution of leadership from administrative scenarios. *Journal of Applied Psychology, 65,* 249–260.

Staw, B.M., & Ross, J. (1985). Stability in the midst of change: A dispositional approach to job attitudes. *Journal of Applied Psychology, 70,* 469–480.

Stier, H., & Grusky, P. (1990). An overlapping persistence model of career mobility. *American Sociological Review, 55,* 736–756.

Strube, M.J. (Ed.). (1990). Type A behavior [Special issue]. *Journal of Social Behavior and Personality, 5*(1).

Van Sell, M., Brief, A.P., & Schuler, R.S. (1981). Role conflict and role ambiguity: Integration of the literature and directions for future research. *Human Relations, 34,* 43–71.

Vecchio, R. (1985). Predicting employee turnover from leader-member exchange: A failure to replicate. *Academy of Management Journal, 23,* 478–485.

Wanous, J. (1980). *Organizational entry: Recruitment, selection and socialization of newcomers.* Reading, MA: Addison-Wesley.

Waters, L., Roach, D., & Waters, C. (1976). Estimates of future tenure, satisfaction, and biographical variables as predictors of termination. *Personnel Psychology, 29,* 57–60.

Wild, R. (1970). Job needs, job satisfaction, and job behavior of women manual laborers. *Journal of Applied Psychology, 54,* 157–162.

Wolfgang, A.P. (1994). Job stress, coping, and dissatisfaction in the health professions: A comparison of nurses and pharmacists. In P.L. Perrewé & R. Crandall (Eds.), *Occupational stress: A handbook* (pp. 193–204). (Original work published 1991)

Zufelt, J. (1988). Use the conquering force within you. *Executive Excellence, 5,* 11–12.

APPENDIX

Below are the seven questions developed by the researchers to measure job burnout. A five-point Likert scale format was used with strongly agree, agree, undecided, disagree, and strongly disagree serving as the possible responses.

1. My job causes me to become much more cold and callous at work.

2. I often try to remove myself emotionally from my job.

3. Since taking this job, I find myself treating people I work with impersonally.

4. By the time I'm ready to leave work, I feel all used up.

5. My job makes me feel burned out.

6. I feel energetic about going to work each day. (R)

7. My job makes me feel emotionally drained.

(R) = Reverse coded.

The Moderating Effects of Self-Esteem on the Work Stress–Employee Health Relationship

Daniel C. Ganster
John Schaubroeck

Organizational psychologists have recently begun to show a renewed interest in the role of personality variables in determining workers' reactions to their jobs and roles. This interest follows a general hiatus from the examination of individual differences factors that arose from the frequent failure of organizational researchers to uncover significant relationships between personality variables and job attitudes and behaviors. Weiss and Adler (1984) argued that personality variables have often been included in organizational studies with little thought as to their theoretical role in the phenomena under study. It is not surprising, then, that researchers have been disappointed with their findings. Unfortunately, small and generally insignificant correlations between personality variables and organizational outcomes, even in situations where there was little justification for positing personality effects, had led theorists to abandon the study of personality in the work setting.

Recent interest in individual differences variables comes from two directions. First, in the study of life stress in general and work stress in particular, several dispositional constructs have been the focus of much attention. Some constructs, such as the Type A behavior pattern and Negative Affectivity, have been examined primarily as main effect predictors of health and well-being (Ganster, Schaubroeck,

Sime, & Mayes, 1991). Others, such as personal hardiness, are of interest mostly because of their interactive, or buffering, role in the stress-health relationship (Kobasa, 1988). There is growing evidence that dispositions can serve either as "protective" or "vulnerability" factors, and determine how individuals appraise, cope, and react to a variety of stressors. The other direction from which interest in personality variables arises is work-based theories of self-esteem. In the late 1960's Korman (1966, 1970) proposed a self-consistency theory of self-esteem. Positing that workers were motivated to adopt attitudes, beliefs, and behaviors in order to maintain their pre-existing levels of self-esteem, Korman predicted that self-esteem would be related to a variety of work outcomes, including job performance and occupational choice. Although Korman's model met with limited empirical success (Dipboye, 1977), Brockner (1988) has developed a convincing case for the study of self-esteem in the workplace that has effectively rekindled interest in this construct. In the present paper we extend Brockner's (1988) work in self-esteem, and in particular, his "plasticity" hypothesis, to the relationship between role stresses and somatic health outcomes.

Since the early work of Kahn, Wolfe, Quinn, Snoek, and Rosenthal (1964), the concept of role stress has had a prominent position in the study of work stress. While the largest body of literature exists for the constructs of role conflict and role ambiguity, many studies have also accumulated data that incorporate concepts of role overload as well. This literature has been subjected to both conceptual (Van Sell, Brief, & Schuler, 1981) and quantitative reviews (Fisher & Gitelson, 1983; Jackson & Schuler, 1985). Jackson and Schuler (1985), in particular, noted that the relationship between role stresses and "job tension" has been well established, but that the impact of role stresses on physiological or explicitly health-related outcomes has not been clearly delineated. A major interpretive problem with the job tension findings, moreover, is that these correlations are likely inflated by the measurement overlap between the scales measuring role stress and those measuring job tension. This problem is particularly likely when the job tension scale of Kahn et al. (1964) is used because items from this index make direct reference to problems of role conflict and role ambiguity. Thus, it is not clear that the two scales are even measuring theoretically distinct constructs. Studies that have examined more conceptually distinct outcome variables such as physiological and health outcomes (e.g., Caplan & Jones, 1975; Fusilier, Ganster, & Mayes, 1987; Ivancevich, Matteson, & Preston, 1982) have not demonstrated a consistent main effect of role stresses. In the present paper we echo Jackson and Schuler's (1985) call for the study of moderator variables in understanding role conflict and ambiguity by examining employee self-esteem.

Self-Esteem as a Moderator of the Role Stress and Health Relationship

Brockner (1988) notes that "the essence of self-esteem is the favorability of individuals' characteristic self-evaluations" (p. 11). One can further distinguish the concept of global self-esteem from self-esteem specific to particular areas of one's life. Our focus in this study is on the former construct. While self-esteem can be viewed as a dependent variable, subject to modification by salient experiences, one's global self-esteem, at least in adults, is generally considered a stable disposition that affects their perceptions and responses to the external environment. Self-esteem has been regarded as a critical element in mental health. Rosenberg's (1965) study of over 5,000 high school students demonstrated moderate to strong correlations between self-esteem and depressive affect ($r = -.30$) and a list of physiological indicators of anxiety that included trembling hands, pounding heart, pressures or pains in the head, sweating hands, and dizziness ($r = -.48$). Beck's (1967) research on depression also indicated the important role of low self-esteem in that disorder. Self-esteem's role as a correlate of physical disease, however, is less established.

While self-esteem (SE) has not been shown to be a reliable predictor of physical health, it might play an important role in determining whether stressful environmental demands exact a toll on health. Brockner (1983, 1988) has advanced the hypothesis that low SEs are generally more susceptible to environmental, and in particular, organizational, events than are high SEs. Brockner (1988) reviewed extensive evidence that this "plasticity hypothesis" explains a number of organizational processes, including the effects of (a) peer group interaction on workers' performance and attitudes, (b) performance appraisal feedback on workers' subsequent performance, (c) feedback on communication processes, (d) socialization on role taking, and (e) modeling on leadership. Finally, and most germane to the present paper, two studies have demonstrated that self-esteem moderates the effects of role stress on work behaviors and attitudes. Specifically, Mossholder, Bedeian, and Armenakis (1981) examined how self-esteem affected how nurses reacted to role conflict and role ambiguity. Surveying nurses in a large southwestern hospital, they found that self-reports of role conflict were negatively related to supervisory ratings of job performance for low SEs but not high SEs. Role ambiguity, in turn, was negatively related to job satisfaction for low SEs but not for high SEs. Overall, these results provide support for the general validity of the plasticity hypothesis, at least insofar as how SE affects the attitudinal and behavioral responses of employees to work demands.

More recently, Pierce, Gardner, Dunham, and Cummings (1993) tested the plasticity hypothesis with workers from an electrical utility. Pierce et al. expanded the investigation of self-esteem moderating effects by examining more role stressors (they added role overload, work environment support, and supervisor support to measures of role conflict and ambiguity) and by using an organizationally-based measure of self-esteem instead of global self-esteem. Their moderated regression analyses indicated general support for the plasticity hypothesis.

While the Mossholder et al. (1981) and Pierce et al. (1993) findings indicate that low SEs are attitudinally and behaviorally more susceptible to role stresses than are high SEs, it has not yet been demonstrated that the plasticity hypothesis can explain the effects of role stress on health outcomes. It would seem that the plasticity hypothesis could be extended to the explanation of health outcomes through an analysis of how self-esteem might determine one's coping activities when encountering role demands. In Hall's (1972) model of coping with role conflict, he posited that individuals (in his case women) could cope with role conflicts by (a) renegotiating structurally imposed demands, (b) personally redefining role demands, or (c) passive role behavior. Using the last strategy, the individual attempts to cope with the role demands by striving to improve her performance so that she can meet the existing role demands rather than actively trying to change the demands themselves. This passive coping approach can lead to a mounting spiral of ever increasing pressures and frustrations. The individual tries harder and harder to meet the conflicting role demands and ambiguous standards that, by their nature, are almost sure to result in frustration and depression. It is this chronic state of frustration and tension that is expected to lead to the development of somatic complaints and ill-health. How likely is it, then, that the low SE individual is more apt to be a passive coper? Brockner (1988), among others, has demonstrated that low SEs are less confident in their ability to influence their social environment, especially in ways that would have esteem-enhancing outcomes. Moreover, Brockner (1988) has speculated that "low SEs' experience of their interpersonal environment in the work setting could interfere dramatically with their ability to accomplish their tasks, especially if the work requires coordination with co-workers" (p. 220).

Given the low self-efficacy of low SE employees when faced with such socially mediated demands as role conflict and ambiguity, we would hypothesize that they are more apt to take a passive role in dealing with these stresses. Consequently, we would hypothesize that the health of low SEs will be more affected by these stresses than that of high SEs. This reasoning translates into the following hypothesis:

Self-esteem will interact with role conflict and role ambiguity in determining the level of somatic health complaints. The relationship between role conflict and role ambiguity and somatic health complaints will be stronger for lower SE workers.

METHOD

Sample

Survey data were obtained from 157 members of a medium-sized city's fire department. All respondents were male, had a mean age of 35.5 years, an average educational attainment of 13.42 years, and had an average job tenure of approximately 12 years. The sample was relatively free from major health problems, but a majority had reported using medical services (other than for a routine medical examination) within the past four months. The sample was recruited at the department's 13 fire houses to participate in a study of work stress and health, and represents 70% of the available workforce. Participants were not compensated for their time, but the researchers were allowed to obtain all measures while the firemen were on duty, resulting in a relatively high response rate.

Measures

Role Stress. Role conflict and role ambiguity were measured with the scales developed by Rizzo, House, and Lirtzman (1970). Role ambiguity was assessed with six 7-point Likert items (e.g., "I feel certain about how much authority I have"), and had an internal consistency reliability of .73 in the present sample. Role conflict was measured with eight 7-point Likert items (e.g., "I receive incompatible requests from two or more people"), and had a reliability of .71. Although these scales have been scrutinized for their item response characteristics (self vs. other wording and positive vs. negative item wording), Jackson and Schuler (1985) concluded in their review that "the Rizzo et al. role ambiguity and role conflict scales have been and are satisfactory measures of two role constructs" (p. 17). Scale scores were computed for both measures by averaging their respective items.

Self-Esteem. SE was assessed with the ten-item 5-point Likert scale of Rosenberg (1965). This scale is perhaps the most commonly used measure of self-esteem, and Rosenberg (1979) reviews the extensive data regarding the construct validity of this scale. In the present sample we calculated SE scores by averaging the scale items. The measure showed a reliability of .81.

Somatic Complaints. Negative health symptoms were measured with a check-list that asked the respondent to denote the frequency of experience in the last month of 17 health complaints, ranging from palpitating heart and chest pains to nausea, headaches, and sleep disorders. An overall scale for somatic complaints was formed by averaging the 5-point Likert-scaled items. The overall scale reliability was .89. This and similar scales have been used in large scale epidemiological studies (cf. Caplan, Cobb, French, Harrison, & Pinneau, 1975) and have been shown to provide valid indications of stress-related somatic health symptoms. Unlike job-related tension scales (cf. Kahn et al. 1964), however, the symptom wording contains no conceptual overlap with respondent reports of role stresses.

Demographic control variables. In order to control for the possible confounding effects of plausible correlates of somatic health, several personal characteristics were also assessed. These included age, education, and cigarette smoking.

Procedure

The initial phase of the research consisted of a qualitative assessment of the firefighter's job and working environment. Members of the research team interviewed individual firefighters, their captains, and the district chiefs and also conducted informal focus groups in several of the fire stations. The investigators also participated in demonstration training exercises with two of the engine companies. The goals of this phase of the research were two-fold. First, we wanted to gain the trust and confidence of the firefighters in order to maximize the participation rate in the study and the quality of the data. Second, we hoped to gain an appreciation of the demands of the occupation and how they interacted with the social environment in the fire stations. To briefly describe the work environment, the job alternates between periods of extreme physical demand and exposure to noxious environmental hazards, and periods of maintenance, study, and training activities. The firefighters work 24-hour shifts and live together in a group setting that is characterized by high cohesiveness and strong norms. Cohesiveness was evident from the consistency in which the men reported the norms prevalent in the fire houses and the pressure that was exerted on them to comply with these norms. A strong social norm concerned cooperation and assistance among the firefighters. Each man was expected to protect the others, especially those in his company, from the dangers of firefighting. This norm even extended to keeping oneself in physical condition so that one could better aid others in emergency situations. We comment more on the qualitative data below. However, from this phase of the research we felt confident that our assessment of role stresses was relevant to this occupation and that we were tapping meaningful demands for this sample.

Following the qualitative phase, we administered questionnaires to the firefighters at their fire stations during the middle of the work day. We administered the measures on two separate occasions, usually about three days apart, for each respondent. On the first administration we collected data regarding perceptions of role stressors and demographic variables. On the second day we obtained responses on the self-esteem and somatic complaints scales. Our rationale here was to reduce the effects of common method variance by temporally separating the questionnaires. This procedure is expected to reduce the consistency and priming artifacts that often inflate the observed relationships between variables obtained from self-reports (Campbell & Fiske, 1959). The correlation between self-esteem and somatic complaints might be expected to be the most susceptible to this artifactual inflation because they were obtained at the same time. However, this relationship was not of central concern to the study, in contrast to the case with the role stress-health relationship and the moderating effect of self-esteem. After data were analyzed, results were fed back to the firefighters in a series of workshops held in the fire stations.

RESULTS

Table 1 displays the descriptive statistics and zero-order correlations among the study variables. Conflict had a significant positive correlation with somatic health complaints ($r = .21$, $p < .05$), but role ambiguity did not. Self-esteem and health complaints had a significant negative, albeit small, correlation ($r = -.19$, $p < .05$). Within this sample, role conflict showed a higher mean level and somewhat more variability than did role ambiguity.

Table 2 displays a summary of a hierarchical multiple regression that tested the main effects of the role stresses and self-esteem, and their interactions, after controlling for the demographic variables. The regression procedure followed Cohen and Cohen's (1983) guidelines: variables entered the equation in a hierar-

TABLE 1   Descriptive Statistics and Correlations among Study Variables

	Mean	SD	1	2	3	4
1. Role Conflict	3.55	0.99	—	.30**	-.12	.21
2. Role Ambiguity	2.56	0.88		—	-.14	.05
3. Self-Esteem	3.95	0.50			—	-.19*
4. Somatic Ill Health	1.65	0.44				—

*p < .05; *p < .01*

chical order so as to partial main effect terms from interaction terms. The demographic variables as a set did not account for any of the variance in somatic health so they were dropped from the regression to conserve degrees of freedom. Table 2 shows the results of a regression in which self-esteem was entered on the first step, followed by role conflict and ambiguity on the second step, and then the two interaction terms between the role stresses and self-esteem on the third step. At each step the incremental R^2 was assessed for significance before considering the significance of the individual variables in that step. This procedure corresponds to the Fisher "protected t" strategy as described by Cohen and Cohen (1983) and serves to protect the overall study-wise alpha level. The unstandardized regression coefficients reported are those obtained at the last step of the regression. The significance values indicated are those obtained at the time each variable entered the equation.

As consistent with the zero-order correlation, self-esteem was significant and explained about 4% of the variance in somatic health. The second step in which the role stress variables were entered was only marginally significant (F(2, 153) = 2.89, p = .06), but the conflict variable itself was significant. Finally, the third step comprising the two interaction terms was highly significant, accounting for 8% of

TABLE 2   Regression Results of Role Stress and Self-Esteem on Health Outcomes

Independent Variable	B	ΔR^2	df	F
Step 1: Self-Esteem	.94*	.04	1, 155	5.68*
Step 2: Role Conflict	.88*	.04	2, 153	2.89
Role Ambiguity	.60			
Step 3: Conflict X SE	-.19**	.08	2, 151	7.44**
Ambiguity X SE	-.16			
Constant	-2.44			

Note: Regression coefficients are those obtained at the final step of the regression. Significance levels are those obtained at the time the variable entered the equation.
*p < .05; **p < .01*

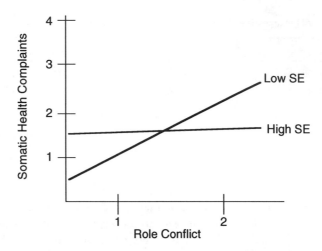

FIGURE 1   Interaction between Role Conflict and Self-Esteem

the somatic health variance ($F_{(2, 151)} = 6.64$, $p < .01$). Within this step only the role conflict by self-esteem interaction term was significant ($F_{(1, 151)} = 7.44$, $p < .01$). This interaction is displayed in Figure 1 by graphing the equations relating role conflict to somatic health at different levels of SE. For illustrative purposes, low and high SE values represent those at approximately one standard deviation above and below the sample mean. As illustrated, role conflict shows a significant positive relationship with somatic health complaints for those respondents low in SE. In contrast, high SEs appear to be unaffected by role conflict. Thus, the SE plasticity hypothesis is supported with respect to role conflict but not role ambiguity.

DISCUSSION

The current findings lend support to the generality of the self-esteem plasticity hypothesis in explaining the effects of role conflict on somatic health outcomes. Only for low SE firefighters was there a significant relationship between their perceptions of conflict and somatic complaints. These findings directly extend those of Mossholder et al. (1981) and Pierce et al. (1993) to the specific realm of occupational stress and health outcomes. No effects were found for role ambiguity, however, either as a main effect predictor of somatic health or as interacting with self-esteem. This raises the question as to whether the plasticity hypothesis applies to work stressors in general, or is very limited in its scope. It would seem that, at least in the current setting, the null findings for role ambiguity might be explained by its general lack of salience in this occupation. In addition to the nonsignificant correlation of role ambiguity with somatic complaints and its lower mean level relative to role conflict, the qualitative data provide some insight into this issue.

The interview data mentioned earlier supplement our understanding of the nature of the occupational demands facing this sample, especially in regard to the salience of the various role stressors that they face. Interestingly, while firefighters acknowledge the physical risks of their occupation, they take great pride in this aspect of their work and almost unanimously report that fighting fires does not constitute the most stressful part of their work. We were struck by how the stressors that the firefighters spontaneously reported stemmed from the nature of the paramilitaristic structure of the fire department and the formalization and lack of discretion that resulted. Many firefighters also reported that the social environment produced a set of stressors that might best be described as intersender role conflict. Given a very flat organization, there was little opportunity for promotion to captain (the next rank and second from the top of the whole department), and thus much competition ensued among firefighters for a spot on the career ladder. Competition for the few promotion slots often put firefighters in conflict with the prevailing norms of cooperation, assistance, and self-sacrifice for one's mates. Many firefighters spoke of the inherent conflict between meeting the expectations of the hierarchy that was required to receive the kind of performance appraisal that would make one promotable, and meeting the norms of the workgroup that stressed within-group loyalty over following standard operating procedures. This inherent conflict perhaps could be seen as spilling over into a certain amount of role ambiguity as well, but in general, individuals reported that their roles as professional firefighters were clear and well-defined. Role overload, in the classic sense of having too many tasks to accomplish, was also not a prevalent concern in this setting. In fact, many firefighters complained of the relative infrequency of real "working fires," and bemoaned their periods of "scut work" around the fire stations. These reports appear to converge with the quantitative data in that they reaffirm the importance of role conflict in this setting and the unimportance of role ambiguity. Thus, we would proffer the explanation that the results reflect more the irrelevance of role ambiguity in this setting as a stressor than they do the inapplicability of the plasticity hypothesis of self-esteem.

Given the non-experimental nature of the design, we are constrained in making causal inferences about the role of stress and self-esteem in determining somatic health. However, we did attempt to reduce the artifactual inflation of observed correlations between stressors and health by temporally separating the administrations of their measures. While this strategy is expected to reduce consistency and priming artifacts, some method variance effects might nevertheless be at work. Artifacts arising out of same-source data collection, however, do not provide a ready explanation for the pattern of interaction observed between role conflict and self-esteem. Thus, we think that the data provide credible evidence about the difference in vulnerability between high and low self-esteem workers to the stressful demands of work.

The most compelling direction for future research would be to explore the processes that theoretically mediate the vulnerability effects of low self-esteem.

Our extension of the SE plasticity hypothesis to health outcomes derived from our reasoning that low SEs were more apt to take a passive coping approach to work stressors, especially those involving the social environment. By not directly confronting problems generated by role conflict, the low SE might attempt to meet the demands of different senders by simply working harder to meet everyone's expectations. This strategy results in the imposition of higher demands that, though self-set, exact an eventual toll on the individual. We would encourage researchers to examine the coping actions that high and low SEs actually employ. This research question would seem amenable to both laboratory experimentation and naturalistic observation in the work setting.

REFERENCES

Beck, A.T. (1967). *Depression: Clinical, experimental, and theoretical aspects.* New York: Hoeber.

Brockner, J. (1983). Low self-esteem and behavioral plasticity: Some implications. In L. Wheeler & P.R. Shaver (Eds.), *Review of personality and social psychology* (Vol. 4, pp. 237–271). Beverly Hills, CA: Sage.

Brockner, J. (1988). *Self-esteem at work.* Lexington, MA: Heath.

Campbell, D.T., & Fiske, D.W. (1959). Convergent and discriminant validation by the multitrait-multimethod matrix. *Psychological Bulletin, 56,* 81–105.

Caplan, R.D., Cobb, S., French, J.R.P., Jr., Van Harrison. R., & Pinneau, S.R., Jr. (1975). *Job demands and worker health.* Ann Arbor, MI: Survey Research Center, University of Michigan.

Caplan, R.D., & Jones, K.W. (1975). Effects of workload, role ambiguity, and Type A personality on anxiety, depression, and heart rate. *Journal of Applied Psychology, 60,* 713–719.

Cohen, J., & Cohen, P. (1983). *Applied multiple regression/correlation analysis for the behavioral sciences* (2nd ed.). Hillsdale, NJ: Erlbaum.

Dipboye, R.L. (1977). A critical review of Korman's self-consistency theory of work motivation and occupational choice. *Organizational Behavior and Human Performance, 18,* 108–126.

Fisher, C.D., & Gitelson, R. (1983). A meta-analysis of the correlates of role conflict and ambiguity. *Journal of Applied Psychology, 68,* 320–333.

Fusilier, M.R., Ganster, D.C., & Mayes, B.T. (1987). Effects of social support, role stress, and locus of control on health. *Journal of Management, 13,* 517–528.

Ganster, D.C., Schaubroeck, J., Sime, W., & Mayes, B.T. (1991). The nomological validity of the Type A personality among employed adults. *Journal of Applied Psychology (Monograph), 76,* 143–168.

Hall, D.T. (1972). A model of coping with role conflict: The role behavior of college educated women. *Administrative Science Quarterly, 17,* 471–486.

Ivancevich, J.M., Matteson, M.T., & Preston, C. (1982). Occupational stress, Type A behavior, and physical well-being. *Academy of Management Journal, 25,* 373–391.

Jackson, S.E., & Schuler, R.S. (1985). A meta-analysis and conceptual critique of research on role ambiguity and role conflict in work settings. *Organizational Behavior and Human Decision Processes, 36,* 16–78.

Kahn, R.L., Wolfe, D.M., Quinn, R.P., Snoek, J.D., & Rosenthal, R.A. (1964). *Organizational stress: Studies in role conflict and ambiguity.* New York: Wiley.

Kobasa, S.C. (1988). Conceptualization and measurement of personality in job stress research. In J.J. Hurrell, Jr., L.R. Murphy, S.L. Sauter, & C.L. Cooper (Eds.), *Occupational stress: Issues and developments in research* (pp. 100–109). London: Taylor & Francis.

Korman, A.K. (1966). The self-esteem variable in vocational choice. *Journal of Applied Psychology, 50,* 479–486.

Korman, A.K. (1970). Toward an hypothesis of work behavior. *Journal of Applied Psychology, 54,* 31–41.

Mossholder, K.W., Bedeian, A.G., & Armenakis, A.A. (1981). Role perceptions, satisfaction, and performance: Moderating effects of self-esteem and organizational level. *Organizational Behavior and Human Performance, 28,* 224–234.

Pierce, J.L., Gardner, D.G., Dunham, R.B., & Cummings, L.L. (1993). Moderation by organization-based self-esteem of role condition-employee response relationships. *Academy of Management Journal, 36*(2), 271–288.

Rizzo, J.R., House, R.J., & Lirtzman, S.I. (1970). Role conflict and ambiguity in complex organizations. *Administrative Science Quarterly, 15, 150–163.*

Rosenberg, M. (1965). *Society and the adolescent self-image.* Princeton, NJ: Princeton University Press.

Van Sell, M., Brief, A.P., & Schuler, R.S. (1981). Role conflict and role ambiguity: Integration of the literature and directions for future research. *Human Relations, 34, 43–71.*

Weiss, H.M., & Adler, S. (1984). Personality and organizational behavior. In B.M. Staw & L.L. Cummings (Eds.), *Research in organizational behavior,* (Vol.6, pp. 1–50). Greenwich, CT: JAI Press.

Coping with Work Stress:
The Influence of Individual Differences

Stephen J. Havlovic
John P. Keenan

Stress is in part the result of a lack of fit between individuals and their environment (Argyle, 1989). Dual careers, downsizing of business organizations, increased competition, and changing technology are some of the phenomena that may have contributed to poor worker-environment fit and associated stress during the past decade. In an article by a *New York Times* reporter titled "Stress Casts Shadow Over Workplace," it is noted that "Paradoxically, the growing interest in workplace stress comes after years of predictions that corporate life in the 1980s would be easier, with shorter workdays because of new technology" (Fowler, 1989, p. 1G).

Organizations can help to reduce stress and its costs, estimated to cost U.S. businesses billions of dollars annually (Matteson & Ivancevich, 1987). Selection and placement procedures and employee training, as well as career counseling can be used more effectively in the 1990s to improve worker-organization fit, and to help reduce or eliminate the strain and related ailments caused by stress. In particular, training can focus on adaptive coping techniques or methods designed to help workers cope with job stressors (e.g., role ambiguity). Effective coping has

Author's Note: The authors appreciate helpful comments made by Janina Latack, Wanda Trahan, and Laura Koppes on an earlier version of this manuscript.

the potential of improving work satisfaction, reducing tension, lowering turnover and absenteeism, in addition to other positive outcomes. Such improvements will clearly benefit both individuals and their employers.

In an effort to further the study of what organizations can do to help lower employee strain, this paper refines the Latack Coping Scale (Latack, 1986; Latack & Aldag, 1986) and utilizes the modified scale to examine the relationship between the common work stressors of role conflict and role ambiguity with the coping strategies used by a sample of Midwestern managers. Individual differences (Type A behavior; Gender; Job Tenure; Managerial Experience) are also considered in our examination of the coping activities utilized.

INFLUENCES ON COPING ACTIVITIES

The coping literature was reviewed in order to gain insight into the influences on individual coping activities. In general the literature indicated that "...we still know relatively little about the specific coping strategies individuals use in dealing with stress, [or] the process by which individuals select and implement these strategies...." (Edwards, 1988, p. 233). However, there were several variables which appeared to trigger or influence coping responses in a number of studies.

Role conflict and role ambiguity are two role stressors widely present in the work place, associated with work tension (Kahn, Wolfe, Quinn, Snoek, & Rosenthal, 1964; Van Sell, Brief, & Schuler, 1981; Cooper, 1981; Fisher & Gitelson, 1983; Jackson & Schuler, 1985) and believed to influence coping activities (Latack, 1986). Empirical studies of how people cope with role conflict and role ambiguity are just beginning to emerge. "Theoretical models of stress...have recognized the importance of coping in mitigating the severity of tension and other deleterious effects of role stressors" (Parasuraman & Cleek, 1984, p. 179).

Evidence exists which suggests that individual differences influence the coping activities used. Baron concludes that, "...Type As seem to employ different strategies for coping with stress than Type Bs....They tend to deny that they are upset by the stress, and often project their own feeling of tension and anxiety onto others" (1986, p. 220). Type A behavior has been described by Matteson and Ivancevich (1987) as being "...compulsively competitive, always in a hurry, often angry and hostile" (p. 220). The converse Type B behavior pattern has been conceived as the relative absence of Type A behaviors in individuals who exhibit a different coping style that is characterized by a relative lack of time urgency, impatience, and hostile responses (Rosenman, 1990). Kirmeyer and Diamond (1985) found that police officers with Type A behavior patterns used coping methods which were more active and narrowly focused when encountering stressful events. A study by Carver, Scheier, and Weintraub (1989) found those with Type A personalities to be positively correlated with active coping and planning coping scales.

Gender has been included in a number coping studies with differences in coping methods sometimes detected, but the findings vary considerably. For example, "Within each type of event, small but consistent gender differences were noted on some of the coping measures. Women were more likely to use avoidance coping...." (Billings & Moos, 1981, p. 154). The opposite was found in a study by Parasuraman and Cleek, "Manager sex [gender] was unrelated to maladaptive coping, but showed a positive relationship to adaptive coping, indicating that female managers engage in more adaptive coping than male managers" (1984, p. 185). In a study by Nelson and Sutton (1990) gender was not correlated with coping methods. As a whole, the influence of gender on coping activities has been inconsistent.

Managerial experience and job tenure are relatively unexplored in terms of their influence on coping activities. Work experience was not related to coping in the Nelson and Sutton (1990) study. Organizational tenure was found to be negatively associated with adaptive and maladaptive coping in the Parasuraman and Cleek study (1984).

Our hypothesized model is shown in Figure 1. The proposed paths are based upon our review of the relevant literature and to a large degree on the research of Parasuraman and Cleek (1984) and Newton and Keenan (1985) who view individual differences and work stressors as important determinants of coping behaviors.

Supervisors are a logical group to examine in terms of the potential interconnections between role ambiguity, role conflict and coping. It has been well documented that supervisors experience inter-sender role conflict by being caught in the middle between superiors and subordinates (Driscoll, Carroll, & Sprecher, 1978; Kahn et al., 1964; Gardner & Whyte, 1945; Roethlisberger, 1945). Supervisors also face potential role ambiguity in their positions, when they lack direction from their superiors, have inadequate job descriptions, are unsure about which aspects of their jobs are most important, or have unclear evaluation procedures (Kahn et al., 1964).

METHOD

Sample

Our sample consists of 135 past graduates of continuing management education programs at a Midwestern university. These programs dealt with a variety of topics, including communication and leadership. All subjects were currently employed in supervisory level positions. From the 135 completed questionnaires, 92 were fully usable (68.1%) having responses on all items (i.e., no missing values). The questionnaire included scales to measure work stressors, coping activities, and Type A behavior. Gender, job tenure, and managerial experience were assessed with single items. In terms of the subjects analyzed 57% were male, and the average age was 37 years with a range of 21 to 62 years.

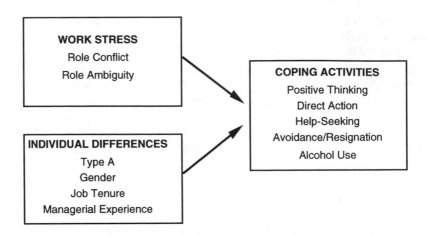

FIGURE 1   Hypothesized Model of Work Stressors, Individual Differences, and Coping Activities

Measurement of Coping Activities

The Aldwin and Revenson (1987) definition of coping is utilized in this study: "Coping encompasses cognitive and behavioral strategies used to manage a stressful situation (problem-focused coping) and the attendant negative emotions (emotion-focused coping)." We believe that the coping and alcohol usage items from the Latack (1986) Coping Scales tap the content of this definition of coping. The alcohol use items were extracted from the separate symptom management scale as other researchers have identified alcohol as a separate coping method (e.g., Violanti, Marshall, & Howe, 1985). Taken as a whole, these coping items are comprehensive in that they measure both focus (problem/emotion) and method (cognitive/behavioral) dimensions as well as social versus solitary and control versus escape components (Latack & Havlovic, 1992).

Table 1 contains the rotated principal axis factor matrix obtained using the coping items with the sample of Midwestern managers. A scree test supported the retention of five coping factors. Consistent with Latack and Aldag (1986) the varimax rotation supported three control oriented (positive thinking, help-seeking, and direct action) and two escape oriented (avoidance/resignation and alcohol use) coping factors. These coping factors fit closely with the notion of adaptive and maladaptive coping as identified by Parasuraman and Cleek (1984). Not included in Table 1 are five coping items which were discarded because they did not achieve a significant loading on any factor and one item which loaded on both avoidance/ resignation and help-seeking factors.

The coping subscales are shown in Table 2. Each item was measured on a five point scale ranging from "Hardly ever do this" to "Almost always do this." The

TABLE 1   Rotated Factor Loading Matrix (Varimax) of Coping Items

Item	Avoidance/ Resignation	Positive Thinking	Direct Action	Help Seeking	Alcohol Use
1	-.015	-.235	-.035	**.380**	-.114
2	**.644**	-.096	-.045	-.012	-.186
3	**.427**	.023	-.147	-.141	-.023
4	-.109	**.620**	-.206	.016	.022
5	.063	**.715**	.171	-.010	-.004
6	-.079	.067	**.389**	.111	-.072
7	-.145	.064	.202	**.442**	-.046
8	**.792**	-.115	.035	-.085	-.168
9	-.078	**.557**	.162	.028	.022
10	-.034	.153	**.603**	.119	-.038
11	**.640**	.084	-.107	.011	.117
12	**.519**	.088	-.080	-.052	.082
13	.101	**.556**	.153	.053	-.059
14	-.139	.043	**.704**	.058	.065
15	.218	**.467**	.239	.044	-.101
16	**.336**	.209	-.087	-.102	-.222
17	-.120	**.607**	.157	.253	-.046
18	-.140	.161	**.531**	.117	-.021
19	-.183	.080	.268	**.442**	-.026
20	-.253	.082	**.397**	-.016	-.054
21	.082	.100	-.083	**.478**	.192
22	-.115	.002	-.012	**.681**	.072
23	-.093	.210	.183	**.402**	.003
24	.048	.019	**.469**	-.282	.257
25	-.024	-.083	.016	.011	**.874**
26	-.026	-.031	-.074	.104	**.768**
Eigenvalue	3.51	2.69	1.85	1.51	1.50
% Variance Explained	11.0	8.4	5.8	4.7	4.7

Note: Factors loading at .30 or greater are boldfaced.

subscale items were summed to achieve factor-based scores. Coefficient alpha values for the subscales ranged from .610 for help-seeking coping activities to .828 for the alcohol use subscale. These alpha values are in the same range as most of the reliabilities found in the coping literature (Latack & Havlovic, 1992).

Measurement of Work Stress and Type A Behavior

The Rizzo, House, and Lirtzman (1970) subscales were used to measure role conflict (e.g., I have to do things that should be done differently) and role

TABLE 2   Coping Subscale Items

Instructions for items 1-24. *In this section, think of stressful situations you have faced* at work. *Then, using the scale below, indicate how often you react in each of the following ways in response to such situations: 1) Hardly ever do this; 2) Seldom do this; 3) Occasionally do this; 4) Frequently do this; 5) Almost always do this.*

Avoidance/Resignation (alpha = .739)
8. Try to keep away from this type of situation.
2. Avoid being in this situation if I can.
11. Separate myself as much as possible from the people who created this situation.
12. Try not to get concerned about it.
3. Tell myself that time takes care of situations like this.
16. Accept this situation because there is nothing I can do to change it.

Positive Thinking (alpha = .758)
5. Think of ways to use this situation to show what I can do.
4. Remind myself that other people have been in this situation and that I can probably do as well as they did.
17. Think about the challenges I can find in this situation.
9. Try to see this situation as an opportunity to learn and develop new skills.
13. Try to think of myself as a winner—as someone who always comes through.
15. Tell myself that I can probably work things out to my advantage.

Direct Action (alpha = .691)
14. Devote more time and energy to doing my job.
10. Put extra attention on planning and scheduling.
18. Try to work harder and more efficiently.
24. Throw myself into my work and work harder, longer hours.
20. Give it my best effort to do what I think is expected of me.
6. Try to be very organized so that I can keep on top of things.

Help-Seeking (alpha = .610)
22. Seek advice from people outside the situation who may not have power but who can help me think of ways to do what is expected of me.
21. Request help from other people who have the power to do something for me.
19. Decide what I think should be done and explain this to the people who are affected.
7. Talk with people (other than my supervisor) who are involved.
23. Work on changing policies which caused this situation.
1. Get together with my supervisor to discuss this.

Instructions for items 25-26. *People may deal with job tension in different ways. How often do you do the following when you feel tense because of your job: 1) Hardly ever do this; 2) Seldom do this; 3) Occasionally do this; 4) Frequently do this; 5)Almost always do this.*

Alcohol Use (alpha = .828)
25. Drink a moderate amount (i.e., 2 drinks) of liquor, beer or wine.
26. Drink more than a moderate amount of liquor, beer or wine.

ambiguity (e.g., I know what my responsibilities are). An article by King and King (1990) has raised questions about the construct validity of the Rizzo et al. measures. The scale has been favorably evaluated in the past (e.g., Schuler et al., 1977). A recent study by Smith, Tisak, and Schmieder (1993) confirmed the factors of the Rizzo et al. subscales across three diverse samples.

During the course of the Western Collaborative Group Study (WCGS) Rosenman et al. (1964) and Bortner and Rosenman (1967) developed a fourteen-item questionnaire for self-report assessment of Type A Behavior Pattern (TABP). This instrument was later shortened by Bortner (1969) to seven items. Predictive accuracy for the Type A concept was found through use of the seven-item Type A scale in the Belgian-French Pooling Project (1984). The Bortner Scale has been shown to have both prospective (French-Belgian Collaborative Group, 1982) and retrospective validity (Koskenvuo, Kaprio, Langinvainio, Romo, & Sarna, 1981; Heller, 1979). The results of analysis suggest that the study of Type A behavior through self-report measures is likely to lead to an underestimation of the true prevalence of this trait in any given research sample (Carmelli, Rosenman, & Swan, 1990). Based on the work of Brief, Schuler, and Van Sell (1981) and Bortner (1969) we used this seven-item scale to measure Type A behavior (e.g., "Never feel rushed even under pressure"). Type B behavior was not assessed.

RESULTS

Table 3 contains the descriptive statistics for the study variables. A wide range of work stressor scores were obtained with median scores equivalent to the means. An approximately normal distribution for Type A behaviors was obtained, but the reliability of the Type A behavior scale was unsatisfactory and therefore any Type A influences will be attenuated. On average the subjects were relatively new to their jobs and had limited managerial experience. The majority of the sample indicated that they infrequently react to stressful situations with alcohol use.

Significant correlations (two tail tests) were obtained between a number of the variables (see Table 4). A significant positive correlation was found between role conflict and role ambiguity (r = .40) which is consistent with prior studies (Fisher & Gitelson, 1983; Jackson & Schuler, 1985). Based on his review of prior empirical research, Burke concluded: "It has been...clearly demonstrated that role conflict and role ambiguity are consistently correlated with each other" (1988, p. 81). Role conflict was higher for males and correlated with alcohol use. Those with higher role ambiguity were less likely to use direct action coping activities.

Type A behavior was found to be associated with positive thinking activities. The males in the sample had greater managerial experience, but there was no relationship between gender and coping activities. As would be expected, managerial experience was positively correlated with job tenure. Those with greater managerial experience and job tenure were more likely to use positive thinking

TABLE 3   Descriptive Statistics (n = 92)

Variable	Mean	Standard Deviation	Range	Coefficient Alpha	#Scale Items
Work Stressors					
Role Conflict	22.45	5.80	8 - 35	.735	8
Role Ambiguity	14.02	4.48	6 - 27	.815	6
Individual Differences					
Type A Behavior	39.73	5.71	29 - 51	.529	7
Gender (Male)	.57	.50	0 - 1	—	1
Job Tenure	4.20	4.72	0 - 23	—	1
Managerial Experience	4.84	6.11	0 - 28	—	1
Coping Methods					
Positive Thinking	20.76	3.70	11 - 28	.758	6
Direct Action	22.61	2.86	15 - 29	.691	6
Help-Seeking	21.22	3.04	12 - 30	.610	6
Avoidance/Resignation	15.53	3.61	6 - 24	.739	6
Alcohol Use	3.62	1.77	2 - 9	.828	2

coping activities. Increased job tenure was related to less utilization of help-seeking coping activities, but this was not the case for managerial experience.

A positive correlation was obtained between the control coping methods of positive thinking and direct action. Those engaging in help-seeking coping activities were less likely to use avoidance/resignation coping approaches. These findings are consistent with the control/escape constructs presented by Latack (1986).

The results of the regression analyses used to test our hypothesized coping model (Figure 1) are contained in Table 5. Both the full and "trimmed" equations (t-values greater than or equal to 1.0) are shown. A small but significant amount of variance (after adjustment for shrinkage 4.6% to 11.8%) is explained in all of the "trimmed" equations.

Those experiencing greater role conflict were found to use more escape coping activities (alcohol use and avoidance/resignation). Direct action coping methods were utilized less frequently by those experiencing role ambiguity. These findings would suggest that these subjects were not coping well with these work stressors. Maladaptive coping strategies were being utilized as opposed to control oriented adaptive coping activities.

Those exhibiting Type A behaviors were more inclined to use positive thinking, and there was some evidence of their also using more direct action. This would be consistent with the tendency of Type As to want to control the situation. Gender was not related to any of the coping activities selected. Job tenure was

TABLE 4 Correlations Among the Variables

	1	2	3	4	5	6	7	8	9	10
1 Role Conflict										
2. Role Ambiguity	.395[a]									
3. Type A Behavior	.170	.133								
4. Gender	.227[b]	.172	.062							
5. Job Tenure	-.020	-.046	.168	.121						
6. Managerial Experience	.084	.027	.132	.208[b]	.387[a]					
7. Positive Thinking	.068	-.030	.238[b]	.134	.269[a]	.280[a]				
8. Direct Action	-.063	-.226[a]	.124	-.105	-.096	-.085	.217[b]			
9. Help-Seeking	-.006	-.065	-.046	-.176	-.218[b]	-.003	.134	.149		
10. Avoidance/Resignation	.179	.147	.072	-.010	.184	.036	.031	-.193	-.208[b]	
11. Alcohol Use	.346[a]	.105	.155	.122	-.122	.024	-.078	.018	.056	-.123

($n = 92$); [a]$p < .01$; [b]$p < .05$

TABLE 5 Coping Activity Regression Analyses (Standardized Solution)

	Positive Thinking		Direct Action		Help Seeking		Avoidance/ Resignation		Alcohol Use	
	(1)	(2)	(1)	(2)	(1)	(2)	(1)	(2)	(1)	(2)
Role Conflict	.040		.020		.045		.163	.183c	.328a	.322a
Role Ambiguity	-.081		-.253b	-.247b	-.067		.108		-.060	
Type A Behavior	.187c	.185c	.183c	.157	-.009		.004		.126	.123
Gender	.071		-.054		-.171	-.173	-.083		.063	
Job Tenure	.157	.164	-.112		-.247b	-.246b	.220c	.187c	-.158	-.136
Managerial Experience	.178	.192c	.020		.127	.128	-.049		.029	
R²	.151	.142	.099	.075	.088	.084	.085	.067	.154	.147
Adjusted R²	.092	.113	.036	.055	.023	.052	.020	.046	.095	.118
F Value	2.53b	4.85a	1.56	3.63b	1.36	2.67c	1.32	3.21b	2.59b	5.06a
Degrees of Freedom Residual	85	88	85	89	85	88	85	89	85	88

(n = 92); [a]p < .01; [b]p < .05; [c]p < .10

188

related to less use of help-seeking activities and greater avoidance/resignation coping. It would appear that employees with lower job tenure more actively seek help from others and perhaps don't sidestep work related problems. Individuals with more managerial experience tended to utilize increased positive thinking.

DISCUSSION AND CONCLUSION

This study extends prior research on the determinants of coping activities. While causality can not be determined from a cross-sectional study and these findings should be considered exploratory, there are none the less some interesting patterns which emerge. First, male and female managers did not exhibit different coping activities. Secondly, Type A behavior, job tenure, and managerial experience were significantly associated with the coping activities used. While Type As used more control oriented coping methods, help-seeking activities were not employed. Clearly Type A behavior as an influence on coping should be included in future studies. Evidence exists suggesting that "…maintaining control is of paramount importance to Type A individuals (Glass, 1977), who may react strongly to their internal demands for maintaining control in their occupational environments by working harder, longer, and faster. Although such coping behaviors may result in high job performance ratings, there is a small probability that such individuals pay a price in somatic distress" (Lee, Ashford, & Bobko, 1990, p. 877).

Those with greater managerial experience used increased positive thinking and at a slightly greater relative magnitude than those exhibiting Type A behaviors. Unfortunately, those with greater job tenure seem to be relying more on escaping the situation than using proactive coping activities. This is not viewed as desirable and organizations experiencing this type of withdrawal should attempt to intervene perhaps by changing the organizational culture to one encouraging better participation between coworkers and subordinates (help-seeking) and increased problem solving. Leiter (1991) has found burnout to decrease with control coping and to increase with escapist coping strategies.

Organizational training implications would appear to surface from this study in light of the maladaptive coping used by those experiencing work stress and having greater job tenure. Training interventions for more tenured individuals to learn how to use more positive thinking, direct action, and help-seeking activities would appear to be warranted. This could particularly benefit those who are experiencing high role ambiguity and are using fewer direct action coping strategies. Constructive alternatives are also clearly needed for those with conflicting roles who are relying on alcohol use and avoidance/resignation activities.

Some evidence exists that coping varies across situations (Newton & Keenan, 1985; Latack & Aldag, 1986). This suggests that organizational surveys could be used to determine the coping activities used by employees to deal with a variety of work stressors while controlling for individual differences. Based on survey results, the need for organizational change or training courses could be determined. If needed, organizational or training interventions could then be developed and initiated to help to increase the use of adaptive coping strategies in light of individual and situational differences. Training which emphasizes supervisory support of subordinates should be considered as this type of social support has been shown to increase coping actions and reduce tension-anxiety (Kirmeyer & Dougherty, 1988).

Studies are needed to replicate and extend the research to date on the determinants of coping activities. Longitudinal field studies of coping activities (pre and post intervention) would be an useful extension in light of organizational culture and training interventions designed to increase the use of adaptive coping. Further evaluation should include the effectiveness of the intervention(s). For example, organizations using training to increase adaptive coping could evaluate the program in terms of effectiveness criteria such as work satisfaction, absenteeism, turnover, illness, somatic tension and other types of strain. Control groups should be used and both short and long term effects examined.

While this cross-sectional study focused on supervisory level managers, future longitudinal studies should also consider using panel data beginning prior to promotion into supervisory positions and tracking stressors, coping, and strain into the post promotion environment. In particular, the impact of prepromotion stress management training on coping with managerial stressors (e.g., role conflict, role ambiguity, role overload) could be analyzed while controlling for individual differences. Future studies should consider larger samples which would allow for additional individual difference variables (e.g., locus of control).

REFERENCES

Aldwin, C.M., & Revenson, T.A. (1987). Does coping help? A reexamination of the relation between coping and mental health. *Journal of Personality and Social Psychology, 53,* 337–348.

Argyle, M. (1989). *The social psychology of work* (2nd ed.). London: Penguin.

Baron, R.A. (1986). *Behavior in organizations* (2nd ed.). Boston: Allyn and Bacon.

Belgian-French Pooling. (1984). Assessment of type A behavior by the Bortner scale and schemic heart disease. *European Heart Journal, 5,* 440–446.

Billings, A.G., & Moos, R.H. (1981). The role of coping responses and social resources in attenuating the impact of stressful life events. *Journal of Behavioral Medicine, 4,* 139–157.

Bortner, R.W., & Rosenman, R.H. (1967). The measurement of pattern A behavior. *Journal of Chronic Disorders, 20,* 525–533.

Bortner, R.W. (1969). A short rating scale as a potential measure of pattern A behavior. *Journal of Chronic Disorders, 22,* 87–91.

Brief, A.P., Schuler, R.S., & Van Sell, M. (1981). *Managing job stress.* Boston: Little, Brown.

Burke, R.J. (1988). Sources of managerial and professional stress in large organizations. In C.L. Cooper & R. Payne (Eds.), *Causes, coping and consequences of stress at work.* Chichester, GB: John Wiley.

Carmelli, D., Swan, G.E., & Rosenman, R.H. (1990). Self-ratings and perceptions of type A traits in adult twins. In M.J Strube (Ed.), Type A behavior [Special issue]. *Journal of Social Behavior and Personality, 5*(1), 263–276.

Carver, C.S., Scheier, M.F., & Weintraub, J.K. (1989). Assessing coping strategies: A theoretically based approach. *Journal of Personality and Social Psychology, 56,* 267–283.

Cooper, C.L. (1981). *The stress check: Coping with stresses of life and work.* Englewood Cliffs, NJ: Prentice Hall.

Driscoll, J.W., Carroll, D.J., & Sprecher, T.A. (1978). The first-level supervisor: Still "the man in the middle." *Sloan Management Review, 19*(2), 25–37.

Edwards, J.R. (1988). The determinants and consequences of coping with stress. In C.L. Cooper & R. Payne (Eds.), *Causes, coping and consequences of stress at work.* Chichester, GB: John Wiley.

Fisher, C.D., & Gitelson, R. (1983). A meta-analysis of the correlates of role conflict and ambiguity. *Journal of Applied Psychology, 68,* 320–333.

Fowler, E.M. (1989, October 1). Stress casts shadow over workplace. *Wisconsin State Journal,* p. 1G.

French-Belgian Collaborative Group. (1982). Ischemic heart disease and psychological patterns. *Advances in Cardiology, 29,* 25–31.

Gardner, B., & Whyte, W.F. (1945). The man in the middle: Position and problems of the foreman. *Applied Anthropology, 4*(2), entire issue.

Heller, R.F. (1979). Type A behavior and coronary heart disease. *British Medical Journal, 11,* 368.

Jackson, S.E., & Schuler, R.S. (1985). A meta-analysis and conceptual critique of research on role ambiguity and role conflict in work settings. *Organizational Behavior and Human Decision Processes, 36,* 16–78.

Kahn, R.L., Wolfe, D.M., Quinn, R.P., Snoek, J.D., & Rosenthal, R.A. (1964). *Organizational stress: Studies in role conflict and ambiguity.* New York: Wiley.

King, L.A., & King, D.W. (1990). Role conflict and role ambiguity: A critical assessment of construct validity. *Psychological Bulletin, 107,* 48–64.

Kirmeyer, S.L., & Diamond, A. (1985). Coping by police officers: A study of role stress and type A and type B behavior patterns. *Journal of Occupational Behavior, 6,* 183–195.

Kirmeyer, S.L., & Dougherty, T.W. (1988). Work load, tension, and coping: Moderating effects of supervisor support. *Personnel Psychology, 41,* 125–139.

Koskenvuo, M., Kaprio, J., Langinvainio, H., Romo, M., & Sarna, S. (1981). Psychosocial and environmental correlates of coronary-prone behavior in Finland. *Journal of Chronic Diseases, 34,* 331–340.

Latack, J.C. (1986). Coping with job stress: Measures and future directions for scale development. *Journal of Applied Psychology, 71,* 377–385.

Latack, J.C., & Aldag, R.J. (1986). *The dynamic constellation of coping: An examination of measures across samples and job situations.* Paper presented at the annual meeting of the Decision Sciences Institute, Honolulu.

Latack, J.C., & Havlovic, S.J. (1992). Coping with job stress: A conceptual evaluation framework for coping measures. *Journal of Organizational Behavior, 13,* 479–508.

Lee, C., Ashford, S.J., & Bobko, P. (1990). Interactive effects of "type A" behavior and perceived control on worker performance, job satisfaction, and somatic complaints. *Academy of Management Journal, 33,* 870–881.

Leiter, M.P. (1991). Coping patterns as predictors of burnout: The function of control and escapist coping patterns. *Journal of Organizational Behavior, 12,* 123–144.

Matteson, M.T., & Ivancevich, J.M. (Eds.). (1987). *Controlling work stress: Effective human resource and management strategies.* San Francisco: Jossey-Bass.

Parasuraman, S., & Cleek, M.A. (1984). Coping behaviors and managers' affective reactions to role stressors. *Journal of Vocational Behavior, 24,* 179–193.

Nelson, D.L., & Sutton, C. (1990). Chronic work stress and coping: A longitudinal study and suggested new directions. *Academy of Management Journal, 33,* 859–869.

Newton, T.J., & Keenan, A. (1985). Coping with work-related stress. *Human Relations, 38,* 107–126.

Rizzo, J.R., House, R.J., & Lirtzman, S.I. (1970). Role conflict and ambiguity in complex organizations. *Administrative Science Quarterly, 15,* 150–163.

Roethlisberger, F.J. (1945). The foreman: Master and victim of double talk. *Harvard Business Review, 23*(3), 283–298.

Rosenman, R.H. (1990). Type A behavior pattern: A personal overview. In M.J Strube (Ed.) Type A behavior [Special issue]. *Journal of Social Behavior and Personality, 5,* 1–24.

Rosenman, R.H., Friedman, M., Straus, R., Wurm, M., Kositchek, R., Hahn, W., & Werthessen, N.T. (1964). A predictive study of coronary disease: The western collaborative group study. *Journal of the American Medical Association, 189,* 15–22.

Schuler, R.S., Aldag, R.J., & Brief, A.P. (1977). Role conflict and ambiguity: A scale analysis. *Organizational Behavior and Human Performance, 20,* 111–128.

Smith, C.S., Tisak, J., & Schmieder, R.A. (1993). The measurement properties of the role conflict and role ambiguity scales: A review and extension of the empirical research. *Journal of Organizational Behavior, 14,* 37–48.

Van Sell, M., Brief, A.P., & Schuler, R.S. (1981). Role conflict and role ambiguity: Integration of the literature and directions for future research. *Human Relations, 34,* 43–71.

Violanti, J.M., Marshall, J.R., & Howe, B. (1985). Stress, coping, and alcohol use: The police connection. *Journal of Police Science and Administration, 13,* 106–110.

Job Stress, Coping, and Dissatisfaction in the Health Professions: A Comparison of Nurses and Pharmacists

Alan P. Wolfgang

Cherniss (1980) has depicted burnout as a process which begins with job stress. It is the effort expended in dealing with that stress which eventually can result in burnout. He also notes that, to prevent burnout, we first must understand the underlying cause of the problem (i.e., stress). Therefore, when considering burnout among health professionals, "job stress is a logical starting point" (p. 43).

Since the early 1980s, numerous studies have been conducted to identify the sources of job stress among health professionals, as well as the effects of that stress. Examples can be found from many individual health professions which were the target of such investigations, including nursing (Ceslowitz, 1989; Dewe, 1989; Fox, Dwyer, & Ganster, 1993; McLaney & Hurrell, 1988), pharmacy (Ciaccio, Jang, Caiola, Pfifferling, & Eckel, 1982; Lahoz & Mason, 1990; Ortmeier & Wolfgang, 1991; Wolfgang, Kirk, & Shepherd, 1985), and medicine (Krakowski, 1982; May, Revicki, & Jones, 1983; Richardson & Burke, 1991; Roeske, 1981). It would be inappropriate, however, to believe that merely identifying sources of stress among health professionals is adequate for truly understanding job stress.

Author's Note: The author acknowledges the editors, anonymous reviewers, and Dr. John A. McMillan; their comments on an earlier version of this manuscript were most helpful.

Matteson and Ivancevich (1987) have defined stress as "an adaptive response, moderated by individual differences, that is a consequence of any action, situation, or event that places special demands upon a person" (p. 10). This means that predicting the amount of stress which results from exposure to given situations, as well as the impact of that stress, requires knowing more than just what potentially stressful situations health professionals face on the job. To gain a greater understanding of stress, it must be thought of as a transaction between people and their environments (Lazarus & Launier, 1978). Faced with a given situation or event, people appraise its significance to their well-being and the availability of coping resources, where coping is defined as "cognitive and behavioral efforts to manage specific external and/or internal demands that are appraised as taxing or exceeding the resources of the person" (Lazarus & Folkman, 1984, p. 141). Thus, if we are interested in predicting the impact of stress, the degree to which coping resources exist must be considered (Osipow & Davis, 1988). Lazarus and Launier (1978), in fact, have stated that "the ways people cope with stress are even more important...than the frequency and severity of episodes of stress themselves" (p. 308).

Recognizing the importance of considering both levels of job stress and coping resources in studying stress among health professionals, the present study attempted to bridge a knowledge gap which exists in the literature. To date, studies of stress in the health professions have focused almost exclusively on single professions. As a result, little is known about interprofession differences in job stress. One study has identified differences in overall stress levels and frequency of exposure to specific potentially stressful situations among pharmacists, nurses, and physicians, but it did not address coping or outcomes of that stress (Wolfgang, 1988b). The present study investigated stress, coping, and one outcome measure, job dissatisfaction, among two groups of health professionals—nurses and pharmacists. It is especially appropriate to consider job stress among nurses and pharmacists because of the occupational similarities between these professionals. While nursing and pharmacy are by no means identical, there are many occupational attributes which they possess in common. For example, both pharmacists and nurses: (a) primarily work as employees, rather than owners of their own businesses; (b) have opportunities to work in a wide variety of settings, but often function as members of large, bureaucratic organizations; and (c) have been termed "dependent practitioners" because of their relative lack of autonomy and reliance on physicians (Jonas, 1990, p. 52). The specific objectives of this study were to:

(1) compare nurses and pharmacists on levels of perceived job stress, job dissatisfaction, and use of various coping methods;

(2) assess the relationship between stress, coping, and job dissatisfaction for nurses and pharmacists; and

(3) develop equations for nurses and pharmacists to explain job dissatisfaction in terms of stress and coping, after controlling for demographic characteristics.

METHOD

Subjects

A cover letter and three-page questionnaire were mailed first class to 600 registered nurses and 600 registered pharmacists. For each group, 150 professionals were selected randomly from mailing lists obtained from four states. Two criteria were used in selecting these states: (a) one state was to be chosen from each of four geographic regions of the nation (i.e., West, South, Midwest, and East); and (b) mailing lists had to be available from the appropriate board of pharmacy or nursing at a reasonable cost. The four states utilized for selection of nurses were Idaho, Louisiana, Missouri, and Maine. Pharmacists were chosen from Idaho, Louisiana, Missouri, and Virginia. Questionnaires for eight pharmacists and 10 nurses were returned as undeliverable, thus leaving 592 pharmacists and 590 nurses as potential respondents.

Two weeks after the initial mailing, a second questionnaire was mailed to increase the response rate. Overall, responses were received from 319 nurses and 327 pharmacists. This yielded response rates of 54.1 percent and 55.2 percent for nurses and pharmacists, respectively.

Since the focus of this study was the means by which health professionals cope with job-related stress, all respondents who were not in active practice at the time of the survey were eliminated from subsequent data analyses. The results reported, therefore, reflect responses from 280 nurses and 279 pharmacists who were in active practice. In the sample of actively practicing nurses, 97.1 percent were female, the median age was 40 years, and almost three-fourths of the nurses indicated that a hospital was their primary work setting. For comparison, recent national data indicated that 96.7 percent of nurses were female, the median age was 39 years, and approximately 68 percent worked in hospitals (American Nurses' Association, 1987). In the present sample, 74.2 percent of the nurses were married. Those citing a nursing diploma, associate degree, and bachelor degree as their highest educational preparation comprised 28.2, 25.0, and 34.6 percent of the sample, respectively.

Among actively practicing pharmacists in the sample, 27.2 percent were female and 73.0 percent of the pharmacists worked in community (retail) settings; comparable national data show that 23.7 percent of pharmacists are female and approximately 65 percent work in community settings (National Association of Boards of Pharmacy, 1989–90). Other demographic characteristics of the pharmacist sample included a median age of 39 years, 81.0 percent being married, and 91.0 percent holding a bachelor of science as their highest educational degree.

Measures

The questionnaire consisted of four sections. First, respondents completed the Health Professions Stress Inventory (HPSI), which consists of 30 job situations that health professionals might be expected to encounter in the performance of their duties. Respondents indicate how often they have found each situation to be stressful in their work environment, on a five-point Likert-type scale (from "never" to "very often"). Each item is scored from zero to four; thus, the possible range of scores on the entire HPSI is zero to 120. Evidence indicates that the HPSI is both a valid and reliable measure of health professionals' job stress. Each job situation on the HPSI was derived from previous studies of stress in the health professions, thus maximizing content validity, and HPSI scores have shown significant correlations with scores on an index of job-related tension (Wolfgang, 1988c). In the present study, the Cronbach's alpha reliability coefficients were .89 for nurses and .87 for pharmacists.

The second section of the questionnaire included a list of 17 strategies which people may utilize when dealing with job-related stress. These strategies were adapted from previous work by Billings and Moos (1981), who also classified the strategies according to the method of coping which they represent (Table 1). Thus, each strategy could be designated as representing one of three methods of coping: (a) active-cognitive, which involves managing one's appraisal of a situation's stressfulness (e.g., trying to see the positive side, considering several alternatives for handling problems); (b) active-behavioral, which is characterized by overt attempts to deal directly with problems and their effects (e.g., taking some positive action, talking with a friend); or (c) avoidance, which attempts to avoid active confrontation of problems or to indirectly reduce emotional tension (e.g., keeping your feelings to yourself, eating more). Respondents were asked to think about recent situations on their jobs which had been especially stressful, and to indicate how often they used each strategy to deal with such stress on a five-point Likert-type scale (from 0 = "never" to 4 = "very often"). For nurses, the Cronbach's alpha coefficients were .69 for the active-cognitive scale (six items), .50 for the active-behavioral scale (six items), and .40 for the avoidance scale (five items). For pharmacists, the alpha coefficients for the active-cognitive, active-behavioral, and avoidance scales were .67, .54, and .45, respectively. While these reliability coefficients may be lower than normally desired, Billings and Moos (1981) have noted that this is not unexpected. They have pointed out that since the use of one coping strategy could effectively reduce stress and thus decrease the need to use other strategies, this may, in effect, place an upper limit on the scales' internal consistency coefficients.

The third part of the questionnaire contained a job dissatisfaction scale, which consisted of four Likert-type items (Caplan, Cobb, French, Harrison, & Pinneau, 1975). In addition to inquiring as to the level of overall satisfaction, these items asked respondents to indicate whether they would recommend their job to a good friend, decide again to take the same job, or try to find a new job in the next year. The scoring of this scale (possible range = 4–13) resulted in higher scores on the scale indicating greater dissatisfaction with one's job; lower scores indicated greater job satisfaction. The Cronbach's alpha coefficients for the scale were .83 for nurses and .81 for pharmacists.

TABLE 1   Coping Methods and Component Coping Strategies

Active-Cognitive	Active-Behavioral	Avoidance
1. Drawing on past experience with similar situations	1. Trying to find out more about the situation	1. Keeping your feelings to yourself
2. Taking things one step at a time	2. Taking some positive action	2. Preparing for the worst
3. Considering several alternatives for handling problems	3. Talking with a friend	3. Taking frustrations out on other people
4. Trying to see the positive side	4. Talking with a spouse or other relative	4. Trying to reduce tension by eating more
5. Trying to step back from the situation and be more objective	5. Exercising more	5. Trying to reduce tension by smoking more
6. Praying for guidance or strength	6. Talking with a professional (e.g., doctor, clergy, lawyer)	

Finally, the questionnaire collected data on several demographic characteristics of respondents, including sex, age, work setting, marital status, average number of hours worked per week, and highest educational degree. Data analyses utilized programs of the Statistical Analysis System (SAS Institute, 1985).

RESULTS

Table 2 contains the mean scores for nurses and pharmacists on the Health Professions Stress Inventory (HPSI), the job dissatisfaction scale, and the three coping methods scales. On the HPSI, nurses reported significantly more job stress than did pharmacists (t(520) = 3.75, p < .001). Pharmacists had higher scores on the job dissatisfaction scale (t(553) = 2.86, p < .01), indicating that nurses were reporting higher levels of job satisfaction. On the coping methods scales, nurses reported more frequent use of active-cognitive (t(552) = 3.87, p < .001), active-behavioral (t(548) = 4.02, p < .001), and avoidance (t(550) = 2.94, p < .01) coping.

Significant intercorrelations of the stress, dissatisfaction, and coping scale scores reveal both similarities and differences between the nurses and pharmacists in the sample (Table 3). For both groups of professionals, HPSI scores were correlated positively with job dissatisfaction scale scores. This indicates that, as perceived stress increased, job dissatisfaction increased. Among both pharmacists and nurses, avoidance coping scores were positively correlated with both HPSI and job dissatisfaction scores. Thus, greater use of avoidance coping was associated with higher levels of stress and lower job satisfaction. For pharmacists, stress scores also were correlated positively with scores on both the active-cognitive and active-behavioral coping scales. Among nurses, scores on the same two coping scales were correlated negatively with job dissatisfaction scale scores, indicating that more frequent use of active-cognitive and active-behavioral coping strategies were associated with greater job satisfaction.

TABLE 2   Mean Scores on the Health Professions Stress Inventory (HPSI),
Job Dissatisfaction Scale, and Coping Methods Scales by Profession

	Nurses		Pharmacists		
Scale	M	(SD)	M	(SD)	t
HPSI	59.48	(14.71)	54.77	(13.99)	3.75**
Job Dissatisfaction	6.35	(2.14)	6.88	(2.22)	2.86*
Active-Cognitive Coping	17.07	(3.03)	16.06	(3.11)	3.87**
Active-Behavioral Coping	14.68	(3.16)	13.57	(3.31)	4.02**
Avoidance Coping	7.19	(2.78)	6.51	(2.65)	2.94*

$*p < .01; **p < .001$

As noted above, a prime objective of this study was to determine the relative contribution of job stress and coping to health professionals' job dissatisfaction, after controlling for demographic characteristics. To examine the relative influence of these independent variables, separate multiple regression analyses were conducted for pharmacists and nurses, with job dissatisfaction as the dependent measure. Dummy coding was utilized for several demographic variables, including sex (male and female), marital status (married, single, and other), practice setting (hospital, nursing home, home health care, and other for nurses; independent retail, chain retail, hospital, and other for pharmacists), and highest degree earned (diploma, associate degree, bachelor degree, and other for nurses; bachelor degree, doctor of pharmacy degree, and other for pharmacists). Respondents with missing data for any of the variables to be entered into the regression equations

TABLE 3   Intercorrelations of Health Professions Stress Inventory (HPSI), Job
Dissatisfaction Scale, and Coping Methods Scales Scores by
Profession

	1	2	3	4
Nurses				
1. HPSI				
2. Job Dissatisfaction	.47**			
3. Active-Cognitive Coping	.17*	-.10		
4. Active-Behavioral Coping	.17*	-.10	.42**	
5. Avoidance Coping	.30**	.24**	.04	.02
Pharmacists				
1. HPSI				
2. Job Dissatisfaction	.46**			
3. Active-Cognitive Coping	-.11	-.21**		
4. Active-Behavioral Coping	-.03	-.22**	.45**	
5. Avoidance Coping	.26**	.16*	-.06	-.02

$*p < .01; **p < .001$

TABLE 4   Results of Multiple Regression Analyses with Job Dissatisfaction as the Dependent Variable[a]

Independent Variable	F	p	Beta Coefficient
Nurses (n = 242)			
HPSI (Stress)	53.06	.0001	0.45
Active-Cognitive Coping	4.65	.0321	-0.14
Active-Behavioral Coping	4.80	.0295	-0.14
Avoidance Coping	1.99	.1596	0.08
Pharmacists (n = 230)			
HPSI (Stress)	44.18	.0001	0.45
Active-Cognitive Coping	0.28	.5991	-0.03
Active-Behavioral Coping	7.02	.0087	-0.17
Avoidance Coping	0.03	.8587	0.10

[a]For the full equations, including stress, coping methods, and demographic variables: R^2 (nurses) = .34 and R^2 (pharmacists) = .33

were deleted, thus leaving data from 230 pharmacists and 242 nurses for use in the regression analyses.

Since there were no hypotheses regarding health professionals' demographic characteristics, this information was used as a control in the regression analyses. For both pharmacists and nurses, demographic variables were entered into the regression equations first, thus partialling out their influence. After controlling for the demographic variables, each of the other independent variables (i.e., stress and coping scores) were entered last into the regression equations in order to identify each of those variables' unique contribution to the dependent variable, job dissatisfaction. Initially, three interaction terms also were entered into the regression equations for both nurses and pharmacists. These terms were the interaction of stress with each of the coping methods (i.e., active-cognitive, active-behavioral, avoidance). For both groups of health professionals, all interaction terms failed to achieve statistical significance ($p > .05$). Thus, the interaction terms were dropped from the regression equations, leaving the main effects for stress and coping in the subsequent analyses.

Insight into the relationship between job dissatisfaction, job stress, and coping can be gained by examination of Table 4. For nurses, job stress, active-cognitive coping, and active-behavioral coping were related significantly to job dissatisfaction ($p < .05$). Analysis of the standardized partial regression or beta coefficients indicates that job stress had the greatest relative influence on nurses' job dissatisfaction, with a standardized coefficient considerably larger than the coefficients for any of the coping methods. The sign of each standardized coefficient also is useful in understanding the independent variables' relationships with job dissatisfaction, with a positive coefficient indicating that higher values on that variable were associated with greater dissatisfaction. Thus, higher levels of job stress were associated with higher levels of job dissatisfaction. By contrast, active-behavioral and active-cognitive coping scores exhibited negative coefficients. Thus, use of those coping methods was associated with improved job satisfaction.

As shown in Table 4, results of the regression analysis for pharmacists in the sample revealed many similarities to those for nurses. Job stress again was related significantly to job dissatisfaction ($p < .05$); the beta coefficient also indicated that job stress had the greatest relative influence on job dissatisfaction. As with nurses, use of active-behavioral coping was related significantly to job dissatisfaction ($p < .05$) and, as indicated by the negative standardized coefficient, was associated with improved job satisfaction. Active-cognitive coping, however, did not show a significant relationship with job dissatisfaction among pharmacists as it had among nurses. For both nurses and pharmacists, avoidance coping exhibited a positive standardized coefficient, but its relationship with job dissatisfaction failed to reach statistical significance.

DISCUSSION

When evaluating research of this type, one must be concerned about the generalizability of the results to the population of nurses and pharmacists. As noted above, the demographic characteristics of the sample were similar to those reported for pharmacists and nurses across the country. Thus, there is little reason to suspect that the respondents in this study were not representative of their respective professional populations. Non-response bias is another concern when mail surveys are used as the method of data collection. To check on the possibility of non-response bias, data for the last 100 nurse and pharmacist respondents (who were assumed to be most like those individuals who did not respond) were compared to data for all other responding nurses and pharmacists. No pattern of differences was found between "early" and "late" responders in either group of professionals so non-response bias was deemed not to present a problem in this study. It is acknowledged, however, that one limitation of this study is the fact that health professionals from all areas of the country could not be sampled. Thus, replication of the study with other samples will be necessary in order to assess the true generalizability of these results.

Because so little is known about interprofession differences in job stress, one of the primary objectives of this study was to compare nurses and pharmacists in terms of perceived job stress, job dissatisfaction, and use of certain methods of coping. For instance, it might have been anticipated that nurses would report higher levels of job stress because they tend to have much more direct patient contact and are more likely than pharmacists to deal with life-and-death emergencies, but there is little evidence available to support such expectations. In the one previous study which compared stress among nurses and pharmacists, nurses did report significantly greater levels of job stress (Wolfgang, 1988b). While job dissatisfaction was not addressed in that previous study, responses to one career satisfaction item revealed that pharmacists were less satisfied than nurses with choice of career (Wolfgang, 1988a).

Although there were some differences in magnitude, intercorrelations between stress, dissatisfaction, and coping (Table 3) indicated similar relationships between those variables for nurses and pharmacists. It must be noted, however, that, although statistically significant, several of these intercorrelations were 0.2 or lower and thus may be of only limited practical significance.

Perhaps the most important objective of this study involved the explanation of nurses' and pharmacists' job dissatisfaction in terms of job stress and the use of various coping strategies. By examining the standardized partial regression coefficients in Table 4, it is possible to determine both the direction and the relative magnitude of the relationships between job dissatisfaction and these independent variables. For nurses in the sample, perceived job stress appeared to be the greatest contributor to job dissatisfaction. It might have been anticipated that higher levels of stress would be associated with greater dissatisfaction. Dissatisfaction with one's job has long been cited as a consequence of excessive job stress (French & Caplan, 1972; Kahn, Wolfe, Quinn, Snoek, & Rosenthal, 1964; Margolis, Kroes, & Quinn, 1974). What may be somewhat more surprising is the fact that the beta coefficient for job stress was considerably larger than the coefficient for any of the coping methods. It has been suggested that coping may be more important than the amount of stress itself in determining the impact of that stress (Lazarus & Launier, 1978), but it appears that, in this sample of health professionals, the amount of perceived stress had the greatest impact on job dissatisfaction. For both pharmacists and nurses, none of the stress X coping interaction terms were related significantly to job dissatisfaction. Thus, no evidence was found to support the role of coping as a moderator between job stress and satisfaction.

Active-behavioral coping and active-cognitive coping also exhibited significant relationships (p < .05) with nurses' job dissatisfaction in the regression equation. The standardized coefficient for each type of coping was negative, indicating that increased use of these coping strategies was associated with improved job satisfaction. These findings are similar to those reported by Ceslowitz (1989), who found that nurses reporting lower burnout scores used coping strategies classified as planful problem solving and seeking social support, which were much like the active-cognitive and active-behavioral methods used in this study. Likewise, in their study of hospice nurses, Chiriboga, Jenkins, and Bailey (1983) reported that those nurses with the most favorable scores on a measure of adaptive status (e.g., work satisfaction, effectiveness) used more cognitive or rational coping strategies. In a more recent study of nurses, Kandolin (1993) found that the use of active coping strategies (e.g., talking about situations with friends) was associated with fewer stress symptoms. Thus, there is strong evidence to indicate that coping which allows nurses to manage their appraisal of the stressfulness of a situation or to deal directly with problems can have a positive impact on outcomes such as job satisfaction. It should be recalled that, in the present study, nurses reported using active-cognitive and active-behavioral coping strategies more often than pharmacists. This may explain why nurses reported greater job satisfaction despite perceiving higher levels of job stress.

Although not found to be related significantly to job dissatisfaction, avoidance coping is of interest because it exhibited a positive coefficient, unlike the other two forms of coping, which indicates that greater use of avoidance coping was associated with increased dissatisfaction among nurses. In their study of hospice nurses, Chiriboga et al. (1983) also reported that the use of "emotional avoidance" as a coping strategy tended to be associated with less favorable job outcomes. Similarly, Kandolin (1993) found that use of passive coping strategies (e.g., alcohol use) was accompanied by more symptoms of burnout and stress.

The regression analysis for pharmacists exhibited both similarities to and differences from that conducted for nurses in the sample. As was the case for nurses, the amount of stress perceived by pharmacists was the greatest contributor to explanation of job dissatisfaction. The positive coefficient again indicated that increasing stress was associated with greater dissatisfaction with one's job. Only one coping method, active-behavioral coping, showed a significant relationship with job dissatisfaction. Similar to the case for nurses, the negative coefficient indicated that increased use of this coping method corresponded to improved job satisfaction.

As is often the case in research projects such as this one, more questions may be raised than answered. For example, although demographic characteristics were used only for control purposes in this study, one obvious difference between nursing and pharmacy is the proportion of males and females in each profession. Nursing has relatively few male practitioners, while pharmacy remains a male-dominated profession. It seems natural to wonder whether the differences between the professions noted in this study actually might be due to gender differences. As a check on this possibility, t-tests were used to compare male and female pharmacists. If job dissatisfaction, stress, and coping are more a function of gender than profession, then significant differences should be found between male and female pharmacists. In this case, only active-behavioral coping exhibited a significant difference ($p < .05$), with female pharmacists reporting more frequent use of such coping strategies. There were no significant male-female differences in terms of pharmacists' job dissatisfaction, stress, or the use of other coping strategies. While this lack of gender differences may indicate that stress and dissatisfaction are more a function of profession than gender, further study of this issue is warranted.

The results of this study certainly can increase our understanding of both the similarities and differences in how the stress process operates within two groups of health professionals. The results of the regression analyses indicated that for both nurses and pharmacists, perceived job stress had a greater influence on job dissatisfaction than did any of the coping methods. It also must be acknowledged, however, that the equations resulting from the regression analyses (i.e., main effects for stress and the three coping methods, plus demographic variables) explained only about one-third of the variance in both nurses' ($R^2 = .34$) and pharmacists' ($R^2 = .33$) dissatisfaction. This implies that there are other variables which must be considered in future studies if a better understanding of job stress and job satisfaction in the health professions is to be achieved.

What other variables should be taken into consideration? While any number of variables could be nominated for inclusion in future research, two broad concepts will be mentioned here. First, this study did not attempt to measure the "psychological resources" which individuals may draw upon to deal with stress. As defined by Pearlin and Schooler (1978), the present study took into account coping responses, "things that people do...to deal with the life-strains they encounter in their different roles." It did not include consideration of psychological resources, which "represent some of the things people are, independent of the particular roles they play" (p. 5). Pearlin and Schooler have noted that, in dealing with job-related stress, these psychological resources are somewhat more effective than the use of specific coping responses. A wide variety of "psychological resources" have been investigated in other studies and might prove beneficial to future studies of health professionals' job stress. In their review of the literature, Ivancevich and Matteson (1980) identified several such variables which seem particularly relevant, including need for achievement, self-esteem, tolerance of ambiguity, locus of control, and Type A personality. Support for the investigation of such variables is provided by the work of Jamal and Baba (1991), who found that Type A nurses experienced significantly greater job stress and overload than did Type B nurses.

The second concept which should be considered for utilization in future studies is that of collective coping. For the most part, the coping responses used in this study represent steps which individuals take to deal with stress. But it may be that many problems associated with one's job "are not responsive to individual coping responses. Coping with these may require interventions by collectivities rather than by individuals" (Pearlin & Schooler, 1978, p. 18). This viewpoint is supported by a recent study of employee pharmacists in which Wolfgang (in press) found that co-worker social support moderated the relationship between stress and job dissatisfaction. Among nurses, Anderson (1991) has reported that support from social networks can buffer the impact of stress on individuals, while George, Reed, Ballard, Colin, and Fielding (1993) found that social support moderated the relationship between exposure to AIDS patients and negative mood. Thus, focusing on health professionals' social support networks and the availability of group coping responses may improve our ability to explain the consequences of job stress.

REFERENCES

American Nurses' Association (1987). *Facts about nursing*. Kansas City, MO: Author.

Anderson, J.G. (1991). Stress and burnout among nurses: A social network approach. In P.L. Perrewé (Ed.), Handbook on job stress [Special issue]. *Journal of Social Behavior and Personality, 6*(7), 251–272.

Billings, A.G., & Moos, R.H. (1981). The role of coping responses and social resources in attenuating the impact of stressful life events. *Journal of Behavioral Medicine, 4*, 139–157.

Caplan, R.D., Cobb, S., French, J.R.P., Jr., Van Harrison, R., & Pinneau, S.R., Jr. (1975). *Job demands and worker health: Main effects and occupational differences* (HEW NIOSH No. 75–160). Washington, DC: U.S. Government Printing Office.

Ceslowitz, S.B. (1989). Burnout and coping strategies among hospital staff nurses. *Journal of Advanced Nursing, 14*, 553–557.

Cherniss, C. (1980). *Staff burnout: Job stress in the human services*. Beverly Hills, CA: Sage.

Chiriboga, D.A., Jenkins, G., & Bailey, J. (1983). Stress and coping among hospice nurses: Test of an analytic model. *Nursing Research, 32,* 294–299.

Ciaccio, E.A., Jang, R., Caiola, S.M., Pfifferling, J.H., & Eckel, F.M. (1982). Well-being: A North Carolina study. *American Pharmacy, NS22,* 244–246.

Dewe, P.J. (1989). Stressor frequency, tension, tiredness and coping: Some measurement issues and a comparison across nursing groups. *Journal of Advanced Nursing, 14,* 308–320.

Fox, M.L., Dwyer, D.J., & Ganster, D.C. (1993). Effects of stressful job demands and control on physiological and attitudinal outcomes in a hospital setting. *Academy of Management Journal, 36*(2), 289–318.

French, J.R.P., Jr., & Caplan, R.D. (1972). Organizational stress and individual strain. In A.J. Morrow (Ed.), *The failure of success* (pp. 30–66). New York: Amacom.

George, J.M., Reed, T.F., Ballard, K.A., Colin, J., & Fielding, J. (1993). Contact with AIDS patients as a source of work-related distress: Moderating effects of support and estrangement. *Academy of Management Journal, 36,* 157–171.

Ivancevich, J.M., & Matteson, M.T. (1980). *Stress and work: A managerial perspective.* Glenview, IL: Scott, Foresman.

Jamal, M., & Baba, V.V. (1991). Type A behavior, its prevalence and consequences among women nurses: An empirical examination. *Human Relations, 44,* 1213–1228.

Jonas, S. (1990). Health manpower: With an emphasis on physicians. In A.R. Kovner (Ed.), *Health care delivery in the United States* (4th ed., pp. 50–86). New York: Springer.

Kahn, R.L., Wolfe, D.M., Quinn, R.P., Snoek, J.D., & Rosenthal, R.A. (1964). *Organizational stress: Studies in role conflict and ambiguity.* New York:Wiley.

Kandolin, I. (1993). Burnout of female and male nurses in shiftwork. *Ergonomics, 36,* 141–147.

Krakowski, A.J. (1982). Stress and the practice of medicine II—Stressors, stresses, and strains. *Psychotherapeutics and Psychosomatics, 38,* 11–23.

Lahoz, M.R., & Mason, H.L. (1990). Burnout among pharmacists. *American Pharmacy, NS30,* 460–464.

Lazarus, R.S., & Folkman, S. (1984). *Stress, appraisal, and coping.* New York: Springer.

Lazarus, R.S., & Launier, R. (1978). Stress-related transactions between person and environment. In L.A. Pervin & M. Lewis (Eds.), *Perspectives in interactional psychology* (pp. 287–327). New York: Plenum.

Margolis, B.L., Kroes, W.H., & Quinn, R.P. (1974). Job stress: An unlisted occupational hazard. *Journal of Occupational Medicine, 16,* 659–661.

Matteson, M.T., & Ivancevich, J.M. (Eds.). (1987). *Controlling work stress: Effective human resource and management strategies.* San Francisco: Jossey-Bass.

May, H.J., Revicki, D.A., & Jones, J.G. (1983). Professional stress and the practicing family physician. *Southern Medical Journal, 76,* 1273–1276.

McLaney, M.A., & Hurrell, J.J. (1988). Control, stress, and job satisfaction in Canadian nurses. *Work & Stress, 2,* 217–224.

National Association of Boards of Pharmacy (1989–90). *Survey of Pharmacy Law.* Park Ridge, IL: Author.

Ortmeier, B.G., & Wolfgang, A.P. (1991). Job-related stress: Perceptions of employee pharmacists. *American Pharmacy, NS31,* 635–639.

Osipow, S.H., & Davis, A.S. (1988). The relationship of coping resources to occupational stress and strain. *Journal of Vocational Behavior, 32,* 1–15.

Pearlin, L.I., & Schooler, C. (1978). The structure of coping. *Journal of Health and Social Behavior, 19,* 2–21.

Richardson, A.M., & Burke, R.J. (1991). Occupational stress and job satisfaction among physicians: Sex differences. *Social Science and Medicine, 33,* 1179–1187.

Roeske, N.C.A. (1981). Stress and the physician. *Psychiatric Annals, 11,* 245–258.

SAS Institute. (1985). *SAS user's guide: Statistics* (Version 5 ed.). Cary, NC: Author.

Wolfgang, A.P. (1988a). Career satisfaction of physicians, nurses, and pharmacists. *Psychological Reports, 62,* 938.

Wolfgang, A.P. (1988b). Job stress in the health professions: A study of physicians, nurses, and pharmacists. *Behavioral Medicine, 14,* 43–47.

Wolfgang, A.P. (1988c). The health professions stress inventory. *Psychological Reports, 62,* 220–222.

Wolfgang, A.P. (in press). Job stress and dissatisfaction among employee pharmacists: The role of coworker social support and powerlessness. *Journal of Pharmaceutical Marketing and Management.*

Wolfgang, A.P., Kirk, K.W., & Shepherd, M. D. (1985). Job stress in pharmacy practice: Implications for managers. *American Pharmacy, NS25,* 446–449.

PART FOUR

An Examination of Burnout

The Purpose of Burnout: A Jungian Interpretation

Anna-Maria Garden

This paper outlines a conceptual framework for burnout and summarizes a series of research studies which had as their objective clarification of the nature of the phenomenon and its unique theoretical content. It sets out an a priori argument for including personality as a variable in burnout research. This is followed by a brief description of the research findings, which provided empirical support for this argument. The implications of these are discussed, firstly, in terms of using personality as a moderator and, hence, as a measure of individual differences. Secondly, the theoretical implications of the findings are elaborated using a Jungian framework of personality.

The working framework I adopt is that of Freudenberger, who defines burnout as: "To deplete oneself. To exhaust one's physical and mental resources. To wear oneself out by excessively striving to reach some unrealistic expectation imposed by one's self or by the values of society" (Freudenberger, 1980, p. 1).

THE ROLE OF PERSONALITY

Much of the confusion and contradictions in the burnout field can be seen to have arisen from an *inappropriate unit of analysis*. The issue of what the individual brings to the burnout process is a largely unexplored one, and empirical research is notably sparse (Shirom, 1989; Keinan & Melamed, 1987). Yet, as Maslach states, "there is general agreement that burnout occurs at an individual level [and that it] is an internal *psychological* experience" (1982b, p. 32). Since the unit of analysis should be appropriate to the phenomenon under scrutiny, one would have expected the individual, if not the individual psyche, to be at the forefront of burnout research.

Yet, the argument has generally been for a social-psychological approach to the study of burnout (Burke, 1993). However, one reason for this is that there have been *few* studies of the personality correlates of burnout and *many* of the situational correlates of burnout.

Where "individual" factors have been included in the research design, they have usually concentrated on demographic indicators (Burke, 1993; Lee & Ashforth, 1993; Pines, 1983), which may not be sufficiently subtle to capture those features of the individual salient to burnout. Alternatively, some authors, by assuming that burnout is a stress response, have assumed that personality traits that influence an individual's response to stress, such as the "Type A" personality, also represent the traits that cause burnout (Cherniss, 1980, p. 127). Whether this can be empirically supported remains an open question (Cordes & Dougherty, 1993).

A few studies have systematically examined the relative importance of situational versus personality factors (Gann, 1979). Dolan and Renaud (1992) found personality factors to be important in the overall process, but that they were secondary to organizational determinants. But there is some question whether the personality indicators used are simply mapping characteristics of the individual that are an outcome of the burnout experience (e.g., not wanting to be involved, and having low self-esteem were discussed as both outcomes of, and antecedents to, burnout).

In general, so far only lip service has been paid to the notion that individual factors play a role in burnout. Yet, as Fischer (1983) states, "since burnout is not a general phenomenon specific to any particular setting, *the sufficient cause must be sought among personal psychological factors*" [italics added] (p. 41).

Not only have the individual causes of burnout not been as systematically investigated as situational causes, but avoidance of the role of the individual has meant that *moderating* psychological factors have also been relatively ignored. Most authors, by focusing on the external "stressors" presumed to cause burnout in a particular setting, are defining those "stressors" as universally applicable. This ignores the critical fact that what is stressful to one person may not be stressful to another (Lazarus & Folkman, 1984). Kobasa has illustrated the potential usefulness of personality in understanding people's differential sensitivity to apparently stressful events (Kobasa, Maddi, & Cowington, 1981; Kobasa & Puccetti, 1983). Similarly, Hochwarter, Perrewé, and Kent (1994) have demonstrated the usefulness of persistence as an individual difference variable in the stressor–strain relationship. The question exists, therefore, whether various situational factors cited in the literature can be considered causes of burnout, or whether they can be only considered causes of burnout for some people and not others.

Similarly, the *form* with which burnout manifests itself is also usually considered independent of the psychology of the individual. This is implicit in definitions and descriptions which are predicated on situational, not individual factors. The unresolved question is whether the individual does affect the manifestation of burnout and whether this influence occurs in a predictable and *systemized* way rather than being wholly unique for each individual.

THE FRAMEWORK FOR PERSONALITY

There are a number of ways in which individual psychological factors can be taken into account. One way, which seems particularly relevant to the burnout field, is the psychological type theory of Carl Jung (1971), since there would appear to be a confounding of occupation and Jungian type in burnout research. This theory (Jung, 1968, p. 48) proposes the existence of distinct personality types.

The Jungian theory of types consists of pairs of polar opposites. The two dimensions critical for the present paper are referred to as "functions." One function-dimension is a perceiving dimension and a person is either a "sensing" type or an "intuitive" type. The second function-dimension is a judging one and a person is either a "thinking" type or a "feeling" type. Those with a preference for feeling (rather than thinking) have been shown to be more tender-minded and to have greater concern and awareness for people (Myers, 1980). In contrast, those with a preference for thinking have been shown to have a higher task orientation and achievement orientation, and to neglect "the art of friendship and of relationship with other people" (Fordham, 1966, p. 37). Those with a preference for intuition (rather than sensing) are described as being more enthusiastic, having new and original ideas, and being alert to the possibilities in situations rather than the present reality. In contrast, the sensing types prefer to concentrate on the actual here-and-now concrete reality of the situation and display a preference for facts (rather than ideas) and a concern to be realistic (Myers, 1980).

Thus, a person has a preferred perceiving function (intuition and sensation) and a preferred judging function (thinking and feeling). A person also tends to further order these two preferred functions such that one of *them* is preferred to the other (e.g., intuition may be preferred to thinking), ending with a four-level hierarchy with the most preferred or dominant function at the top, then the second function, referred to as the auxiliary, next the third or tertiary, and finally the least preferred or inferior function at the bottom.

The dominant function is well developed, or specialized, through receiving a great deal of the available psychic energy at the expense of the others, particularly the inferior (or fourth) function which remains in the unconscious along with the third (tertiary) and possibly also the second (auxiliary). One uses all the functions at one time or another, but the more "preferred" functions are more reliable, since functions that are developed are considered to be in the conscious sphere (rather than the unconscious). The preferred conscious functions, particularly the dominant, become the means of adapting to the world as well as the basis with which to organize one's personality. The ego tends to become *identified* with the nature of the preferred functions. Feeling types may become identified with being the sort of person who is caring and concerned about other people and ignore the fact that the opposite (the thinking function) is also present in their psychic make-ups. An intuitive may become identified with being creative, insightful and enthusiastic and, often, repress that side of them that is concerned with mundane reality. People tend to choose environments or jobs which are compatible with their most preferred functions (Myers & McCaulley, 1985). In such environments or jobs, these functions can, of course, be developed further.

CONFOUNDING OF OCCUPATION AND PERSONALITY

It would appear, a priori, that a confounding of occupation and these two Jungian types has occurred. It should be noted that most research on burnout, at least initially, occurred in the human services (Cordes & Dougherty, 1993). To a secondary extent it has occurred in business, particularly with managers. (However, even studies or theories of "managerial burnout" have looked at human service managers, e.g., Lee & Ashforth, 1993.) This is significant for two reasons. First, there is a considerable amount of evidence to support the notion that different occupations attract different Jungian types. Keen (1982), as well as Myers and McCaulley (1985), report data indicating that, in the health-related, counselling, and education fields, the proportion of feeling types to thinking types is at least 80/20. In other words, the "feeling" type in terms of the Jungian theory, as opposed to the "thinking" type, is associated with those human service workers studied in the burnout area. Other occupations, such as managers, engineers, or students, tend to be predominantly "thinking" rather than "feeling," again, in a proportion of around 80/20 (Keen, 1982; Myers & McCaulley, 1985). Thus, in the different occupational spheres studied in burnout research, there is over-representation of one type relative to another.

The second reason for confounding to have occurred is that the distinctions between feeling and thinking types also seem to reflect the differences in descriptions of burnout in the human services and other, non-human service spheres. A primary way of distinguishing between feeling and thinking types is a personal versus an impersonal stance toward others, a priority given to people rather than to work or achievement (Myers, 1980, p. 25). Yet, in the human services literature, attention is focused on attitudes towards other people during burnout (Leiter & Maslach, 1988; Maslach, 1982a), whilst in the managerial literature attention is focused on one's attitude towards success at work and achievement (Levinson, 1981; Ginsburg, 1974). Similarly, in the former, burnout is assumed to derive from emotionally-demanding work (Leiter & Maslach, 1988; Maslach & Jackson, 1982,

p. 228). Yet, in managerial settings it is the demands to achieve and mental pressures which are implicated (Levinson, 1981; Ginsburg, 1974). A recent study (Dolan & Renaud, 1992), suggesting that social support played a very marginal role in buffering executive burnout, could be due to its lack of relevance to thinking types, as opposed to feeling types. The different descriptions may be due only to the different situational demands.[1] Yet, the differences in personality composition, combined with the consistent importance of personality as a moderator in the psychological stress literature (Lazarus & Folkman, 1984), leads to the empirical question of whether the different descriptions of the causes and characteristics of burnout in the two occupations are also affected by the different personalities involved. And, if that is the case, then our understanding of the burnout phenomenon may be clarified.

SUMMARY OF FINDINGS

The series of studies I have pursued has been designed to unravel this potential confounding of occupation and personality (Garden, 1985, 1987, 1988, 1989). They involved comparison of a human services sample of 81 occupational health nurses and a group of 196 mid-career management students. Each group was sampled with a questionnaire assessing burnout and the other variables described in the following sections.

The measure of burnout used in the research studies focused on energy depletion as its key indicator. This is in keeping with Shirom's (1989) argument that "the major conclusion which may be drawn from past validation efforts is that the *unique* content of burnout has to do with the depletion of an individual's energetic resources" (p. 33, my emphasis). Energy depletion (or exhaustion) is the dimension of burnout for which there is most definitional agreement irrespective of occupational setting (Cordes & Dougherty, 1993; Lee & Ashforth, 1993).

The derivation of the measure used in these studies has been described in detail elsewhere (Garden, 1985, 1987, 1988, 1989). To summarize, factor analysis was performed on items relating to energy depletion and exhaustion. These had been rated on a frequency scale from 1 to 5 (Never to Always). Items in the measure were selected using the criterion of having a factor loading above .50 on that factor and not above .30 on any other. The two simple items in the measure are "feeling exhausted" and "severe energy fluctuations." Two composite items included are "at limit," referring to being unable to perform a range of specific activities because one is too tired, e.g., reading, talking to others; and "non-renewal," comprising a number of questions gauging whether sleeping, rest, or vacation relieved sensa-

---

[1] *It is also instructive to consider that the conventional wisdom about what burnout is, as derived from the human services sphere, may not be generalisable to other occupations. A different understanding of burnout would have emerged if the initial formal differentiation of the phenomenon had occurred outside the human services.*

tions of tiredness. Cronbach's alpha was .87 for the non-human services management students and .79 for the occupational health nurses.

The measure of personality was the Myers Briggs Type Indicator (MBTI; Myers & McCaulley, 1985). This is an operationalization of the Jungian theory of types and is a self-report questionnaire of 126 forced-choice items. The MBTI attempts to measure the dichotomous pairs of polar opposites in Jung's original theory. The MBTI was administered 9–12 months prior to the main questionnaire, since it is important to map personality factors that precede burnout and are not merely an outcome of it (Corrigan et al., 1994). The MBTI has been shown to have adequate reliability and validity (Carlyn, 1977; McCaulley, 1981).

The findings were similar in each of the two occupational settings, and can be summarized as follows:

(i) A process of *reversal*. Increasing energy depletion was accompanied by a decrease in the characteristics normally associated with that function or type. Specifically, hostility and lack of concern for others was associated with higher levels of burnout only for feeling types who, by definition, are supposed to have a higher concern for others. In contrast, thinking types showed, on balance, a more positive attitude toward others the higher the level of burnout. Similarly, for thinking types only, higher levels of burnout were consistently and significantly associated with lower levels of ambition or will to achieve. For sensing types only, burnout was positively associated with "losing touch with reality," and for intuitives only, a significant negative association occurred between enthusiasm and originality, and the measure of burnout. Significant relationships only occurred for that personality type dimension for which there was some a priori theoretical relevance (e.g., the sensing-intuitive dichotomy did not moderate the relationship between burnout and negative feelings toward others).

(ii) A process of *convergence*. T-test comparison of means indicated that, at low levels of burnout, the differences between types that one would theoretically predict, were obtained. For example, concern for others (will to achieve) was significantly higher for feeling (thinking) types than for thinking (feeling) types. However, when burnout was high, no significant differences occur between any of the types for any of the characteristics. Put simply, the opposite types differed when they were low in energy depletion and were similar when they were high in energy depletion.

(iii) Burnout was related to *"fit"* between personality and environmental demands, not *"lack-of-fit."* Theoretically, emotional demands should be easier for feeling types to deal with and mental demands should be easier for thinking types to deal with. Yet, in both occupations, emotional demands predicted energy depletion for feeling types but not for thinking types. Mental demands predicted energy depletion for thinking types but not for feeling types. In other words, burnout was most strongly associated with the kind of demand each type is naturally adapted to deal with.

With a measure of experienced stressfulness of their working situation used as the dependent variable instead of energy depletion, the results showed almost the reverse. In general, the pattern was for emotional demands to be a stronger predictor for thinking types, and mental demands to be a stronger predictor for feeling types.

An overall map of the above findings is represented in Figure 1. These findings indicated that various "characteristics" and "causes" of burnout could be personality-specific. There was systematic variation depending on Jungian type, irrespective of occupation. One could expect different manifestations of burnout for different psychological types irrespective of occupation.

This suggests that the conventional wisdom that depersonalization is a part of burnout may derive solely from the nature of the human services and confounding with the predominant psychological type (feeling) existing in such occupational settings. The symptom of negative reactions to people may not apply to thinking types even in a human services occupation and, therefore, may not be applicable to all individuals even within that occupation. Similarly, emotional demands may not be a generalizable notion of what causes burnout, even in the human services. Other types of environmental demand, such as mental demands, may be relevant. The symptom of a decline in the will to achieve described in the burnout literature derived from managerial settings (where there is usually a predominance of thinking types), may derive from the underlying personality-type composition in that setting. Some authors (notably Edelwich & Brodsky, 1980) have emphasized the decline in enthusiasm and "spark" in someone suffering from burnout. From the findings reported here, one would predict that their observational settings might have a predominance of the "intuitive" types.

This argues for the *systematic* incorporation of personality as a variable in burnout research. This may be a way forward in the field and a means of bringing order out of confusion. It would then be possible to determine which symptoms are common across all individuals, and which are different. Measures of burnout could then be derived which do not contain type-specific symptoms.

The above sets out the *methodological* implications of the findings. However, the findings referred to as a process of reversal indicated not simply that burnout may be manifested in symptoms particular to different psychological types, but that that symptom would be a decrease in the attribute normally characteristic of that type. Similarly, the process of convergence suggested that what is occurring in burnout is not just a lessening of the characteristic orientation but a loss of what previously distinguished one type from its opposite. These *conceptual* implications are represented by the type-specific symptoms set out in Figure 1. For feeling types, the characteristic symptom would be a lessening in concern for others, for thinking types a lessening of ambitiousness, for intuitives a lessening of enthusiasm, and for sensing types a lessening of concern for facts and reality. These notions are testable in future research, bearing in mind that the reversal and convergence processes, according to this framework, would only occur with characteristics with some a priori theoretical relevance to the type dichotomy concerned. Thus, changes such as an increasing tendency to explosiveness, or

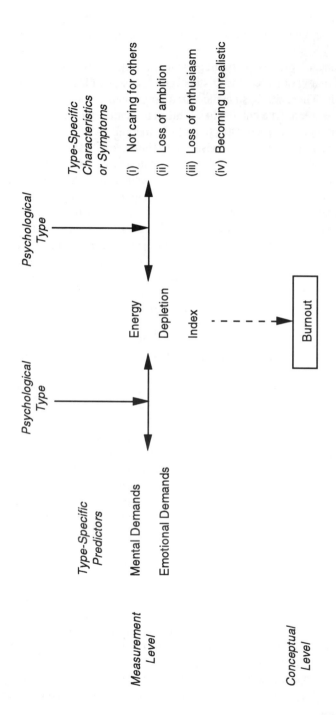

FIGURE 1 Overall Map of the Findings of this Study

emotional eruptions would not be categorized as potentially type-specific since these would be expected to occur for all types. The most intriguing, counter-intuitive finding concerned the fact that burnout was related to what was suitable to a particular type rather than to what was unsuitable to a particular type. One would have expected any moderating effect of type to be in exactly the opposite direction. Why aren't thinking types more depleted by emotional demands than by mental demands? Burnout is sometimes thought of as a result of lack-of-fit between person and environment. Yet these findings seem to imply that fit is a predictor. The general trend was for demands that are unsuitable to a type to be predictors of experienced stress whereas demands suitable to a particular type to predict burn-out. This contrasts with the general view that burnout is a stress syndrome (Cordes & Dougherty, 1993).

The implications of these findings could rest there, and the preceding sections could be read on their own. What follows is a discussion which extends the findings into their deeper theoretical implications. This account uses the Jungian theory of the *dynamics* of the psyche to explain the content of the previous findings, and not simply the implications of a measure of individual differences. Looking at types within a dynamic or developmental perspective helps to explain the content of the counter-intuitive results discussed above.

A DEVELOPMENTAL PERSPECTIVE

The immediate explanation for the empirical findings presented earlier would be that burnout is accompanied by the lack of use, or inability to use, the conscious function(s) upon which one has to come to rely as a primary tool to deal with one's world, as well as for self-identification. The conscious or preferred functions may no longer be available for use purely as a simple *consequence* of energy depletion. The conscious functions are more likely to *drop into the unconscious* if one is fatigued. This would not mean that it was not in operation at all, but its use would be expected to be more sporadic and less disciplined, forcing the individual to rely on another, previously non-conscious, function (the thinking types' increase in concern for others, representing a surfacing of the feeling function).

The consequence of a feeling function sinking back into the unconscious would be an inability to relate to the world in the usual concerned pleasant way. The unconscious thinking function, to the degree to which it had previously been actively repressed, may emerge in a more disruptive, more hostile or generally negative way. Previously negative feelings towards others, for example, which could not be allowed to be released because of the feeling function's desire to create harmony may, with that feeling function no longer in control, erupt or otherwise be released.

Similarly, if a thinking type loses the same degree of facility with the thinking function, the high value placed on achievement (or competence, rationality or logic, etc.) would not be maintained if the thinking function ceases to be fully in the conscious sphere.

Inability to rely on the function which one has been accustomed to use for adaptation to the environment, and for self-identification, would in itself be a distressing experience, evoking feelings of helplessness. Getting through each day would be a struggle; one would feel unable to perform as well as previously, would lose one's sense of priorities, etc. All these feelings are frequently described as intrinsic to the burnout experience. In other words, the loss of use of one's conscious functions would create exactly the same experience as that used to describe burnout. One then has a catastrophe model of the psyche involving a switch in kind, not just degree, in how the psyche operates. During burnout, eruptions of one's opposite functions would be more frequent and represent the explosiveness and touchiness which characterize descriptions of burnout. The contents of the unconscious, by definition not subject to the control of the will, would more readily spill out into the behavior of the person.

ENANTIADROMIA

The above theoretical explanation describing the effects of conscious functions "dropping" into the unconscious because of energy depletion accounts for the processes named reversal and convergence. However, tiredness, *in itself,* would not make one more sensitive to the type of demands one is adapted to while not being sensitive to the type of demands one is not adapted to (whether emotional or mental demands). Nor does it explain why this pattern is reversed for the measure of experienced stress. To explain those findings we need to draw on another theoretical concept. This is referred to as *"enantiadromia"* by Jung. This is "the emergence of the unconscious opposite in the course of time. This characteristic phenomenon practically always occurs when an extreme one-sided tendency dominates conscious life" (Jung, 1960, p. 262). The functions in the unconscious may act simply as a counterweight (Jung, 1960, p. 419), not *necessarily* acting in a contrary way to the conscious functions. The contents of the conscious sphere can, however, become so developed relative to the unconscious that the latter no longer "balances" the former but begins to act in opposition to the aims of the conscious ego. The more one's ego becomes attached to that *one* way of being, and the more one identifies only with the first and/or second functions, the more one-sided the individual becomes, since the conscious sphere needs to be balanced by its opposite in the unconscious.

In a sense, psychic health is maintained by not living only out of one's conscious functions. The psychic process, called individuation in Jungian theory, requires as an *internal* demand for adaptation and growth that balance and wholeness are developed. If the ego thwarts this psychic demand, the unconscious may retaliate. As Jacobi (1962, p. 31) explains, "when the conscious portion of the psyche develops too one-sidedly...it may easily land itself in a kind of hypertrophy." When this occurs for too long, it will call into action the "self-regulating mechanism of the psyche" (Jacobi, 1962, p. 38). If one had been relying too much on, or identifying too much with, one function and creating an imbalance in the psyche, the unconscious may *purposively "pull" the conscious functions down.* The finding that the two types differed from each other at low levels of energy depletion but, with increasing energy depletion, became more similar to each other, would be consistent with the idea of enantiadromia (as a balancing-out process) playing a part.

This explanation also accounts for the "fit" findings. Too much use of the thinking function in a thinking type would cause the enantiadromia process. This over-use would only be possible with too many *mental* demands when the thinking function is conscious. The over-use of the thinking function in a thinking type would not be as likely to occur with *emotional* demands. Similarly, with a feeling type, over-use of the conscious (feeling) function would be most likely to occur with too many emotional demands. It is also possible that, if the psyche is trying to redress a one-sided orientation, one becomes particularly sensitive to those environmental demands which use the conscious functions.[2]

Note that it is possible that the causal direction assumed above is reversed. When an individual of a particular type is operating in a way which is not consistent with being that type, this could lead to high energy depletion. When being consistent with that type, low energy depletion results. However, this explanation has difficulty encompassing the findings concerning the measure of experienced stress, since it would predict a thinking type who, being conflicted about or not acting true to that type, would be more comfortable or compatible with emotional demands than with mental demands.

DISCUSSION

The implications of the above relate primarily to the notions of psychic balance and adaptation to inner demands. Enantiadromia, which is set off when one way of being or functioning in the world is overemphasized, seems to be a part of the burnout process. This would imply that burnout has a positive purpose, i.e., psychic balance, and that the latter may be the very reason it occurs.

Freudenberger states that a burnout experience usually has its roots in "the area of a person's life that seemed to hold the most promise" (1980, p. 13). It tends to involve overidentification with the part of themselves which previously brought rewards. This is, of course, the dynamic that occurs if one over-relies on

---

[2] *Other findings (Garden, 1985) support these notions. For example, burnout was inversely associated with hours of leisure, but not related to hours of work, or with people. Since it is often in leisure that one uses one's opposite psychological functions, the less time spent on leisure activities the more a one-sided conscious orientation would become reinforced.*

one's conscious functions. One does so precisely because they have been the means of survival, adaptation, success and self-gratification. Overidentification or over-reliance on a function will be encouraged by being rewarded for that way of being (the most "caring" nurse, the most "insightful" psychologist). In other words, when dealing with the most developed function, we are dealing with "that part of a person's life which seemed to hold the most promise" in Freudenberger's terms.

The explanation of these findings hinges on the theoretical explanation of enantiadromia. This, combined with the findings of "reversal" described earlier, provides a potential means for distinguishing the phenomenon of burnout from that of other similar phenomena, whether anxiety or depression. With burnout, but not with the latter two, one could predict a lessening only of that characteristic which previously defined that particular type. Further, it would be associated primarily with environmental demands which the individual is compatible with, rather than with those demands one would theoretically predict would be incompatible.

From the standpoint of the conscious sphere, and the centre of that sphere, the ego, enantiadromia, or the withdrawal of conscious functions, is experienced as dysfunctional and "bad." From the standpoint of the psyche as a whole, the process is a functional one designed to bring about greater balance to the whole psyche, not simply to satisfy the orientation of the conscious ego. Viewed in this way, the enantiadromia and the convergence which occurs may be the *purpose of burnout.*

A few authors have, in fact, suggested that something akin to this may be occurring. Khalsa (1978) argued that burnout may function as a call to growth and used Dabrowki's theory of positive disintegration to explain the purpose of burnout. In this theory, the personality develops primarily through dissatisfaction with, and fragmentation of, the existing psychic structure. "This development occurs through periods of psychic disintegration and subsequent periods of secondary integration at a higher level of functioning" (Khalsa, 1978, p. 27). Much of what occurs in burnout may, according to Khalsa, be seen as the workings of this "positive disintegration" stage. Freudenberger (1983, p. 26) has also argued that it may "be useful...to consider the homeostatic function of burnout, that is, to consider how the process of burnout provides signals for us to monitor and alter maladaptive personal and social systems."

This "homeostatic" function may be considered, at one level, to stimulate a search for balance as a desirable outcome of the otherwise-dysfunctional experience of burnout. In this sense, homeostasis would reflect a return to a system of equilibrium. In Jungian terms this would refer to a return to a psychic state where over-emphasis on the conscious function(s) did not occur, but the relations between the conscious and unconscious functions was that of a mutual compensation.

The more radical view is that stimulating the psyche to seek greater balance in the use of one's functions, one's psychic orientation and lifestyle, may be the very *reason for burnout* to occur. Thus, in contrast with Selye's conception of stress being the "common denominator of all the adaptive reactions of the body" (Selye,

1956, p. 64), the conceptualization developed in this paper would argue that burnout is a common denominator of the adaptive reactions of the *psyche.*

Further, in contrast to the view of Lazarus and Folkman (1984, p. 19) that stress is a "particular relationship between the person *and the environment* that is appraised by the person as taxing or exceeding his or her resources and endangering his or her well-being" (my emphasis), burnout would refer more to *inner adaptation,* adaptation to the demands of the psyche for growth. External events or situations may be a catalyst for, or be a parallel to, the inner adaptive demands. But there may also be endogenous needs for adaptation. This is a different notion than homeostasis, for here one could not return to an original state even if that had formerly been one of equilibrium between one's conscious and unconscious functions. The Jungian notion of psychic growth ("individuation") would require a new equilibrium, only obtained after the requisite growth in the unconscious function had occurred.

One implication of the preceding is the possible need to allow for the role of the unconscious in one's research design. In contrast, it has usually been assumed in previous research that burnout is both observable and a conscious phenomenon. There has been a tendency to equate what burnout "is" with what it is seen or felt to be. (Similarly, measures constructed from factor analysis are equated with the definition of what burnout is). Another author to discuss the role of the unconscious in burnout is Meyer (1979) who used a phenomenological approach. He distinguished between "descriptive observable symptoms of burnout: 'feeling overwhelmed,' 'anxious' or 'fatigued'" (p. 101), and a stage preceding these reactions, which is an unconscious "incubation period." This unconscious incubation period, involving a form of inner psychic conflict or a re-socialization process of the individual's personality vis-a-vis the organization, results in a "reaction," the second stage of the burnout process. Others working in a more clinical way (notably Fischer, 1983; and Freudenberger, 1980, 1985) have also highlighted the importance of the unconscious in burnout.

Determining the precise nature of any unconscious processes involved should be one aim of future burnout research. Freudenberger (1980) has argued that burnout has to do with internalizing unrealistic expectations, usually early on in one's life. Meyer (1979) asserts that it arises from an inner conflict mirrored in the socialization process to a new organization. Fischer (1983) argues that a narcissistic trauma is involved but that future research is needed to determine the source of such a trauma. Whatever theoretical framework one chooses to use, it would appear that the distinctive qualities of the nebulous burnout phenomenon may be accurately discerned, or at least illuminated, by examining the unconscious, a far cry from limiting the study of burnout to its social-psychological features. Further, it suggests a return to a more clinical qualitative approach, at least as a supplement of the standardized quantitative one advocated in the burnout field in the 1980s.

CONCLUSION

This paper has attempted to illustrate the usefulness in burnout research of incorporating personality as a variable. Both the a priori arguments as well as the empirical findings of the series of research studies conducted, suggested that Jungian type may be being confounded with occupation in burnout research. This argued for the systematic inclusion of personality type as a moderating variable. A framework for what those type-specific causes and characteristics might be was set out.

In addition to the notion of the usefulness of a measure of individual differences, the theoretical account of the findings from a Jungian perspective suggested that the unique and distinctive content of burnout might be properly understood from a developmental perspective emphasizing the role of the unconscious.

From the preceding, a working definition of burnout would be that it is a process which is *experienced as* a chronic unrelenting depletion of energy, not easily renewed by activities such as sleep, rest or vacation. Other symptoms which emerge are personality-specific and reflect a loss of precisely those attributes which previously defined or characterized the individual. These symptoms are the manifestation of an underlying process involving the dynamic of enantiadromia. Specifically, an individual becomes overidentified with one way of being in the world, and over-relies on a one-sided conscious orientation. This means one does not live out the opposite sides of one's being. The unconscious will purposely withdraw the established means of both identification and survival from continued use, in order to propel individuals to live more completely from all sides of their nature. This dynamic is not merely for the purpose of maintaining a homeostatic equilibrium but to propel the individual to a deeper and enlarged integration of personality than previously. Burnout would be intensified the more the individual resisted the development required and clung to the previously useful and successful, but now out-dated and over-used, modus operandi. This perspective suggests an approach for future research which may enable us to determine how burnout may be distinguished from other related phenomena. This is likely to require us to pay attention to the subtleties of the psyche, specifically the unconscious.

REFERENCES

Burke, R.J. (1993). Toward an understanding of psychological burnout among police officers. *Journal of Social Behavior and Personality, 8*(3), 425–438.

Carlyn, M. (1977). An assessment of the Myers Briggs Type Indicator. *Journal of Personality Assessment, 41*(5), 461–473.

Cherniss, C. (1980). *Staff burnout: Job stress in the human services.* Beverly Hills, CA: Sage.

Cordes, C.L., & Dougherty, T.W. (1993). A review and an integration of research on job burnout. *Academy of Management Review, 18*(4), 621–656.

Corrigan, P.W., Holmes, E.P., Luchins, D., Buican, B., Basit, A., & Parks, J.J. (1994). Staff burnout in a psychiatric hospital: A cross-lagged panel design. *Journal of Organizational Behavior, 15,* 65–74.

Dolan, S.L. (1994). Individual, organizational and social determinants of managerial burnout: Theoretical and empirical update. In P.L. Perrewé & R. Crandall (Eds.), *Occupational stress: A handbook* (pp. 223–238). Washington, DC: Taylor & Francis. (Original work published 1992)

Edelwich, J., & Brodsky, A. (1980). *Burnout: Stages of disillusionment in the helping professions.* New York: Human Services Press.

Fischer, H.J. (1983). A psychoanalytic view of burnout. In B.A. Farber (Ed.), *Stress and burnout in the human service professions* (pp. 40–45). New York: Pergamon.

Freudenberger, H.J. (1980). *Burnout: How to beat the high cost of success.* New York: Bantam.

Freudenberger, H.J. (1983). Burnout: Contemporary issues, trends and concerns. In B.A. Farber (Ed.), *Stress and burnout in the human service professions* (pp. 23–28). New York: Pergamon.

Freudenberger, H.J. (1985). *Women's burnout.* New York: Doubleday.

Gann, M.L. (1979). *The role of personality factors and job characteristics in burnout: A study of social service workers.* Unpublished doctoral dissertation, University of California, Berkeley.

Garden, A.M. (1985). *Burnout: The effect of personality.* Unpublished doctoral dissertation, Massachusetts Institute of Technology, 0373736.

Garden, A.M. (1987). Depersonalization: A valid dimension of burnout? *Human Relations, 40,* 545–560.

Garden, A.M. (1988). Jungian type, occupation and burnout: An elaboration of earlier study. *Journal of Psychological Type, 14,* 2–14.

Garden, A.M. (1989). Burnout: The effect of psychological type on research findings. *Journal of Occupational Psychology, 62,* 223–234.

Ginsburg, S.G. (1974, August). The problem of the burned out executive. *Personnel Journal,* pp. 598–600.

Hochwarter, W.A., Perrewé, P.L., & Kent, R.L. (1994). The impact of persistence on the stressor-strain and strain-intentions to leave relationships: A field examination. In P.L. Perrewé & R. Crandall (Eds.), *Occupational stress: A handbook* (pp. 153–166). Washington, DC: Taylor & Francis. (Original work published 1993)

Jacobi, J. (1962). *The psychology of C. G. Jung.* London: Routledge & Kegan Paul.

Jung, C.G. (1960). On the nature of the psyche. *Collected Works, Vol. 18.* Bollinger Series XX. Princeton University Press.

Jung, C.G. (1968). *Man and his symbols.* New York: Dell.

Jung, C.G. (1971). Psychological types. *Collected Works, Vol. 6.* Bollinger Series XX. Princeton University Press.

Keen, P.G. (1982). *Cognitive style research: A perspective for integration.* Unpublished paper, Sloan School of Management.

Keinan, G., & Melamed, S. (1987). Personality characteristics and proneness to burnout: A study among internists. *Stress Medicine, 3,* 307–315.

Khalsa, R.K. (1978). *Healing the healer.* Los Angeles: Centre for Health and Healing.

Kobasa, S.C., Maddi, S.R., & Cowington, S. (1981). Personality and constitution as mediators in the stress-illness relationship. *Journal of Health and Social Behavior, 22,* 368–378.

Kobasa, S.C., & Puccetti, M.C. (1983). Personality and social resources in stress resistance. *Journal of Personality and Social Psychology, 45,* 839–850.

Lazarus, R.S., & Folkman, S. (1984). *Stress, appraisal, and coping.* New York: Springer.

Lee, R.T., & Ashforth, B.E. (1993). A longitudinal study of burnout among supervisors and managers: Comparisons between the Leiter and Maslach (1988) and Golembiewski et al. (1986) models. *Organizational Behavior and Human Decision Processes, 54,* 369–398.

Leiter, M.P., & Maslach, C. (1988). The impact of interpersonal environment on burnout and organizational commitment. *Journal of Organizational Behavior, 9*(1), 297–308.

Levinson, H. (1981, May/June). When executives burn out. *Harvard Business Review,* pp. 73–81.

Maslach, C. (1982a). *Burnout: The cost of caring.* Englewood Cliffs, NJ: Prentice-Hall.

Maslach, C. (1982b). Understanding burnout: Definitional issues in analyzing a complex phenomenon. In W.S. Paine (Ed.), *Job stress and burnout: Research, theory, and intervention perspectives* (pp. 29–41). Beverly Hills, CA: Sage.

Maslach, C., & Jackson, S.E. (1982). Burnout in health professions: A social psychological analysis. In G.S. Sanders & J. Suls (Eds.), *The social psychology of health and illness* (pp. 227–251). Hillsdale, NJ: Erlbaum.

McCaulley, M.H. (1981). Jung's theory of psychological types and the MBTI. In P. McReynolds (Ed.), *Advances in personality assessment* (Vol. V, pp. 294–352). San Francisco: Jossey Bass.

Meyer, J. (1979). *Social construction of burnout: An emergent theory.* Unpublished doctoral dissertation, Boston University.

Myers, I. (1980). *Gifts differing.* Palo Alto, CA: Consulting Psychologists Press.

Myers, I., & McCaulley, M.H. (1985). *Manual: Myers Briggs Type Indicator.* Palo Alto, CA: Consulting Psychologists Press.

Paine, W.S. (1982). *Job stress and burnout.* Beverly Hills, CA: Sage.

Pines, A.M. (1983). On burnout and the buffering effects of social support. In B.A. Farber (Ed.), *Stress and burnout in the human service professions* (pp. 155–174). New York: Pergamon.

Selye, H. (1976). *The stress of life* (rev. ed.). New York: McGraw-Hill.

Shirom, A. (1989). Burnout in work organizations. In C.L. Cooper & I. Robertson (Eds.), *International review of industrial and organizational psychology* (pp. 25–48). London: Wiley.

Individual, Organizational and Social Determinants of Managerial Burnout: Theoretical and Empirical Update

Shimon L. Dolan

In the past two decades, research on aspects of occupational demands and their deleterious consequences on the psychological well-being of employees has been flourishing. While psychological manifestations of inability to cope with organizational life vary significantly, by and large, typical measures used in organizational research include assessments of emotional state such as anxiety, depression, irritation and even job dissatisfaction. In the past few years, however, we have witnessed a surge of write-ups in the stress arena using the concept of burnout (see Cordes & Dougherty, 1993). While the burnout literature originally concentrated on the helping/caring professions such as nursing, medicine, social work, etc., it is discussed today in other contexts. In broad terms, burnout refers to the "syndrome of physical and emotional exhaustion involving the development of negative self-concept, negative job attitudes and loss of concern and feeling for clients" (Pines & Maslach, 1978, p. 233).

The burnout concept, coupled with the well established research tradition in occupational stress, leads to broadening the burnout syndrome into many other

Author's Note: This is an updated version of the article published in 1992 in the *Journal of Social Behavior and Personality,* 7(1), 95–110 (with Stéphane Renaud), and entitled Individual, Organizational and Social Determinants of Managerial Burnout: A Multivariate Approach.

occupational categories. In spite of these contemporary trends and the rise in popularity of burnout research, findings should be treated with caution due to contradicting conceptual and theoretical views of this phenomenon. On one hand, there are those who assert that burnout is not significantly different from other stress related psychological outcomes (Paine, 1982; Payne, 1984; Shirom, 1989); on the other hand, there are researchers who argue that, if proper attention is paid to conceptualization and level of analysis, burnout is clearly a distinct phenomenon worthy of independent research (Garden, 1994; Cordes & Dougherty, 1993).

Perhaps due to the above mentioned controversy, burnout has not been recognized as a defined "mental disorder" in the Diagnostic and Statistical Manual of Mental Disorders (DSM-III; American Psychiatric Association, 1987). Notwithstanding that, "burnout" has been recently recognized as a compensable occupational disease by numerous workers' compensation boards and tribunals both in the U.S. and in Canada (Lippell, 1988). The critical questions discussed by these tribunals mirror the theoretical arguments found in academic research. Namely: 1) How valid is the diagnosis of burnout and what is the level of confidence in its measuring device (i.e., issues of construct and predictive validity)? And 2), assuming that a case of burnout has been established, what are the plausible causes and/or antecedents to it? Customarily, employers retain the thesis that weaknesses in the individual's personality and/or other personal attributes are the prime cause of the manifested condition. Unions and other employee associations argue the contrary; namely, that the psychological and affective work environment are the provoking agents.

Ample research on burnout among organizational psychologists emphasizes the fact that the concept is an outcome of complex interaction or multiplication of individual needs and resources and the various demands, constraints, and facilitators within the individual work environment (Handy, 1988). Empirical examination of these interactions is made difficult due to lack of conceptual and theoretical clarity as well as methodological problems (Garden, 1994; Shirom, 1989). For instance, a popular research design is to measure the relationships between job/organizational stressors and burnout by relying almost exclusively on subjects' self-reporting via questionnaire for both the job conditions and the affective reactions to them (Spector, Dwyer, & Jex, 1988; Burke, Brief, & George, 1993). This may lead to a problem commonly referred to as "method-variance" or "nuisance." On the other hand, few alternative methodologies are suggested. Nonetheless, a logical way to capture the complex interactions between the numerous variables involved in the etiology of burnout is through a conceptually multivariate approach. Burnout may result from attempts to cope with various environmental-organizational stressors which are perceived as such by people characterized by certain personality profiles. The magnitude of burnout can be also attributed to a host of moderating variables (Barone, 1994; Harris, 1994). Thus, the ultimate effect on burnout is represented by a mosaic of individual traits and job/environment factors (Carroll & White, 1982).

Psychological literature on occupational stress and burnout has focused on several categories of job and organizational stressors. Among these are role problems (conflicts and ambiguities), job content demands (workload and responsibility), work organization (lack of communication and/or commitment), professional perspectives (career ambiguities, skill underutilization), and physical environment (noise, temperature, safety) (see, for example, Beehr & Newman, 1978; Caplan, Cobb, French, Harrison, & Pinneau, 1975; Cooper & Marshall, 1976; Dolan & Balkin, 1987; Dolan, van Ameringen, Corbin, & Arsenault, 1992). Relationships have also been found between qualitative and quantitative overload and burnout (Himle, Jayaratne, & Thyness, 1989; Kaufmann & Beehr, 1989) and work-family conflicts (Izraeli, 1993). All these factors are also cited in research pertaining to executive burnout (Lee & Ashforth, 1993a; Quick, Nelson, & Quick, 1990; Quick & Quick, 1984).

In examining the literature concerning individual differences, personality is often suggested as the key construct in the etiology of burnout (Ganster & Schaubroeck, 1994). The reason for this suggestion stems from both conceptual and methodological issues. For one, most studies on burnout use an underlying assumption about cognitive appraisal by the individual worker of the job demands which become stressful only when perceived as threatening (Lazarus, 1994; Lazarus & Folkman, 1984; Payne, Jabri, & Pearson, 1988).

In addition, a number of variables have been suggested as moderators in the etiology of stress manifestations and burnout. Social support is one such variable (Dolan, van Ameringen, & Arsenault, 1992). Social support is of growing interest as a potential approach to alleviate job stress and burnout. Although it might seem evident that better support improves coping, the study of social support is a complex undertaking (Cobb, 1976; Cohen & Syme, 1985). There is wide disagreement on how to both define and measure social support. For example, certain definitions are more structural in character, pertaining to the number and frequency of relationships with others (Hammer, 1981). Others are more subjective in form and pertain to an individual's perceptions of the supportive quality of his/her social environment (House, 1981). In their definition of social support, LaRocco, House, and French (1980) noted that there are many types of social support: emotional, empathy and understanding, instrumental assistance and provisions of information. They have found social support to operate in two ways. First, it can enhance employee responses on the job because it meets important needs (such as security, approval, belonging and affection); in other words, the positive effects of social support can offset the negative effects of job demands and burnout. Second, social support has been thought to buffer the impact of stress on employee responses (Greenglass, 1991). Thus, the controversy regarding the effects of social support could be summarized as the following: While the vast majority of published research indicates that social support reduces, or buffers, the adverse psychological impact of exposure to organizational/job stress, recent findings have found support to counter-buffer (i.e., Barrera, 1988; Beehr, King, & King, 1990) and amplify the stress response (Dolan & Zeilig, 1994). Clearly, the debate surrounding this issue is far from over.

This paper focuses on several aspects of the proposed relationships between job/organizational demands, personal characteristics, social support and burnout for managerial settings as depicted in Figure 1. The paper explores the moderating role of self-reported cognitions of social support in buffering burnout. It is assumed that the measured level of burnout can be partially accounted for by organizational factors and job demands, that personality traits are also important determinants of it, and that the state of burnout can be moderated through the degree of social support provided. Consequently, by testing social support as a moderator, it is argued that the interactive influence of the latter can be explored.

In summary, the purpose of this study is to: (a) ascertain the relative impact of the job environment (i.e., organizational factors) vs. individual differences (i.e., personality traits) in predicting managerial burnout, and (b) empirically test the role of social support in buffering managerial burnout.

METHOD

Subjects

Two hundred twenty-four senior executives from different private-sector organizations who had attended an executive health clinic participated in the study. The clinic provided an annual medical examination to various executives in many organizations. Only those who had visited the clinic for the first time (i.e., recent clients) were included in the study, due to the fact that the mental assessment was not part of the service rendered by the clinic in the past, and the latter was part of a one-year trial experience. Consequently, one may conclude that the participants in the study happened to be recent recruits in their respective organizations who benefited from this type of service. These executives constituted a normal popula-tion, and none were aware of the content and nature of the medical assessment prior to arriving at the clinic. Data were collected over a 9-month period. Breakdown by gender showed a predominance of male executives (91.5%), typical of these occupations.

Materials and Procedures

A multiple-item computer-assisted questionnaire was administered on site and comprised organization/job assessment, personality trait assessment, social support assessment and the Maslach Burnout Inventory (MBI; Maslach & Jackson, 1986). Because of the computer-aided interactive data gathering, and due to the fact that each manager was requested to answer one question at a time as it appeared on the screen, it is suggested that the method variance bias was minimal (see, for example, Spector, 1987; Williams, Cote, & Buckley, 1989).

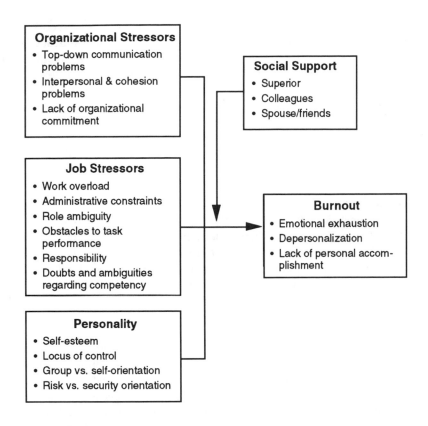

FIGURE 1   A Conceptual Model of Managerial Burnout

Organizational/Job Stressors

Three organizational stress clusters (out of 20 initial items) were extracted following factor analysis. The 20 items required the executives to assert whether or not a particular stressor was present in their organizational environment. In addition, six job stress clusters pertinent to managerial work were extracted from 30 items which used a 7-point Likert-type scale. These job stressors were initially identified in the literature and were further refined through the factor analysis results. Internal reliability coefficients for all scales ranged between .58 and .86. Table 1 presents the alphas, means, standard deviations and observed ranges of all scales. The descriptive statistics showed normal distributions for all independent variables.

Dimensions of Personality

Four personality dimensions were retained. A decision had been made to avoid using elaborate and lengthy measures of personality in order to save time.

Since prior experience indicates that senior executives would not cooperate if requested to spend too much time with the computer, a short assessment of some common personality traits was used. Of the 30 initial items, using a 4-point Likert-type scale and following a factorial analysis, only 11 items were retained to denote the following dimensions: Self-esteem, Locus of control, Group vs. self-orientation and Risk vs. security orientation. These are commonly cited by researchers to denote predisposition or proneness to burnout (see, for example, Arsenault, Dolan, & van Ameringen, 1991; Dolan, van Ameringen, Corbin, & Arsenault, 1992; Ganster & Schaubroeck, 1994). Table 1 provides a summary for the alphas, means, standard deviations, and ranges for these personality dimensions.

Social Support

Three dimensions of social support—superior, colleagues, and off-work (i.e., family or friends)—were used here. The same measures have been used extensively by the authors in previous research (see, for example, Arsenault, Dolan, & van Ameringen, 1988; Dolan, van Ameringen, & Arsenault, 1992). While internal reliability in other populations for these variables was significantly higher, Table 1 shows alphas ranging from .69 to .85 in this sample.

Burnout

Burnout was measured with the Maslach Burnout Inventory (MBI; Maslach & Jackson, 1986). The MBI is a 22-item measure which produces three scores: emotional exhaustion, depersonalization and lack of personal accomplishment. The measures and their respective reliabilities, means, standard deviations and ranges are described in Table 1. It should be noted that the alpha calculated for our managerial population, particularly for depersonalization, is significantly lower in this sample ($\alpha = .55$) than those originally reported by Maslach and Jackson (1986) and also Dolan and van Ameringen (1989) for other populations.

Statistical Treatment

Direct effects: Stepwise multiple regression procedures were used for each dimension of burnout. The situational and personality variables were regressed in combination in order to test the overall effect.

Moderator effects: As mentioned earlier, controversy about social support as a moderator variable does exist, perhaps due to divergent methodologies and statistical treatments (Aneshensel & Stone, 1982). The buffering effect of each source of social support for each predictor was assessed in this study through a two-step hierarchical regression procedure. The first step was pursuant to the results of the "Direct Effect" regression analyses. Only predictors that contributed to the explained variance were retained. Subsequently, each of these job/organizational variables and each source of social support was forced into the hierarchical regression procedure. If the results per each equation were significant, a second step followed. In this second step, the interaction term (i.e., Predictor x Social

TABLE 1   Descriptive Statistics for All Variables Used in the Study

	No. of items	alpha	Mean	SD	Range Min	Max
Organizational Stressors						
Top-down communication problems	3	.64	.28	.34	0	1
Interpersonal and cohesion problems	4	.62	.29	.31	0	1
Lack of organizational commitment	4	.58	.13	.22	0	1
Job Stressors						
Work overload	7	.86	4.0	1.48	1	7
Administrative constraints	3	.72	2.3	1.2	1	7
Role ambiguity	4	.71	2.4	1.3	1	7
Obstacles to task performance	4	.66	2.1	1.0	1	7
Responsibility	3	.66	2.4	1.4	1	7
Doubts and ambiguities regarding competency	3	.58	2.5	1.2	1	7
Social Support						
Superiors	4	.85	2.9	.81	1	4
Colleagues	4	.69	3.0	.60	1.3	4
Spouse/Friends	4	.74	3.5	.58	1.3	4
Personality						
Self-esteem	3	.73	2.7	.81	1	5
Locus of control	3	.60	2.3	.68	1	4
Group vs. self-orientation	3	.64	2.4	.81	1	5
Risk vs. security orientation	2	n.a.	2.6	.80	1	5
Burnout						
Emotional exhaustion	9	.90	15.1	10.5	0	46
Depersonalization	5	.55	5.8	4.2	0	23
Personal accomplishment	8	.70	35.7	7.5	0	48

Note: All scores on scales, except for burnout, were calculated by linear addition of scores for all items and subsequent division by the number of items in the scale. Scores on the burnout scales, however, were not converted in order to make comparisons with other published reports as well as with the authors' previous work.

TABLE 2   Results of Product Moment Correlation Analyses

	Emotional Exhaustion	Depersonalization	Personal Accomplishment
Organizational Stressors			
Top-down communication problems	.23	.14	-.03[ns]
Interpersonal and cohesion problems	.30	.13	-.12
Lack of organizational commitment	.31	.26	-.03[ns]
Job Stressors			
Work overload	.51	.22	-.02[ns]
Administrative constraints	.37	.20	-.04[ns]
Role ambiguity	.36	.22	-.22
Obstacles to task performance	.25	.23	-.05[ns]
Responsibility	.53	.30	-.14
Doubts and ambiguities regarding competency	.43	.23	-.10[ns]
Personality			
Self-esteem	.48	.30	-.19
Locus of control	.25	.13	-.15
Group vs. self-orientation	.45	.35	-.17
Risk vs. security orientation	.44	.26	-.21

Note: All correlations are significant at the $p < .05$ level, except for those marked with [ns].

support) was forced into the equation as another independent variable. According to recent write-ups, the buffering effect is evident only when the interaction term is significant (Arnold 1982; Evans, 1991). In other words, if the interaction is adding to the R^2 beyond the contribution of the main effects (i.e., step 1), then one can conclude that social support plays a moderating role (see, also, Cohen & Cohen, 1975). In a similar manner, the moderating effects of social support as a function of the various personality dimensions was tested (i.e., the main effects of personality were removed first). All in all, 39 equations/combinations were tested (i.e., 8 x 3 for Emotional exhaustion; 3 x 3 for Depersonalization; and 2 x 3 for Personal accomplishment; see Table 3).

RESULTS

Table 2 lists the bivariate product moment correlations for all variables in the study. Results show that all predictors (situational and personal) are correlated with the first two

TABLE 3   Stepwise Multiple Regression Analyses

		Emotional Exhaustion		Depersonali- zation		Personal Accom- plishment	
		ΔR^2	R^2	ΔR^2	R^2	ΔR^2	R^2
1	Responsibility	.29	.29				
2	Self-esteem	.11	.40				
3	Work overload	.06	.46				
4	Group vs. self orientation	.05	.51				
5	Lack of org. commitment	.02	.53				
6	Doubts and ambiguities regarding competency	.01	.54				
7	Risk vs. security orientation	.01	.55				
8	Locus of control	.01	.56				
	(F = 83.16; p < .01)						
1	Group vs. self orientation			.11	.11		
2	Self-esteem			.06	.17		
3	Lack of org. commitment			.04	.21		
	(F = 26.00; p < .01)						
1	Role ambiguity					.05	.05
2	Risk vs. security orientation					.02	.07
	(F = 8.35; p < .01)						

dimensions of burnout, namely emotional exhaustion and depersonalization. With regard to personal accomplishment, it seems that there is a moderate/low correlation with all personality traits, yet almost no correlations were found with the situational variables (except for three variables). In relative terms, the highest significant correlations amongst all variables are those associated with emotional exhaustion. For instance, three situational variables, responsibility (r = .53), work overload (r = .51) and doubts and ambiguities regarding competency (r = .43), all have correlations surpassing .40. Similarly, three personality dimensions, low self-esteem (r = .48), orientation towards self (r = .45) and orientation for security (r = .44) also have moderate/high correlations with emotional exhaustion. In analyzing the depersonalization scale, the magnitude of the correlations is significantly lower, notwithstanding the fact that they are all significant at the p < .05 level. And finally, personal accomplishment is correlated with only three of the nine situational stressors: interpersonal problems (r = -.12), role ambiguity (r = -.22), and responsibility (r = -.14), yet it is correlated with personality traits.

TABLE 4   Hierarchical Regression Analyses

		Emotional Exhaustion	
Step	Variables in the Equation	ΔR^2	R^2
Moderated Case I			
1	Responsibility Superior Social Support	.308	.308
2	Responsibility **x** Superior SS	.014	.322
	(F = 33.27; p < .01)		
Moderated Case II			
1	Work Overload Spouse/Friends Social Support	.295	.295
2	Work Overload **x** Spouse/Friends SS	.015	.300
	(F = 30.43; p < .01)		

Table 3 presents the complete/saturated multiple regression model whereby the situational and the personality factors combined were introduced into the regression equations for each of the dimensions of burnout. The explained variance found for the emotional exhaustion measure ($R^2 = .56$) is considered to be extremely high considering the individual level of analysis. Altogether, 8 variables entered as predictors, 4 of which are situational and 4 of which represent personality traits. Nonetheless, in relative terms, the situational factors explain most of the variance of emotional exhaustion (36%). For the depersonalization scale, its explained variance is more modest ($R^2 = .21$). It is the personality traits that first enter the equation, and also explain most of the variance (15%). Predicting personal accomplishments, the third dimension of burnout is more limited, since only 7% of the variance is explained and only two variables entered the equation which could be significant by chance alone.

In order to test the role of social support as a moderator, 39 separate hierarchical regression analyses were performed. Yet, out of all these analyses, in only two cases was social support found to play a moderating role (Table 4). Although problems related to responsibility are the prime contributor to emotional exhaustion (Table 3), support and understanding on behalf of a superior seems to buffer it. Similarly, social support out of work (i.e., family and friends) acts as a buffer for emotional exhaustion resulting from work overload. Given the multiple tests, these results could be chance.

DISCUSSION

The results provide substantial support to analyzing managerial burnout from a multivariate perspective. Both personal and job/organizational characteristics relate to executive burnout. The distinctive pattern of the relationships found suggests that, in relative terms, the job/organizational factors are more significant in predicting two sub-scales of burnout, emotional exhaustion and personal accomplishment. To the contrary, personality factors have higher predictive validity for the depersonalization scale. Consistent with previously reported findings, of the three burnout sub-scales, the one which was best explained throughout the different analyses is the emotional exhaustion subscale. Even when considering the limitations of the study, namely that it is cross-sectional in nature and that some potential problems connected with "method-variance" might apparently be present, the extremely high R^2 (.56) reported for predicting emotional exhaustion, especially when data were analyzed at an individual level of aggregation, lends further support to the validity of the approach.

In more specific terms, one can discuss the findings pertaining to each of the sub-scales of burnout. For example, the results for emotional exhaustion are not surprising. It has already been reported that executives view job responsibility, and the accountability that goes with it, as the single most demanding aspect of their work. Other studies confirm the direct link between the latter and burnout (i.e., Clark & Vaccaro, 1987; Lee & Ashforth, 1993a; Rosch, 1987). Energy is depleted and emotional exhaustion becomes even more severe under conditions where an executive is embroiled in the following circumstances: low self-esteem (Golembiewski & Byong-Seob, 1989; Fimian, 1988); high work load (Himle et al., 1989; Kaufmann & Beehr, 1989); a personality trait of being self-centered/oriented (Josefowitz, 1985); low organizational commitment (Leiter, 1988; Leiter & Maslach, 1988; Perrewé, Rotondo, & Morton, 1993); doubts and ambiguities regarding self-competency (Meir, Melamed, & Abu-Freha, 1990); a security oriented trait (Harris, 1984); and leaning towards an external locus of control (Fimian, 1988). The fact that a mix of 8 variables, organizational and personal, predicts emotional exhaustion shows the complexity of this phenomenon. It also suggests that no single factor can be attributed as the sole agent causing the burnout syndrome. This multi-facet view contradicts the various workers' compensation boards' models which adhere to a classical medical model whereby a single agent/cause is responsible for the development of a disease or a condition. The time has come for them to realize that we need to develop a more sophisticated view and corresponding analysis in order to understand the full complexity of a mental health phenomenon such as burnout. Moreover, it should be acknowledged that the variables identified for a managerial population might differ for other occupations without rendering the model less valid.

The findings for personal accomplishment, although consistent with previously published results, are much more limited. First, only 7% of the variance is accounted for, 5% of which is explained by role ambiguity. The latter has been previously reported as a predictor of burnout for many other professions (Burke, 1982; Dolan & Zeilig, 1994; Himle et al., 1989). For managers, Quick and Quick (1984) argue that vague job demands are a major de-motivating force. Yet, the higher the position the more difficult it becomes to clearly define it. In fact, very few job descriptions exist for CEOs and presidents of organizations due to the complexity of their jobs. Whether or not the job becomes stressful depends on one key personality trait of the manager: whether he/she is able or willing to take risks. If the answer is negative, scores on this sub-scale (i.e., personal accomplishment) will be significantly lower (i.e., this is an inverse scale).

The findings for depersonalization are somewhat problematic. The fact that the two critical determinants of it are individual traits (i.e., high self-centered and low self-esteem) is made more difficult to interpret due to the fact that the depersonalization scale is not always pertinent for managerial populations. This sub-scale also has the lowest empirical internal consistency amongst all dimensions of burnout studied here (alpha = .55). Thus, results are confounded with methodological and conceptual problems and are not easy to interpret.

The results pertaining to the test of social support as a buffer/moderator to burnout seem to be one of the cornerstone findings for this study. The results provide some limited clarifications to the relationships between personal attributes, organizational determinants and burnout, and the role of social support for executives. All in all, social support plays a marginal role in buffering burnout. Of all possible interactions where social support has been tested, none seem to emerge for depersonalization and personal accomplishment, and in only two particular paths/cases does social support play a moderating role. Similar findings, though for a different population (i.e., female accountants) have been recently reported (Dolan & Zeilig, 1994). Evidently, these findings contradict the prevailing concept in the literature for the role attributed to social support. On the other hand, the observations might only characterize managers and executives who may act as if social support is not needed in their case. A common stereotypical perception among managers is that the mere search for support will be interpreted by colleagues, subordinates and/or superiors as a weakness. Thus, in our culture, managers will rather keep a facade of a "macho" (superman) image, rather than admit that support could benefit them. Quick and Quick (1990) have labelled this phenomenon as "the paradox of the successful executive," where, in spite of the fact that they are surrounded by people and have multiple contacts, they eventually have to make decisions alone and ultimately have to face the responsibility stemming from those decisions. It should also be fair to say that, in many instances, executives feel very isolated and have no one to turn to in seeking support.

While high responsibility has been found to be significantly correlated with emotional exhaustion, Table 4 suggests that superior social support may buffer it. Thus, executives might be strongly advised to actively seek support from their respective superiors. Such support might diminish the feeling of emotional exhaustion, and, in the long run, might prove instrumental in maintaining mental as well as physical health. Work overload is another case in point. While the vast majority of managers will not admit their incapabilities of handling excessive work load within the work settings, it was found that friends and family might help in alleviating some of the burden. There are two reasons for executives to turn to family/friends for support and understanding: 1) it is considered to be non-threatening; an admission of not being able to handle workload to outsiders may not cause any harm and may not reduce the image of the executive; 2) obtaining support and gaining sympathy on the part of a spouse, for instance, can justify the remaining of an extra hour (or more) at work in order to complete unfinished business. The contrary is also true to the extent that many executives complain that their spouse and friends do not understand their need to spend much more time at the office. This has been previously reported to increase incidents of role conflict as well as work-family conflict, and eventually contribute to greater emotional exhaustion (Burke, 1993; Izraeli, 1993).

CONCLUSION

Although burnout is a unique type of stress syndrome, this study shows that the most valid indicator for managerial populations is the emotional exhaustion component. The determinants of emotional exhaustion reflect both organizational and personal demands placed on the manager, and the best approach to understanding it is via a multivariate model. The model assumes that a complex mix of organizational and personal factors results in burnout, although reverse causality should be investigated. The paper further suggests that instant assumptions regarding the role of social support in moderating managerial burnout are not accurate. The study of social support is a complex undertaking due to the multiple ways in which it can affect the availability of coping resources. Nonetheless, the findings reported here should be treated with caution due to a number of reasons: (a) the restricted methodology used (i.e., measures of self-report and the cross-sectional nature); (b) the choice of personality traits and their respective measurement (i.e., very short inventories); and (c) the omission of other potential antecedents of burnout (i.e., job, organizational and personal). Studies on the same or similar subjects, using slightly different conceptual and operational measures, have not entirely supported the findings reported here. In one example where supervisory personnel were studied, the authors show that work autonomy and social support were the two most significant variables explaining exhaustion through role stress. However, the same study found that exhaustion can be treated as a moderator variable between these three antecedents and the two other subscales of burnout (Lee & Ashforth, 1993a, b). Furthermore, in order to demonstrate the importance of demographic antecedents of supervisory burnout, the authors also report a significant variation in the results when controlling for the supervisors' experience, life satisfaction and time spent with clients and subordinates. This is consistent with other studies which report significant relationships between demographic variables such as gender, age, family status, experience, and the burnout components.

REFERENCES

American Psychiatric Association. (1987). *Diagnostic and statistical manual of mental disorders* (3rd ed., rev.). Washington, DC: Author.

Aneshensel, C.S., & Stone, J.D. (1982). Stress and Depression: A test of the buffering model of social support. *Archives of General Psychiatry, 39,* 1392–1396.

Arnold, H.G. (1982). Moderator variables: A classification of conceptual analytic and psychometric issues. *Organizational Behavior and Human Performance, 2A,* 143–174.

Arsenault, A., Dolan, S.L., & van Ameringen, M.R. (1988). *An empirical examination of the buffering effects of social support on the relationships between job demands and psychological strain* (pp. 16–18). Proceedings of the Eastern Academy of Management.

Arsenault, A., Dolan, S.L., & van Ameringen, M.R. (1991). Stress and mental strain in hospital work: Exploring the relationship beyond personality. *Journal of Organizational Behavior, 12,* 483–493.

Barone, D.F. (1994). Work stress conceived and researched transactionally. In P.L. Perrewé & R. Crandall (Eds.), *Occupational stress: A Handbook* (pp. 29–37). Washington, DC: Taylor & Francis. (Original work published 1991)

Barrera, M., Jr. (1988). Models of social support and life stress: Beyond the buffering hypothesis. In L.H. Cohen (Ed.), *Life events and psychological functioning: Theoretical and methodological issues.* Newbury Park, CA: Sage.

Beehr, T.A., King, L.A., & King, D.W. (1990). Social support and occupational stress: Talking to supervisors. *Journal of Vocational Behavior, 36,* 61–81.

Beehr, T.A., & Newman, J.E. (1978). Job stress, employee health, and organizational effectiveness: A facet analysis, model, and literature review. *Personnel Psychology, 31,* 655–699.

Burke, M.J., Brief, A.P., & George, J.M. (1993). The role of negative affectivity in understanding relationships between self-reports of stressors and strains: A comment on the applied psychology literature. *Journal of Applied Psychology, 78*(3), 402–412.

Burke, R.J. (1982). Impact of occupational demands on nonwork experiences of senior administrators. *Journal of Psychology, 112*(20), 195–211.

Burke, R.J. (1993). Toward an understanding of psychological burnout among police officers. *Journal of Social Behavior and Personality, 8*(3), 425–438.

Caplan, R.D., Cobb, S., French, J.R.P., Jr., Van Harrison, R., & Pinneau, S.R., Jr. (1975). *Job demands and worker health: Main effects and occupational differences* (HEW NIOSH No. 75–160). Washington, DC: U.S. Government Printing Office.

Carroll, J.F.X., & White, W.L. (1982). Theory building: Integrating individual and environmental factors within an ecological framework. In W.S. Paine (Ed.), *Job stress and burnout* (pp. 41–60). Newbury Park, CA: Sage.

Clark, G.H., & Vaccaro, J.V. (1987). Burnout among CMHC psychiatrists and the struggle to survive. *Hospital and Community Psychiatry, 38*(8), 843–847.

Cobb, S. (1976). Social support as a moderator of life stress. *Psychosomatic Medicine, 38*(1), 300–314.

Cohen, J., & Cohen, P. (1975). *Applied multiple regression/correlation analysis for the behavioral sciences.* Hillsdale, NJ: Erlbaum.

Cohen, S., & Syme, S.L. (1985). Issues in the study and application of social support. In S. Cohen & S.L. Syme (Eds.), *Social support and health.* New York: Academic Press.

Cooper, C.L., & Marshall, J. (1976). Occupational sources of stress: A review of the literature relating to coronary heart disease and mental health. *Journal of Occupational Psychology, 49,* 11–28.

Cordes, C.L., & Dougherty, T.W. (1993). A review and an integration of research on job burnout. *Academy of Management Review, 18*(4), 621–656.

Dolan, S.L., & Balkin, D. (1987). A contingency model of occupational stress. *International Journal of Management, 4*(3), 328–340.

Dolan, S.L., & van Ameringen, M.R. (1989). *Etude sur le stress et la qualité de vie au travail chez les substituts du procureur général du Québec.* Unpublished research report. Montreal: MDS Management.

Dolan, S.L., van Ameringen, M.R., & Arsenault, A. (1992). The role of personality and social support in the etiology of workers' stress and psychological strain. *Industrial Relations, 47*(1), 125–139.

Dolan, S.L., van Ameringen, M.R., Corbin, S., & Arsenault, A. (1992). Lack of professional latitude and role problems as correlates of propensity to quit amongst nursing staff. *Journal of Advanced Nursing, 17,* 1455–1459.

Dolan, S.L., & Zeilig, P. (1994, June 25–28). *Occupational stress, emotional exhaustion and propensity to quit amongst female accountants: The moderating role of mentoring.* Paper presented at the Annual Meeting of the Administrative Sciences Association of Canada, Halifax, N.S.

Evans, M.G. (1991). On the use of moderated regression. *Canadian Psychology, 32*(2), 116–119.

Fimian, M.J. (1988). Predictors of classroom stress and burnout experienced by gifted and talented students. *Psychology in the Schools, 25*(4), 392–405.

Ganster, D.C., & Schaubroeck, J. (1994). The moderating effects of self-esteem on the work stress–employee health relationship. In P.L. Perrewé & R. Crandall (Eds.), *Occupational stress: A handbook* (pp. 167–177). Washington, DC: Taylor & Francis. (Original work published 1991)

Garden, A.M. (1994). The purpose of burnout: A Jungian interpretation. In P.L. Perrewé & R. Crandall (Eds.), *Occupational stress: A handbook* (pp. 207–222). Washington, DC: Taylor & Francis. (Original work published 1991)

Golembiewski, R.T., & Byong-Seob, K. (1989). Self-esteem and phases of burnout. *Organization Development Journal, 7*(1), 51–58.

Greenglass, E.R. (1991). Burnout and gender: Theoretical and organizational implications. *Canadian Psychology, 32*(4), 562–572.

Hammer, M. (1981). Social supports, social networks, and schizophrenia. *Schizophrenia Bulletin, 17,* 45–57.

Handy, J.A. (1988). Theoretical and methodological problems within occupational stress and burnout research. *Human Relations, 41*(5), 351–369.

Harris, J.R. (1994). An examination of the transaction approach in occupational stress research. In P.L. Perrewé & R. Crandall (Eds.), *Occupational stress: A handbook* (pp. 21–28). Washington, DC: Taylor & Francis. (Original work published 1991)

Harris, P.L. (1984). Assessing burnout: The organizational and individual perspective. *Family and Community Health, 6*(4), 32–43.

Himle, D.P., Jayaratne, S., & Thyness, P.A. (1989). The buffering effects of four types of supervisory support on work stress. *Administration in Social Work, 13*(1), 19–34.

House, G.S. (1981). *Work stress and social support.* Reading, MA: Addison-Wesley.

Izraeli, D.N. (1993). Work/family conflict among women and men managers in dual-career couples in Israel. *Journal of Social Behavior and Personality, 8*(3), 371–388.

Josefowitz, N. (1985). *You're the boss!* New York: Warner.

Kaufmann, G.M., & Beehr, T.A. (1989). Occupational stressors, individual strains, and social supports among police officers. *Human Relations, 42*(2), 185–197.

Lazarus, R.S. (1994). Psychological stress in the workplace. In P.L. Perrewé & R. Crandall (Eds.), *Occupational stress: A handbook* (pp. 3–14). Washington, DC: Taylor & Francis. (Original work published 1991)

Lazarus, R.S., & Folkman, S. (1984). *Stress, appraisal, and coping.* New York: Springer.

Lee, R.T., & Ashforth, B.E. (1993a). A further examination of managerial burnout: Toward an integrated model. *Journal of Organizational Behavior, 14,* 3–20.

Lee, R.T., & Ashforth, B.E. (1993b). A longitudinal study of burnout among supervisors and managers: Comparisons between the Leiter and Maslach (1988) and Golembiewski et al. (1986) models. *Organizational Behavior and Human Decision Processes, 54,* 369–398.

Leiter, M.P. (1988). Commitment as a function of stress reactions among nurses: A model of psychological evaluations of work settings. *Canadian Journal of Community Mental Health, 7*(1), 117–133.

Leiter, M.P., & Maslach, C. (1988). The impact of interpersonal environment on burnout and organizational commitment. *Journal of Organizational Behavior, 9*(1), 297–308.

Lippell, K. (1988). *Guide d'orientation sur l'indemnisation des dommages reliés au stress au travail.* IRSST, Montreal.

Maslach, C., & Jackson, S.E. (1986). *Maslach Burnout Inventory Manual* (2nd ed.). Palo Alto, CA: Consulting Psychologists Press.

Meir, E.I., Melamed, S., & Abu-Freha, A. (1990). Vocational, avocational, and skill utilization congruences and their relationship with well-being in two cultures. *Journal of Vocational Behavior, 36*(2), 153–165.

Paine, W.S. (Ed.). (1982). *Job stress and burnout.* Newbury Park, CA: Sage.

Payne, R. (1984). Job stress and burnout: Research theory and intervention perspectives. *Journal of Occupational Psychology, 57,* 175–176.

Payne, R., Jabri, M.M., & Pearson, A.W. (1988). On the importance of knowing the affective meaning of job demands. *Journal of Organizational Behavior, 9,* 149–158.

Perrewé, P.L., Rotondo, F.D., & Morton, K.S. (1993). An experimental examination of implicit stress theory. *Journal of Organizational Behavior, 14,* 677–686.

Pines, A.M., & Maslach, C. (1978). Characteristics of staff burnout in mental health setting. *Hospital and Community Psychiatry, 29,* 233–237.

Quick, J.C., Nelson, D.L., & Quick, J.D. (1990). *Stress and challenge at the top: The paradox of the successful executive.* New York: John Wiley & Sons.

Quick, J.C., & Quick, J.D. (1984). *Organizational stress and preventive management.* New York: McGraw-Hill.

Rosch, P.J. (1987). Dealing with physician stress. *Medical Aspects of Human Sexuality, 21*(4), 73–93.

Shirom, A. (1989). Burnout in work organizations. In C.L. Cooper & I. Robertson (Eds.), *International review of industrial and organizational psychology* (pp. 25–48). London: Wiley.

Spector, P.E. (1987). Method variance as an artifact in self-reported affect and perceptions at work: Myth or significant problem? *Journal of Applied Psychology, 72,* 438–443.

Spector, P.E., Dwyer, D.J., & Jex, S.M. (1988). Relation of job stressors to affective, health, and performance outcomes: A comparison of multiple data sources. *Journal of Applied Psychology, 73,* 11–19.

Williams, L.J., Cote, J.A., & Buckley, M.R. (1989). The lack of method variance in self-report affect and perceptions at work: Reality or artifact. *Journal of Applied Psychology, 74,* 462–468.

The Relationship Between Social Support and Burnout Over Time in Teachers

Esther R. Greenglass
Lisa Fiksenbaum
Ronald J. Burke

Research evidence to date suggests that teacher burnout in several Western industrialized countries has increased and affects a significant proportion of teachers (Shirom, 1989). Burnout may be defined as a state of physical, emotional and mental exhaustion that results from long-term involvement with people in situations that are emotionally demanding (Pines & Aronson, 1981). In line with Maslach and Jackson (1981), we view burnout as a syndrome of emotional exhaustion, depersonalization, and reduced personal accomplishment occurring among individuals who work with people. The dominant source of stress for teachers seems to be the quality of their interpersonal relations, especially with their pupils. Interpersonal relationships with supervisors and colleagues can also be factors (Kyriacou & Sutcliffe, 1977; Wallius, 1982).

Research (Jackson, Schwab, & Schuler, 1986) has yielded a comprehensive model of both sources and consequences of psychological burnout in elementary and secondary school teachers. Sources of burnout in this research were found to

Authors' Notes: Paper presented at the annual meeting of the Canadian Psychological Association, Montreal, May 27-29, 1993.

Grateful acknowledgment is due to York University, Faculties of Arts and Administrative Studies for their support and to Mirka Ondrack for assistance in the data analyses.

include a combination of the individual's unmet expectations and job conditions such as low participation in decision making, high levels of role conflict, a lack of freedom and autonomy and absence of social support networks. According to the model put forth by Cherniss (1980), sources of stress can also function as antecedents of burnout. For Cherniss, these factors include doubts about competence, problems with clients or pupils, lack of stimulation and lack of collegiality. As Cherniss suggests in his model, the level of burnout experienced by individuals is a function of their demands and supports.

Burnout as a Process

Rather than viewing burnout as a reaction to a specific stressful event, researchers increasingly are viewing burnout as a process which emerges gradually over a period of time in reaction to several stressful events (Burke & Greenglass, 1991; Capel, 1991; Cherniss, 1980; Greenglass & Burke, 1990; Leiter, 1990; Wade, Cooley, & Savicki, 1986). Such a model assumes a causal sequence of antecedent conditions between work demands and psychological reactions, i.e., burnout. At the same time, it is expected that certain resources, such as social support, can exert a long-term effect on ameliorating or lowering burnout at a later date.

A process model of burnout should make it possible to study factors that predict burnout over time. Data from Greenglass and Burke (1990) lend support to studying burnout as a process, since results indicated that certain stresses associated with one's job as a teacher were significant contributors to burnout. Wade, Cooley, and Savicki (1986) conducted a one year longitudinal study of burnout to determine how burnout changed over time. Burnout was found to be relatively stable and consistent over time. Factors measured at the original testing which were found to predict higher burnout were less social support from peers and supervisors at work, a more controlling work setting and an attitudinal style in which the individual was more likely to accept personal responsibility for the outcome of clients and to blame oneself when clients did not improve as expected. The relative stability of burnout scores over time has been reported in at least two samples of teachers studied over one school year (Capel, 1991; DePaepe, French, & Lavay, 1985). Burke and Greenglass (1991) also report on the relatively high stability of burnout over time as did Golembiewski and Munzenrider (1988).

Social Support

A variety of studies have found that higher levels of available support from supervisors and administration are associated with lower levels of burnout (Barad, 1979; Dignam, Barrera, & West, 1986; Jayaratne & Chess, 1984). Support from co-workers has been shown to be correlated inversely with burnout (Eisenstat, Felner, Kennedy, & Blank, 1981; Jayaratne & Chess, 1984; Maslach, 1976). Leiter (1990) reported that family resources complemented professionally based resources to alleviate burnout in a sample of mental health workers. It would be expected that social support provided over time should contribute to the lowering of burnout by providing resources for the effective handling of stressful situations which otherwise would cumulatively lead to high burnout levels. This support may be provided by one's supervisor, co-workers or friends and family.

The importance of incorporating family support into burnout research is underlined by the increasing recognition of the reinforcing roles of occupational and familial spheres (Burke & Greenglass, 1986). Marital satisfaction may be regarded as one indicator of the extent to which a person experiences family support and thus may be a predictor of burnout level. Maslach and Jackson (1982), in their research with human service workers, reported that low marital satisfaction was associated with higher burnout.

The Present Study

The present research employs a process model of burnout in order to assess factors related to burnout over time in teachers. It is expected that various sources of stress associated with the classroom should significantly predict a teacher's burnout level. If burnout is a process that develops over time, it is also expected that the level of social support experienced during the previous year should negatively predict the teacher's burnout level. Such support may be from one's supervisor, co-workers, friends, or family members.

METHOD

Respondents

Respondents were 179 women and 182 men employed within a single school board in a large Canadian city. All respondents were employed full-time. While the majority of both the men and women were full-time teachers, 10% of the men and less than 1% of the women were principals of their schools. Approximately one-third of the sample worked in an elementary school, one-quarter to one-third were from the junior high school level and approximately one-third were from second-ary schools. Two-thirds of the men and one-half of the women had graduate or professional degrees; 37% and 27% of the women and men, respectively, had only an undergraduate university degree. The remainder of the sample had some university education or teachers' college.

More men than women were married and living with their spouses (79.7% versus 59.2%, respectively); twice as many women (39.7%) as men (20.3%) were single, separated/divorced or widowed. More men had children than women, and a greater proportion of the men (22.3%) reported that they had children under five years of age than did women (13.6%). A t-test on mean age between men and women showed no significant difference in age (t = -0.80, df = 357, p > .05). The average age was 44 years for the sample.

Procedure

Data were collected using a mail-out questionnaire. There was a total of 361 respondents. In the second year, 833 questionnaires were sent and 473 were returned, giving a response rate of 57%. Of these, 361 respondents had also returned a questionnaire one year earlier.

Measures

Independent variables. Work setting stress was one of the independent variables included in year 2 of this study. Cherniss (1980) identified five sources of stress as antecedents of burnout. These were measured by Burke, Shearer, and Deszca (1984) and include Doubts about Competence (based on three items), Problems with Clients (based on five items), Lack of Stimulation and Fulfillment (three items), Bureaucratic Interference (five items), and Lack of Collegiality (four items). For each of the items, respondents were asked to indicate their degree of agreement on a 7-point scale ranging from 1 (strongly agree) to 7 (strongly disagree). A sample item is "When I began my present job there was a time I felt I was falling short." Burke et al. (1984) provide a more detailed description of these measures.

Additional measures of independent variables include social support (from year 1) and marital satisfaction (in year 2). Social support measures were modified versions of those employed by Caplan, Cobb, French, Van Harrison, and Pinneau (1975). They examined the availability of three types of social support collapsed in one measure: Practical, e.g., "How much can [people][1] be relied on to provide you with assistance when you really need it the most?"; Emotional, e.g., "How much do [people][1] boost your spirits when you feel low?"; and Informational, e.g., "How much useful information does each of these [people][1] provide you with when you really need it most?" Three items assessing each of the above types of support were used to measure social support from one's boss, from co-workers, and from family and friends.

Marital satisfaction in year 2 was measured by a scale developed by Orden and Bradburn (1968) who report that the marital happiness score is positively related to overall self-reported happiness ratings. They asked respondents to indicate how frequently they had undertaken each of nine positive or pleasant activities with their spouse (or partner) as well as how frequently they had disagreed with their spouse (or partner) in nine different areas during the past few weeks. Marital satisfaction scores include a weighing of the positive minus the negative scores.

Dependent variable. The dependent variable, burnout, was measured by the Maslach Burnout Inventory (MBI; Maslach & Jackson, 1986) in the second year. This measure yields scores on three scales: Emotional Exhaustion, Depersonalization, and Lack of Personal Accomplishment. Maslach and Jackson (1981)

---

[1] *For "people" the words, "your immediate supervisor", "other people at work", or "your spouse, friends and relatives" were substituted, depending on the source of the support being assessed.*

TABLE 1   Composite Variables, No. of Items and Alphas (α)

Variable	No. of Items	α
Burnout (year 1)	22	.70
Supervisor Support (year 1)	3	.95
Co-worker Support (year 1)	3	.92
Family and Friend Support (year 1)	3	.89
Burnout (year 2)	22	.68
Cherniss-Sources of Stress (year 2)	5	.65
Marital Satisfaction (year 2)	11	.80

report evidence for the validity of the scale. The inventory provides for measures of frequency and intensity. However, only frequency was used in this study since Gaines and Jermier (1983) found that frequency and intensity were highly intercorrelated. They suggest that only the frequency measure be used in future research. Burnout scores in the first year were also recorded. See Table 1 for the reliability coefficients (α) for the composite variables.

Demographic Variables

Demographic variables assessed included job-related variables such as current position, years in current position, years teaching, hours worked per week, level of school, and number of teachers and students in the school. Additional demographic variables included age, marital status, and children.

RESULTS

Demographic Characteristics

T-tests on work-related variables between women and men showed significant differences on only two variables: number of years teaching ($t = -3.32$, $df = 356$, $p < .01$) and hours worked/per week ($t = -2.80$, $df = 351$, $p < .01$)—men were significantly higher on both variables.

A multiple regression analysis was performed in which the criterion was year 2 burnout. Predictors were year 2 marital satisfaction, year 2 sources of stress, year 1 family and friend support, support x marital satisfaction interaction, and support x sources of stress. Results indicated no significant main effects. However, a significant effect was found due to the interaction between support provided by families and friends at year 1 and sources of stress at year 2 (see Table 2).

Figure 1 presents the significant 2-way interaction in graph form. As sources of stress increased, burnout scores increased. However, with low support, respondents' burnout scores appeared to increase more rapidly as a function of stress than among respondents reporting moderate or higher social support. Further multiple

TABLE 2   Multiple Regression Analysis for Total Sample. Criterion: Overall
MBI Scores (Year 2)

Predictors	R Square	B	t
Model	0.42		
Marital Satisfaction (year 2)		-2.35	-0.98
Sources of Stress (year 2)		-0.09	-0.19
Family and Friend Support (year 1)		-1.35	-1.43
Family and Friend Support x Marital Satisfaction		0.03	0.43
Family and Friend Support x Sources of Stress		0.03	2.02*

*$p < .05$

regression analyses showed a nonsignificant 3-way interaction in year 2 burnout between respondents' sex, sources of stress, and year 1 support from family and friends. Thus, the significant 2-way interaction observed between year 1 support and year 2 sources of stress is not differentially affected by the sex of the teacher.

Social Support, Marital Satisfaction, and Sources of Stress
Intercorrelations were computed separately for women and men among the various measures. Co-worker support was found to be positively and significantly correlated with family and friend support in both women and men. Marital satisfaction was positively and significantly correlated with family and friend support in both women and men. Sources of stress were found to be negatively and significantly correlated with supervisor support and co-worker support. These results were found in both women and men. Significant and negative correlations were found between sources of stress and with family and friend support and with marital satisfaction in women only (see Tables 3 and 4).

Burnout, Support, and Stress
Correlations were computed separately for women and men between year 2 total burnout scores and year 2 sources of stress, supervisor support, co-worker support, family and friend support, marital satisfaction, and year 1 burnout. Significant and positive correlations were found between burnout and sources of stress. These results were found in both women and men. Supervisor support correlated significantly and negatively with total burnout scores in women and men. Co-worker support correlated significantly and negatively with total burnout scores in men only. Marital satisfaction correlated negatively with burnout in both women and men. Highly significant positive correlations were observed in males and females between year 1 and year 2 burnout scores (see Table 5).

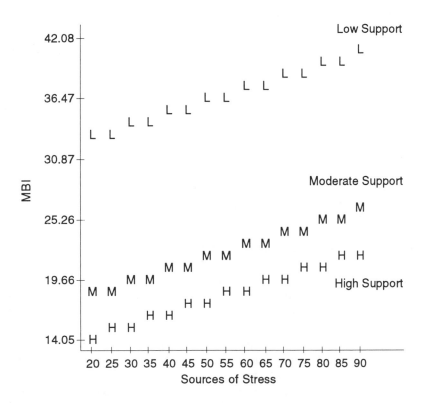

FIGURE 1   Two-way Interaction: Sources of Stress (year 2) x Friend & Family
Support (year 1). Criterion: MBI Scores (year 2)

DISCUSSION

The data from the present study suggest that burnout levels in teachers are
related to perceived work stress. However, additional findings from this study
indicate that a teacher's burnout level depends, as well, on the extent to which he/
she reports receiving social support from friends and relatives one year earlier.
Thus, the data support the idea of conceptualizing burnout as a process. At the same
time, the present data suggest that burnout scores are relatively stable over the time
period of one year, which parallels previous findings (Capel, 1991; DePaepe,
French, & Lavay, 1985). Moreover, the data indicate that there is a significant
interaction between family and friend support (as reported one year earlier) and
stress at work, and its effects on burnout. In the absence of an ongoing supportive
network, burnout increases with stressors in the work environment. With moderate
and higher levels of family and friend support, burnout levels go up less as stress
increases. Groups with more support also have lower overall burnout levels.

TABLE 3   Correlations Between Variables in Female Teachers

	Year 1 Supervisor Support	Year 1 Co-worker Support	Year 1 Family and Friend Support	Year 2 Marital Satisfaction
Year 1 Co-worker Support	.48***			
Year 1 Family and Friend Support	.31***	.44***		
Year 2 Marital Satisfaction	.18	.14	.42***	
Year 2 Cherniss Stress	-.38***	-.31***	-.23**	-.21*

*p < .05; **p < .01; ***p < .001

These data suggest that social support may be exerting an indirect effect on burnout by inoculating respondents against the deleterious effects of work stress. Moreover, these results suggest that social support is functioning here as a buffer. Under conditions of increased support, stressful work experiences are less likely to produce teacher burnout. This explanation parallels results reported by Dignam et al. (1986) suggesting an indirect effect of social support. It will be recalled that social support, as defined in the present study, consists of three dimensions, informational, practical and emotional support. It is possible that information from close others provides the means by which the respondent can cope with stressors in the workplace, in this case, the classroom. In a similar way, practical advice and emotional support can function to provide useful resources which can be employed to cope with stress at work.

The findings of this study build on existing coping theory. According to Folkman and Lazarus (1988), cognitive appraisal plays a central role in coping.

TABLE 4   Correlations Between Variables in Male Teachers

	Year 1 Supervisor Support	Year 1 Co-worker Support	Year 1 Family and Friend Friend Support	Year 2 Marital Satisfaction
Year 1 Co-worker Support	.44***			
Year 1 Family and Friend Support	.14	.26***		
Year 2 Marital Satisfaction	.02	.08	.32***	
Year 2 Cherniss Stress	-.36***	-.45***	-.03	-.13

***p < .001

TABLE 5   Correlations Between Variables and Burnout in Male and Female Teachers

	Year 2 Burnout		
	Total Sample	Females	Males
Year 1 Supervisor Support	-.19**	-.17*	-.21**
Year 1 Co-worker Support	-.15**	-.07	-.23**
Year 1 Family and Friend Support	.02	-.05	.08
Year 2 Marital Satisfaction	-.23***	-.26**	-.21*
Year 2 Cherniss Stress	.59***	.46***	.71***
Year 1 Burnout	.70***	.69***	.73***

*p < .05; **p < .01; ***p < .001*

Two types of cognitive appraisal of potentially stressful encounters are put forth here. First, a primary appraisal is made of the situation's personal significance for well-being. This involves an examination of what is at stake in the situation. Next, a secondary appraisal is made as the individual evaluates the resources available to deal with the situation. According to Folkman, Lazarus, Dunkel-Schetter, DeLongis, and Gruen (1986), these cognitive appraisals are a key determinant of the particular coping strategies that will be used.

The present data support a buffering effect of social support on burnout from family and friends over a period of a year. At the same time, while correlational analyses indicated negative relationships between co-worker and supervisor support and burnout, there was no evidence in this study that this support buffered the individual teacher from work stress to the same extent as family and friend support. Finally, the data did not provide evidence for any gender differences in the effects observed. One may conclude, therefore, that family and friend support provided to teachers one year earlier interacts with work stress on burnout in the same way in men and women teachers.

REFERENCES

Barad, C.B. (1979). *Study of burnout syndrome among social security administration field public contact employees.* Unpublished manuscript, Social Security Administration.

Burke, R.J., & Greenglass, E.R. (1986). Work and family conflict. In C.L. Cooper & I. Robertson (Eds.), *International review of industrial and organizational psychology.* New York: Wiley.

Burke, R.J., & Greenglass, E.R. (1991). A longitudinal study of progressive phases of psychological burnout. *Journal of Health and Human Resources Administration, 13,* 390–408.

Burke, R.J., Shearer, J., & Deszca, E. (1984). Burnout among men and women in police work: An examination of the Cherniss model. *Journal of Health and Human Resources Administration, 7,* 162–188.

Capel, S.A. (1991). A longitudinal study of burnout in teachers. *British Journal of Educational Psychology, 61,* 36–45.

Caplan, R.D., Cobb, S., French, J.R.P., Jr., Van Harrison, R., & Pinneau, S.R., Jr. (1975). *Job demands and worker health: Main effects and occupational differences* (HEW NIOSH No. 75–160). Washington, DC: U.S. Government Printing Office.

Cherniss, C. (1980). *Professional burnout in human service organizations.* New York: Praeger.

DePaepe, J., French, R., & Lavay, B. (1985). Burnout symptoms experienced among special physical educators: A descriptive longitudinal study. *Adapted Physical Activity Quarterly, 2,* 189–196.

Dignam, J.T., Barrera, M., Jr., & West, S.G. (1986). Occupational stress, social support and burnout among correctional officers. *American Journal of Community Psychology, 14,* 177–193.

Eisenstat, R.A., Felner, R.D., Kennedy, M., & Blank, M. (1981). *Job involvement and the quality of care in human service settings.* Paper presented at the 89th annual meeting of the American Psychological Association, Los Angeles.

Folkman, S., & Lazarus, R.S. (1988). The relationship between coping and emotion: Implications for theory and research. *Social Science in Medicine, 26,* 309–317.

Folkman, S., Lazarus, R.S., Dunkel-Schetter, C., DeLongis, A., & Gruen, R.J. (1986). The dynamics of a stressful encounter: Cognitive appraisal, coping, and encounter outcomes. *Journal of Personality and Social Psychology, 50,* 992–1003.

Gaines, J., & Jermier, J.M. (1983). Emotional exhaustion in a high stress organization. *Academy of Management Journal, 26,* 567–586.

Golembiewski, R.T., & Munzenrider, R.F. (1988). *Phases of burnout: Developments in concepts and applications.* New York: Praeger.

Greenglass, E.R., & Burke, R.J. (1990). Burnout over time. *Journal of Health and Human Resources Administration, 13,* 192–204.

Jackson, S.E., Schwab, R.L., & Schuler, R.S. (1986). Toward an understanding of the burnout phenomenon. *Journal of Applied Psychology, 71*(4), 630–640.

Jayaratne, S., & Chess, W.A. (1984). The effects of emotional support on perceived job stress and strain. *Journal of Applied Behavioral Science, 20,* 141–153.

Kyriacou, C., & Sutcliffe, J. (1977). Teacher stress: A review. *Educational Review, 29,* 299–306.

Leiter, M.P. (1990). The impact of family resources, control coping, and skill utilization on the development of burnout: A longitudinal study. *Human Relations, 43,* 1076–1083.

Maslach, C. (1976). Burned out. *Human Behavior, 5,* 16–22.

Maslach, C., & Jackson, S.E. (1981). The measurement of experienced burnout. *Journal of Occupational Behavior, 2,* 99–113.

Maslach, C., & Jackson, S.E. (1982). Burnout in health professions: A social psychological analysis. In G.S. Sanders & J. Suls (Eds.), *The social psychology of health and illness* (pp. 227–251). Hillsdale, NJ: Erlbaum.

Maslach, C., & Jackson, S.E. (1986). *Maslach Burnout Inventory Manual* (2nd ed.). Palo Alto, CA: Consulting Psychologists Press.

Orden, S., & Bradburn, N.M. (1968). Dimensions of marriage happiness. *American Journal of Sociology, 73,* 715–731.

Pines, A.M., & Aronson, E. (with Kafry, D.). (1981). *Burnout: From tedium to personal growth.* New York: Free Press.

Shirom, A. (1989). Burnout in work organizations. In C.L. Cooper & I. Robertson (Eds.), *International review of industrial and organizational psychology* (pp. 25–48). New York: Wiley.

Wade, D.C., Cooley, E., & Savicki, V. (1986). A longitudinal study of burnout. *Children and Youth Services Review, 8,* 161–173.

Wallius, E. (1982). *Work, health and well-being for teachers in Swedish comprehensive schools: X: Teacher stress: Theory and research* (Report No. 161). Laboratory for Clinical Stress Research, Stockholm.

Burnout and Coping Strategies:
A Comparative Study of Ward Nurses

E. Dara Ogus

Burnout in large organizations has been examined in a number of studies (Burke, 1993; Corrigan et al., 1994; Greenglass, Fiksenbaum, & Burke, 1994; Perrewé, 1991). There has been less study of burnout in nurses and hospital settings. Maslach and Jackson (1981) define burnout as a syndrome of emotional exhaustion, reduced personal accomplishment, and depersonalization. Emotional exhaustion refers to feelings of being emotionally overextended and drained by others. Reduced personal accomplishment involves a decline in one's feelings of competence and achievement, and depersonalization refers to a callous response toward people who are the recipient of one's services.

Most research in the past has looked at burnout as an overall concept. More recent studies have concentrated on burnout as involving the three conceptually distinct components. These subscales may share common hypothesized relationships; however, the value of distinguishing among them is shown by findings of differential patterns of correlations between each component and other variables (Leiter, 1991; Ogus, Greenglass, & Burke, 1990). For this reason, both total burnout and its subscales were used in this study.

Author's Notes: I would like to thank Esther R. Greenglass for her assistance in the development of this research.

Burnout is distinct from occupational stress in that it involves three components. In contrast, definitions of occupational stress focus primarily on the experience of strain (Marshall & Barnett, 1993). Thus, burnout is depicted as a process gradually worsening over time, and resulting from a buildup of chronic stress from emotionally demanding situations (Pines & Aronson, 1981).

Though every occupation is susceptible to burnout, the nursing profession seems to be particularly vulnerable (Glass, McKnight, & Valdimarsdottir, 1993). In recent research conducted among hospital nurses, burnout has been found to be significantly associated with lowered morale and reduced performance, as well as a variety of other negative outcomes. These include drug and alcohol abuse, tardiness, absenteeism, and increased job turnover (Goldenberg & Waddell, 1990; Kunkler & Whittick, 1991).

Coping

Some work has investigated nurses' efforts to cope with stress (Schaefer & Moos, 1993). Lazarus and Folkman (1984) have suggested two strategies for coping with stress: direct action and palliation. Direct coping is aimed at solving an existing problem by either changing the situation, one's behavior, or both. Palliation is applied to regulating emotional reactions or making one feel better without actually solving the problem.

In some instances, palliation can act as a positive mechanism in dealing with stress (through, for example, listening to soft music, engaging in yoga, or meditation). However, this study focuses on the more negative forms of palliative coping (i.e., wishful thinking, self-blame, and denial/escape). Counterproductive palliation such as drinking alcohol, overeating, or denial tactics only "put off" dealing with a problem through symptom reduction. They reduce anxiety for the moment but do not help with the issue at hand. Koeske, Kirk, and Koeske (1993) suggest that coping strategies involving confronting or attempting to change the source of stress are more effective than palliative coping strategies involving avoidance or denial tactics.

Ward Units

Less experienced younger nurses tend to start in general rather than specialized units. Based on the theory of adaptation (Indik, Seashore, & Slesinger, 1964), younger medical nurses have not had the experience or time to develop effective coping strategies. In a study of 102 nurses, Kelly and Cross (1985) found that the coping strategies used most often by specialized intensive care nurses tended to focus on problem-solving techniques including drawing upon past experiences, talking over a problem with others, and basing an action on understanding a situation. Conversely, general medical ward nurses tended to use more counterproductive methods for dealing with stress, including crying, sleeping less, and eating more. Frequent use of maladaptive coping behaviours combined with a highly stressed ward environment may be leading ward nurses into a spiral that places them at a higher risk for burnout (Kelly & Cross, 1985).

The present study was designed to examine burnout and coping strategies among medical and surgical ward nurses. It was hypothesized that nurses using high levels of negative palliative coping would experience higher burnout than nurses using palliative coping minimally. Nurses focusing on positive coping methods (i.e., preventive, existential, and internal control) would experience less burnout than nurses using little or no positive coping forms to deal with stress. As well, it was predicted that medical nurses would experience greater burnout than surgical nurses, and would use more negative coping strategies. Surgical nurses would use more positive coping techniques for dealing with stress.

METHOD

Respondents

The sample consisted of 128 female registered nurses from three major community hospitals in a large Canadian city. Respondents represented two clinical areas, medical and surgical wards. Within these wards, respondents were sampled only from full-time nurses working 8-hour shifts (day, evening, night, or continually rotating) and included, in total, 62 medical ward nurses and 66 surgical ward nurses. Most of the sample were unmarried (49% single, 41% married, 10% divorced or separated). The median age was 26 years, with age ranging from 20 to 53 years. Ninety percent of the nurses were diploma graduates (R.N.), 9% had bachelor of science degrees in nursing (B.S.N.), and 1% had master's level degrees (M.S.N.).

Medical units included patients with diseases ranging from cancer to strokes. Many patients were elderly and chronically ill, requiring constant attention. In contrast, surgical units dealt with younger patients in a pre- or post-operative stage requiring much shorter durations of treatment. Medical and surgical units were chosen for this study because their patients represent a wide range of illnesses requiring different types of nursing care that may differentially influence nurses' susceptibility to burnout. It is important, as well, to direct research efforts to compare nursing specialties that have been neglected in the past.

Procedure

The investigator visited each hospital to explain the study, obtain informed consent from each respondent, and leave self-contained research packets with each head nurse to distribute to the nurses. Each packet included a cover letter explaining the nature of the study, methods for ensuring confidentiality, and assurance that participation was voluntary. It was stressed that questions were to be answered without assistance.

Once the nurses completed the survey, they placed the envelopes in sealed boxes to be collected seven days later by the investigator. Each envelope had the word "confidential" written on it, and nurses were informed that only the investigator would have access to the box. Of 237 questionnaires administered, 128 were returned, yielding a response rate of 54%.

MEASURES

Burnout

The Maslach Burnout Inventory (MBI; Maslach & Jackson, 1986) was used to measure burnout. This 22-item inventory yields scores on three subscales: Emotional Exhaustion, Lack of Personal Accomplishment, and Depersonalization, as well as a total burnout score based on the three subscales.

Cronbach alpha coefficients for internal consistency are .90 for Emotional Exhaustion, .79 for Lack of Personal Accomplishment, and .71 for Depersonalization (Maslach & Jackson, 1981). Test-retest reliability, external validity, and absence of social desirability bias have been demonstrated (Maslach & Jackson, 1981, 1986).

Perceived Job Stress

Job stress was measured using parts of the Nursing Stress Inventory (NSI) developed by Numerof and Abrams (1984). The sections used consist of 42 items dealing with five different areas of stress: Organizational Environment (13 items, $\alpha = .88$), Work Demands (6 items, $\alpha = .89$), Death-Related Issues (8 items, $\alpha = .89$), and Lack of Procedural/Administrative Support (7 items, $\alpha = .86$). Items range from conflict with physicians and shortstaffing to caring for a terminal patient. Stress ratings are scored from 1 (no stress) to 5 (very much stress). Total stress scores range from 42 to 210.

Coping

Coping strategies were assessed using four of the scales from the Coping Inventory (Wong & Reker, 1984): Palliative Coping, Internal Control, Preventive Coping and Existential Coping. Palliative Coping includes three sub-categories: Wishful Thinking (four items) involves entertaining unrealistic wishes or fantasies; Self-blame (five items) blames one's own behavior or character for the problem; Denial/escape (five items) regulates one's emotional reaction or makes one feel better without actually solving the problem. Internal Control, based on nine items, is a coping strategy which depends on one's own efforts to change the situation. Preventive Coping is based on six items and includes direct coping techniques aimed at promoting one's well-being and reducing the likelihood of anticipated problems. Existential Coping, based on 11 items, includes ways of coping that attempt to maintain a sense of meaning and coherence or an attitude of acceptance in dealing with life. On a 5-point Likert scale ranging from "not at all" to "a great amount," respondents were asked how likely they were to do each of the things described in the coping items in response to the demands and pressures of their jobs. Sample items include: "Blame myself," "Draw on past experiences in

TABLE 1   Correlations Between Stress, Burnout, and Burnout Subscales

	Stress	Total Burnout	Emotional Exhaustion	Deperson-alization	Lack of Personal Accomplish-ment
Stress	—	.55***	.59***	.47***	.13
Total Burnout	—	—	.91***	.74***	.57***
Emotional Exhaustion	—	—	—	.61***	.28***
Depersonalization	—	—	—	—	.10
Lack of Personal Accomplishment	—	—	—	—	—

*** $p < .001$

trying to change the situation," and "Maintain a sense of purpose, optimism, and zest for life."

RESULTS

Internal consistency reliability coefficients (alphas) of the composite measures (work stress, total burnout, burnout subscales, coping scales) ranged from 0.69 to 0.94.

Correlations between work stress, burnout and burnout subscales were computed. Results showed significant positive correlations between stress and burnout on all burnout scales, except lack of personal accomplishment. That is, the higher the work stress, the greater the level of total burnout, emotional exhaustion and depersonalization (see Table 1).

Correlations were also conducted between burnout (and subscales), palliative coping, internal control, preventive coping, and existential coping. All coping forms were found to be significantly related to burnout, except internal control. That is, the higher the palliative coping, the higher the level of burnout, and the higher the levels of preventive and existential coping, the lower the levels of burnout (see Table 2).

Since significant relationships between the predictors and outcomes were found, it was felt that multiple regressions should be employed. Correlations were first computed between the different coping strategies (see Table 3).

Because preventive coping was highly correlated with all other coping forms, it was felt that a multiple regression including all coping styles would suppress the significant contribution of preventive coping as an important predictor of burnout. Therefore, a multiple regression was conducted for palliative coping, internal control and existential coping with total burnout and the burnout subscales as the

TABLE 2   Correlations Between Burnout and Coping Measures

	Total Burnout	Emotional Exhaustion	Deperson- alization	Lack of Personal Accomplishment
Palliative Coping	.53***	.49***	.34***	.33***
Internal Control	-.22	-.15	-.16	-.22
Preventive Coping	-.35***	-.33***	-.17	-.25***
Existential Coping	-.23*	-.14	-.08	-.32***

*$p < .05$; ***$p < .001$

criteria. Then a second regression was conducted for preventive coping with burnout and its subscales as the criteria.

The four strategies were considered as separate and distinct forms in order to address the contrast between positive and maladaptive forms of coping, and to determine the unique contribution of each type of coping to burnout. Mean scores and ranges for each of the four coping measures were as follows: for palliative coping, M = 29, range = 6 to 30; for preventive coping, M = 20, range = 6 to 30; for existential coping, M = 33, range = 12 to 46, and for internal control, M = 31, range = 18.5 to 45.5.

Palliative Coping

Results indicated main effects for palliative coping with total burnout, as well as all burnout subscales. That is, nurses self-reporting high use of palliative coping

TABLE 3   Correlations Between Palliative Coping, Internal Control, Preventive Coping, and Existential Coping

	Palliative Coping	Internal Control	Preventive Coping	Existential Coping
Palliative Coping	—	-.15	-.33***	-.03
Internal Control	—	—	.69***	.49***
Preventive Coping	—	—	—	.68***
Existential Coping	—	—	—	—

***$p < .001$

TABLE 4   Multiple Regression Equations: Significant Main Effects for
Palliative Coping, Existential Coping, and Internal Control

PREDICTORS	R^2	B	BETA	T	F
		Total Burnout			
Model	.33				20.60
Palliative Coping			0.52	6.92***	
Existential Coping			-0.22	-2.58*	
Internal Control			-0.04	-0.43	
Constant		67.07			
		Emotional Exhaustion			
Model	.26				14.27
Palliative Coping			0.49	6.19***	
Existential Coping			-0.12	-1.30	
Internal Control			- 0.02	-0.18	
Constant		26.24			
		Depersonalization			
Model	.13				6.23
Palliative Coping			0.32	3.82***	
Existential Coping			-0.07	-0.77	
Internal Control			-0.07	-0.81	
Constant		12.20			
		Lack of Personal Accomplishment			
Model	.22				11.81
Palliative Coping			-0.32	3.92***	
Existential Coping			-0.34	-3.75***	
Internal Control			-0.00	-0.01	
Constant		28.63			

*$p < .05$; ***$p < .001$

self-reported more total burnout, more emotional exhaustion, more depersonalization, and a weaker sense of personal accomplishment than nurses reporting low usage of palliative coping (see Table 4).

Existential Coping

Results indicated a main effect for existential coping with total burnout and lack of personal accomplishment, indicating that nurses who used existential coping experienced less total burnout and a greater sense of personal accomplishment than nurses not using existential coping (see Table 4).

Internal Control

No main effects were found for internal control.

TABLE 5   Regression Equations: Significant Main Effects for Preventive
Coping

PREDICTORS	R^2	B	BETA	T	F
		Total Burnout			
Model	.12				17.64
Preventive Coping			-0.35	-4.20***	
Constant		108.43			
		Emotional Exhaustion			
Model	.11				15.90
Preventive Coping			-0.33	-3.99***	
Constant		57.19			
		Depersonalization			
Model	.03				3.82
Preventive Coping			-0.17	-1.95	
Constant		19.54			
		Lack of Personal Accomplishment			
Model	.06				8.63
Preventive Coping			-0.25	-2.94**	
Constant		31.69			

**p < .005; ***p < .001

Preventive Coping

Results indicated main effects for preventive coping with total burnout, emotional exhaustion, and lack of personal accomplishment. That is, nurses using preventive coping techniques reported less total burnout, less emotional exhaustion, and a stronger sense of personal accomplishment than nurses using little or no preventive coping (see Table 5).

Further results revealed that medical ward nurses experienced greater burnout (both total and subscales) than surgical ward nurses (see Table 6). In addition, medical nurses used significantly more negative palliative forms of coping than surgical nurses, while surgical nurses used significantly more preventive coping and existential coping to deal with stress (see Table 7).

DISCUSSION

Coping and Burnout

The data in this study suggest significant relationships between coping and burnout levels. Specifically, nurses who relied heavily on negative palliative coping reported higher burnout than nurses using low levels of palliative coping.

TABLE 6   T-Tests on Total Burnout and Burnout Subscales : Medical Versus Surgical Ward Nurses

Burnout	Medical			Surgical			DF	T
	X	SD	N	X	SD	N		
Total Burnout	89.34	12.44	62	60.24	12.95	66	126	12.95***
Emotional Exhaustion	45.90	7.49	62	29.74	9.15	66	126	10.89***
Depersonal-ization	18.00	6.14	62	10.18	3.99	66	126	8.48***
Lack of Personal Accomplish-ment	25.44	6.61	62	20.32	6.42	66	126	4.44***

***p < .001

TABLE 7   T-Tests on Coping Measures: Medical Versus Surgical Ward Nurses

Coping	Medical			Surgical			DF	T
	X	SD	N	X	SD	N		
Palliative Coping	32.57	8.39	61	25.86	6.72	66	125	4.95**
Internal Control	30.23	4.89	61	31.47	6.12	66	125	-1.27
Preventive Coping	19.06	3.77	62	21.59	3.94	66	126	-3.71**
Existential Coping	31.71	6.15	62	34.77	5.98	66	126	-2.85*

* p = .005; **p < .001

Nurses using high levels of preventive and existential coping reported lower burnout than those using these strategies minimally or not at all. These findings are in line with Milne and Watkins (1986) who found that nurses relying on more productive coping forms (e.g., drawing on past experience, taking positive action) were able to deal more effectively with work stress, and thus improve individual functioning and work performance. In a study of 473 teachers, Greenglass, Burke, and Ondrack (1990) similarly found productive coping forms, specifically preventive and existential coping, to be negatively related to burnout.

For nurses, it is customary to think in terms of reducing stress (e.g., fewer beds or more nurses) when stress reaches unacceptable levels. In contrast, Milne and Watkins (1986) suggest that positive ways of thinking about and reacting to increased stress (developing problem-solving skills, rational re-labelling of stressors and realistic expectancies) may be equally effective in helping to combat burnout. Thus, coping style may play an important role in lowering levels of burnout for nurses and other human service professionals.

Internal Control and Burnout

With regard to internal control and burnout, no significant effects were found. Internal control is a strategy which depends on one's own efforts to change the situation. In nursing, it is understandable why this strategy may not be as helpful for the following reason:

> It seems management never asks themselves why nurses keep leaving. They look only at the immediate problem—finding another nurse. They preach about how we should work smarter, not harder [which] doesn't help... and if you complain or try to suggest changes, they label you a trouble-maker (Carey, Doan-Johnson, & Kelley, 1988, p. 38).

If nurses are criticized or ignored repeatedly, it is not surprising that they may resort to other forms of coping to deal with stress.

Burnout Components

Of the three burnout components, studies, including this one, suggest that emotional exhaustion is most closely related to total burnout (Lee & Ashforth, 1993). Depersonalization is a distinctly new construct only recently being examined in job stress literature. In this study, depersonalization was found to be related to palliative coping. Some have proposed that depersonalization is a coping strategy in itself. It has also been suggested that depersonalization could be the result of an inability to cope (Ogus et al., 1990). If depersonalization is viewed as a maladaptive form of coping, it is not surprising that a relationship was found between depersonalization and palliative coping. When nurses are frustrated or having difficulty dealing with overwhelming stress, it makes sense that they would resort to separating themselves emotionally from the painful aspects of the job in order to cope, albeit maladaptively. Perhaps a future study could investigate in more detail the mechanisms that link these two variables.

Regarding personal accomplishment, psychologists have long been interested in similar concepts including self-esteem, self-efficacy, and expectancy (MacNeil & Weisz, 1987). These self-evaluative concepts are central to experienced job stress. Thus, personal accomplishment must also continue to be examined as a unique component of burnout in stress outcomes.

Ward Units and Coping

With regard to coping styles across wards, it was found that medical nurses used more counterproductive methods for dealing with stress. Surgical nurses relied on more positive strategies. Medical nurses experienced significantly more burnout than their surgical counterparts. These results confirm previous findings (Kelly & Cross, 1985).

As suggested by Dewe (1993), work setting or type of nursing unit may be pertinent to coping style and level of burnout. Generally, surgical nurses receive patients with specific ailments, with many scheduled for elective surgery. In contrast, medical ward nurses are often faced with a high percentage of patients having multiple medical problems. Without a definite diagnosis, it is difficult for medical nurses to plan structured treatment schedules. With less immediate feedback, nurses may feel less competent, and greater difficulty in coping may be experienced (Gray-Toft & Anderson, 1981).

Patient load may also be a factor contributing to coping style and burnout levels (Jamal & Baba, 1992). When surgical beds are filled, the overflow is often placed in medical units until appropriate space can be found. Under these circumstances, medical nurses need to have an additional working knowledge of surgical issues, while shifting back and forth from active treatment to chronic care.

Cronin-Stubbs and Rooks (1985) suggest that, since medical nurses manage multiple variables, stress may be provoked more easily compared to surgical nurses who have more defined roles and lower nurse-patient ratios. With a more specialized patient load, surgical nurses may be experiencing greater autonomy in decision-making and direction-setting which can be related to coping techniques. These issues have not been given enough attention, particularly with regard to unit comparisons.

As a potential solution, coping skills could be improved by providing relaxation training, systematic desensitization, and cognitive restructuring aimed at altering individuals' perceptions of stressful circumstances (Parkes, 1985). As well, nursing applicants could be screened for their coping skills and offered supplementary coping strategies or workshops, and self-help manuals as preventive measures for future stress (Tyler & Cushway, 1992). Techniques could be aimed at promoting nurses' well-being through use of positive self-statements, pinpointing typical stressful situations, and discussing the pros and cons of potential solutions, and providing an avenue for nurses to voice their grievances and work them through in a supportive setting.

A complementary solution might be to alter the work environment to make it easier for nurses to deal with daily stressors as they arise (Robinson et al., 1991). One recommendation would be to give nurses more control and flexibility through greater involvement in policy decision-making. In this way, they may experience lower levels of burnout than nurses who view their positions as merely jobs. Other suggestions may be to encourage nurses to lead more healthful lifestyles (e.g., more exercise, yoga, reduced consumption of cigarettes, coffee and alcohol, more cultural activities), and to provide reliable child care to nurses during working hours. Issues of causality could not be determined in this study. However, the important possibility of coping strategies mitigating the effects of stress and reducing burnout opens up a major option to nurses whether they are involved in direct patient care, nurse management or nurse education.

REFERENCES

Burke, R.J. (1993). Toward an understanding of psychological burnout among police officers. *Journal of Social Behavior and Personality, 8*(3), 425–438.

Carey, K.W., Doan-Johnson, S., & Kelley, W.J. (1988, February). Nursing shortage poll report. *Nursing,* pp. 33–41.

Corrigan, P.W., Holmes, E.P., Luchins, D., Buican, B., Basit, A., & Parks, J.J. (1994). Staff burnout in a psychiatric hospital: A cross-lagged panel design. *Journal of Organizational Behavior, 15,* 65–74.

Cronin-Stubbs, D., & Rooks, C.A. (1985). The stress, social support, and burnout of critical care nurses: The results of research. *Heart and Lung, 14*(1), 31–39.

Dewe, P.J. (1993). Coping and the intensity of nursing stressors. *Journal of Community and Applied Social Psychology, 3,* 299–311.

Glass, D.C., McKnight, J.D., & Valdimarsdottir, H. (1993). Depression, burnout and perceptions of control in hospital nurses. *Journal of Consulting and Clinical Psychology, 61*(1), 147–155.

Goldenberg, D., & Waddell, J. (1990). Occupational stress and coping strategies among female baccalaureate nursing faculty. *Journal of Advanced Nursing, 15,* 531–543.

Gray-Toft, P.A., & Anderson, J.G. (1981). Stress among hospital nursing staff: Its causes and effects. *Social Science and Medicine, 15A,* 639–647.

Greenglass, E.R., Burke, R.J., & Ondrack, M. (1990). A gender-role perspective of coping and burnout. *Applied Psychology : An International Review, 39*(1), 5–27.

Greenglass, E.R., Fiksenbaum, L., & Burke, R.J. (1994). The relationship between social support and burnout over time in teachers. In P.L. Perrewé & R. Crandall (Eds.), *Occupational stress: A handbook* (pp. 239–248). Washington, DC: Taylor & Francis.

Indik, B., Seashore, S.E., & Slesinger, J. (1964). Demographic correlates of psychological strain. *Journal of Abnormal and Social Psychology, 69,* 26–38.

Jamal, M., & Baba, V.V. (1992). Shiftwork and department type related to job stress, work attitudes and behavioral intentions: A study of nurses. *Journal of Organizational Behavior, 13,* 449–464.

Kelly, J.G., & Cross, D.G. (1985). Stress, coping behaviors, and recommendations for intensive care and medical surgical ward registered nurses. *Research in Nursing and Health, 8,* 321–328.

Koeske, G.F., Kirk, S.A., & Koeske, R.D. (1993). Coping with job stress: Which strategies work best? *Journal of Occupational and Organizational Psychology, 66,* 319–335.

Kunkler, J., & Whittick, J. (1991). Stress management groups for nurses: Practical problems and possible solutions. *Journal of Advanced Nursing, 16,* 172–176.

Lazarus, R.S., & Folkman, S. (1984). *Stress, appraisal, and coping.* New York: Springer.

Lee, R.T., & Ashforth, B.E. (1993). A further examination of managerial burnout: Toward an integrated model. *Journal of Organizational Behavior, 14,* 3–20.

Leiter, M.P. (1991). Coping patterns as predictors of burnout: The function of control and escapist coping patterns. *Journal of Organizational Behavior, 12,* 123–144.

MacNeil, J.M., & Weisz, G.M. (1987). Critical care nursing stress: Another look. *Heart and Lung, 16*(3), 274–277.

Marshall, N.L., & Barnett, R.C. (1993). Variations in job strain across nursing and social work specialties. *Journal of Community and Applied Social Psychology, 3,* 261–271.

Maslach, C., & Jackson, S.E. (1981). The measurement of experienced burnout. *Journal of Occupational Behavior, 2,* 99–113.

Maslach, C., & Jackson, S.E. (1986). *Maslach Burnout Inventory Manual* (2nd ed.). Palo Alto, CA: Consulting Psychologists Press.

Milne, D., & Watkins, F. (1986). An evaluation of the effects of shift rotation on nurses' stress, coping and strain. *International Journal of Nursing Studies, 23*(2), 139–146.

Numerof, R.E., & Abrams, M.N. (1984). Sources of stress among nurses: An empirical investigation. *Journal of Human Stress, 10*(2), 88–100.

Ogus, E.D., Greenglass, E.R., & Burke, R.J. (1990). Gender-role differences, work stress and depersonalization. *Journal of Social Behavior and Personality, 5*(5), 387–398.

Parkes, K.R. (1985). Stressful episodes reported by first-year student nurses: A descriptive account. *Social Science and Medicine, 20*(9), 945–953.

Perrewé, P.L. (Ed.). (1991). Handbook on job stress [Special issue]. *Journal of Social Behavior and Personality, 6*(7).

Pines, A.M., & Aronson, E. (with Kafry, D.). (1981). *Burnout: From tedium to personal growth.* New York: Free Press.

Robinson, S.E., Roth, S.L., Keim, J., Levenson, M., Flentje, J.R., & Bashor, K. (1991). Nurse burnout: Work related and demographic factors as culprits. *Research in Nursing and Health, 14,* 223–228.

Schaefer, J.A., & Moos, R.H. (1993). Work stressors in health care: Context and outcomes. *Journal of Community and Applied Social Psychology, 3,* 235–242.

Tyler, P., & Cushway, D. (1992). Stress, coping and mental well-being in hospital nurses. *Stress Medicine, 8,* 91–98.

Wong, P.T.P., & Reker, G.T. (1984). *Coping behaviours of successful agers.* Paper presented at the annual meeting of the Canadian Psychological Association, Ottawa.

Measuring Burnout: An Updated Reliability and Convergent Validity Study

Kevin Corcoran

Burnout in the social services has gained much popularity since the term was first introduced (Freudenberger, 1971). Research has also examined this drain on social service workers' effectiveness (e.g., Corcoran, 1987; Daley, 1979; Hagen; 1989; Harrison, 1983; Paine, 1982).

Accompanying the research has been the publication of measurement devices. Currently, there are numerous instruments to assess burnout, most which have some evidence of reliability and validity (Shinn, 1982). The most widely used instrument is the Maslach Burnout Inventory (MBI; Maslach & Jackson, 1981), which defines burnout as a progressing from emotional exhaustion to depersonalization to a lack of personal accomplishment. This instrument has very good evidence of reliability, stability, and validity (e.g., Maslach & Jackson, 1981; Poulin & Walter, 1993). Recently, however, research has questioned the factor structure such that depersonalization may be either a symptom of burnout or a coping mechanism to prevent burnout (Wallace & Brinkerhoff, 1991). This finding is by no means consistent across the literature (e.g., Lee & Ashforth, 1993a). Moreover, no data are available which examine the interrelationships of the instruments. Consequently, this article examines the reliability and convergent validity of the two most prominent measurement devices using a sample of masters level social workers.

METHOD

Research Participants and Procedures

Three hundred female social workers were randomly sampled from the Texas chapter of the National Association of Social Workers. Participants were mailed a test booklet which included the burnout scales, additional measures such as demographics, and a self-addressed postage-paid return envelope. Three to four weeks later, a follow-up postcard was sent to all subjects. A total of 147 responses were received, of which eight were unusable, producing an analytic sample of 139. This represents a 46.3% response rate, which is acceptable for this population (Kirk & Fischer, 1976).

The average age of the sample was 40.1 (SD = 10.85). The majority were either married or living with a permanent partner (58.3%), were employed full-time (84.2%), and had an average of 12.93 (SD = 9.95) years of experience. The sample spent an average of 55.2% of the work week in direct client service, averaging 17.02 (SD = 11.55) hours of client contact.

Instrumentation

Maslach Burnout Inventory. The first published instrument was the Maslach Burnout Inventory (MBI; Maslach & Jackson, 1981), which conceptualizes burnout as a "syndrome of emotional exhaustion and cynicism" (p. 99). The authors consider burnout to occur in phases, starting with feelings of emotional exhaustion, which produce depersonalized attitudes toward clients and then decrease a worker's professional accomplishment.

The symptom of emotional exhaustion is considered an essential characteristic of burnout (Maslach, 1982), while depersonalization is the most important symptom for social workers (Harrison, 1980). Previous research has questioned the measurement devices of the remaining phases (Iwanicki & Schwab, 1981; Krowinski, 1981). Consequently, for the current study only the Emotional Exhaustion and Depersonalization subscales were included with scores restricted to the frequency of the experiences.

Occupational Tedium. A similar perspective considers burnout as occupational tedium (Pines & Aronson, 1981). This view includes the syndrome of exhaustion, although it differentiates burnout from tedium in terms of etiology. Tedium results from any prolonged chronic pressure, while burnout is seen as due to the specific emotional pressures found in work with people. Pines and her associates argue that tedium is "almost always part and parcel of the burnout syndrome" (Pines & Aronson, 1981, p. 15), while Krowinski (1981, p. 50) observes the distinction "may serve to further muddy the conceptual waters instead of clearing them."

Pines & Aronson (1981) propose three dimensions of occupational tedium: physical exhaustion, emotional exhaustion and mental exhaustion. Each dimension is operationalized with a 7-item scale. Two important features must be considered when using the Occupational Tedium (OT) scale. First, the instruments have no explicit association with work, but simply ask how often one has had certain burnout experiences. Secondly, 30 of the 36 samples used to estimate the instrument's psychometrics were participants in the workshops on burnout. Even with lenient standards, a threat of selectivity must be considered.

Statistical Procedures

Reliability was estimated by internal consistency and standard error of measurement. The former method assesses reliability in terms of the average correlation among items. The conventional internal consistency statistic is coefficient alpha. Since reliability is affected by the number of items as well as their inter-association, a low alpha coefficient suggests either the items have little in common or too few items compose the instrument. A high coefficient indicates that, on the average, the items are tapping the same construct domain (Nunnally, 1978).

As a consequence of being based on correlations, the alpha coefficient can be affected by sample and population variances. Alpha also has little practical value to test users (Nunnally, 1978). An equally useful and recommended statistic is the Standard Error of Measurement (SEM). This procedure not only enables one to establish a confidence interval of true scores, but is a more stable estimate of reliability across samples.

Principal component factor analyses were performed on both measurement devices. This procedure was used to determine if the rotated factors produced simple structures reflecting each instrument's proposed dimensionality. For the purposes of this study, consideration was given to whether an item had adequate loading exceeding .40 on the hypothesized factor (Nunnally, 1978).

Validation estimates were determined through convergent validity correlations. With this procedure, if instruments are indeed assessing the same construct domain, the scores will be significantly correlated. Strong coefficients are needed in order to suggest the instruments are tapping the same construct domain (Fiske, 1973).

RESULTS AND DISCUSSION

Reliability Estimates

Maslach Burnout Inventory. The internal consistency coefficients for the subscales of the MBI were quite acceptable, as displayed in Table 1. These data tend to suggest the subscales are reliable and contain little error. The latter finding is seen in the scale score standard error of measurement which is also displayed as the percentage of error.

The results of the principal component factor analysis appear to support the proposed dimensionality of the MBI with two factors having eigenvalues of 6.1

TABLE 1   Reliability Estimates for the Proposed Subscales

Instruments	Coefficient alpha	SEM	% of Error
MBI:E	.88	3.05	5.64
MBI:D	.81	2.05	5.86
OT:EE	.82	2.30	5.48
OT:PE	.84	2.14	5.09
OT:ME	.86	1.94	4.62

Key: MBI:E   Maslach Burnout Inventory, Emotional Exhaustion
     MBI:D   Maslach Burnout Inventory, Depersonalization
     OT:EE   Occupational Tedium, Emotional Exhaustion
     OT:PE   Occupational Tedium, Physical Exhaustion
     OT:ME   Occupational Tedium, Mental Exhaustion

and 1.98. However, three of the 14 items either loaded on the incorrect factor or did not have simple structure. The remaining 11 items all loaded on the first factor.

These findings do not support the subscale structure, and suggest the MBI may be unidimensional. In fact, the two subscales were highly correlated, as displayed in Table 2. Consequently, the subscales were collapsed and the reliability re-estimated. As displayed in Table 3, the reliability for the 14-item instrument was high enough to suggest the measure is unidimensional (Nunnally, 1978).

Occupational Tedium. The proposed subscales of the OT were found to be internally consistent (see Table 1), with small amounts of standard error. The factor analytic procedures revealed four significant factors, one with an Eigenvalue of 9.25. The remaining Eigenvalues were less than 2.0, all accounting for less than 10% of the variance. Of the 21 items, nine either loaded on incorrect factors or did not have simple structure. These results do not support the proposed dimensionality, and, in fact, the subscales were quite correlated (see Table 2).

These findings warranted examining the OT as a unidimensional instrument. The results (see Table 3) suggest that, as a 21-item instrument, the OT is highly reliable.

Convergent Validity

Estimates of convergent validity of the originally proposed subscales are displayed in Table 2. The correlations between subscale scores were all statistically significant, accounting for a minimum of 27% shared variance.

Since the subscale structures have been questioned, convergent validity also examined the MBI and OT as unidimensional measurement devices. Here the two measures were highly correlated ($r = .75$, $p < .001$). These two measures have 56.25% shared variance, and appear to be tapping the same construct domain. These findings are consistent with the conceptual definitions of burnout, which

TABLE 2   Convergent Validity Coefficients of Proposed Subscales of Burnout

		a	b	c	d	e
MBI:E	(a)	—				
MBI:D	(b)	.68	—			
OT:EE	(c)	.72	.54	—		
OT:PE	(d)	.72	.58	.78	—	
OT:ME	(e)	.72	.58	.78	1.0	—

Key: MBI:E   Maslach Burnout Inventory, Emotional Exhaustion
    MBI:D   Maslach Burnout Inventory, Depersonalization
    OT:EE   Occupational Tedium, Emotional Exhaustion
    OT:PE   Occupational Tedium, Physical Exhaustion
    OT:ME   Occupational Tedium, Mental Exhaustion

assert it is a manifestation of exhaustion resulting from the demands of one's work in the human services.

CONCLUSIONS

The findings of this study provide support for the psychometric properties of these two operational definitions of burnout. The instruments' proposed subscale structure was fairly reliable, although the data suggest each scale is an internally consistent, unidimensional measure. This finding is consistent with other studies which question if depersonalization is actually burnout (Wallace & Brinkerhoff, 1991; Lee & Ashforth, 1993b).

As for which scale is recommended for future research, the results support the use of either as a psychometrically sound assessment device of the exhaustion component of burnout. The OT may appear slightly more reliable, although

TABLE 3   Reliability Estimates for Revised Measures of Burnout

Instrument	Coefficient alpha	SEM	% of Error
MBI	.90	3.84	4.57
OT	.93	3.87	3.23

researchers will most likely prefer the MBI because it is essentially as reliable as the OT, has the benefit of being a much shorter measurement, and is the most frequently used instrument.

REFERENCES

Corcoran, K. (1987). The association of burnout and social work practitioners' impressions of their clients: Empirical evidence. *Journal of Social Service Research, 10,* 57–66.

Daley, M.R (1979). Burnout: Smothering problems in protective services. *Social Work, 24,* 375–379.

Fiske, D.W. (1973). Can a personality construct be validated? *Psychological Bulletin, 80,* 89–92.

Freudenberger, H.J. (1971). The professional in the free clinic: New problems, new views, new goals. In D.E. Smith, D.J. Bental, & J.L. Schwartz (Eds.), *The free clinic: A community approach to health care and drug abuse.* Beloit, WI: Stash Press.

Hagen, J.L. (1989). Income maintenance workers: Burned-out, dissatisfied, and leaving. *Journal of Social Service Research, 13,* 47–63.

Harrison, W.D. (1983). A social competence model of burnout. In B.A. Farber (Ed.), *Stress and burnout in the human service professions.* New York: Pergamon.

Harrison, W.D. (1980). Role strain and burnout in child-protection service workers. *Social Service Review, 54,* 31–43.

Iwanicki, E.F., & Schwab, R.L. (1981). A cross validation of the Maslach Burnout Inventory. *Educational and Psychological Measurement, 41,* 1167–1174.

Kirk, S.A., & Fischer, J. (1976). Do social workers understand research? *Journal of Education for Social Work, 12,* 63–70.

Krowinski, W.J. (1981). *A construct validation study of the Maslach Burnout Inventory.* A competency paper submitted to the School of Social Work, University of Pittsburgh.

Lee, R.T., & Ashforth, B.E. (1993a). A further examination of managerial burnout: Toward an integrated model. *Journal of Organizational Behavior, 14,* 3–20.

Lee, R.T., & Ashforth, B.E. (1993b). A longitudinal study of burnout among supervisors and managers: Comparisons between the Leiter and Maslach (1988) and Golembiewski et al. (1986) models. *Organizational Behavior and Human Decision Processes, 54,* 369–398.

Maslach, C. (1982). *Burnout: The cost of caring.* Englewood Cliffs, NJ: Prentice Hall.

Maslach, C., & Jackson, S.E. (1981). The measurement of experienced burnout. *Journal of Occupational Behavior, 2,* 99–113.

Nunnally, J. (1978). *Psychometric theory.* New York: McGraw Hill.

Paine, W.S. (1982). Overview: Burnout stress syndromes of the 1980s. In W.S. Paine (Ed.), *Job stress and burnout: Research, theory and intervention perspectives* (pp. 11–29). Beverly Hills, CA: Sage.

Pines, A.M., & Aronson, E. (with Kafry, D.). (1981). *Burnout: From tedium to personal growth.* New York: Free Press.

Poulin, J., & Walter, C. (1993). Social worker burnout: A longitudinal study. *Social Work Research and Abstract, 29,* 5–11.

Shinn, M. (1982). Methodological issues: Evaluating and using information. In W.S. Paine (Ed.), *Job stress and burnout: Research, theory, and intervention perspectives* (pp. 61–79). Beverly Hills, CA: Sage.

Wallace, J.E., & Brinkerhoff, M.B. (1991). The measurement of burnout revisited. *Journal of Social Service Research, 14*(5), 85–111.

PART FIVE

Interventions Aimed at Occupational Strain Reduction

The Impact of Stress Counseling at Work

Cary L. Cooper
Golnaz Sadri

Much of the occupational stress research over the last two decades has explored the causes and consequences of stress (Ivancevich, 1986; Cooper & Payne, 1988; Sauter, Hurrell, & Cooper, 1989). More recently, researchers have turned their attention to strategies that might be used to remedy the problem (Cooper, 1987; Murphy, 1988). A number of different approaches to conceptualizing stress management interventions have been highlighted in the literature. Matteson and Ivancevich (1987) draw a distinction between preventive and curative strategies. Defrank and Cooper (1987) suggest that interventions can focus on the individual, the organization or the individual/organizational interface. Murphy (1988) emphasizes three levels of intervention: primary (stressor reduction), secondary (stress management) and tertiary (Employee Assistance Programs).

The great bulk of the stress intervention research has focused on secondary interventions or stress management techniques (Highley & Cooper, 1993; Marshall & Cooper, 1979). These studies were recently reviewed by Murphy (1988), who concluded that stress management programes appear to be effective in reducing subjective distress (e.g., anxiety) and some psycho-physiological indicators (e.g., blood pressure). Studies examining primary or organizational level (or stressor reduction) strategies are extremely rare (Murphy, 1988, highlights only three), as are studies examining the effectiveness of Employee Assistance Programs (EAPs) or counseling services (Shapiro, Cheesman, & Wall, 1993). This might partly be explained by a reluctance on the part of organizations to implement stressor reduction programs (in the case of the former), which would involve OD or job analysis interventions, and by an unwillingness on the part of the prescribers of counseling services to be evaluated (in the latter case). Walsh and Hingson (1985) provide some evidence of this.

Studies which have examined a counseling intervention have generally considered Employee Assistance Programs (EAPs), and have found such programs to be effective in terms of the following criteria: percent of employees who go to treatment; percent who return to work after treatment; changes in behavior after treatment (work performance and problem behavior such as excessive alcohol consumption); and cost-savings to companies (Swanson & Murphy, 1991; Orlans, 1991). However, Murphy (1988) offers some methodological criticisms of existing studies, in terms of a lack of control groups and lack of more objective measures (e.g., sickness absence), and cross-sectional as opposed to longitudinal designs.

The purpose of this paper is to highlight an empirical evaluation of an in-house stress counseling service provided for employees in the U.K. postal service. It attempts to minimize, as far as is practical in a field study, the methodological problems highlighted by Murphy (1988).

METHOD

Overview

The study was conducted within the U.K. postal service. In the early 1980s, the U.K. Post Office discovered that psychiatric and psychological disturbance ranked as the second highest reason for medical retirement, preceded only by muscular skeletal disorders. As a solution, stress counselors were employed in the North-West and North-East Regions of England (and Northern Ireland), to join the Post Office Occupational Health Service on a three year trial experiment.

Subjects

Two hundred fifty employees attending the counseling service formed the client (or experimental group) in the study. The control group consisted of a comparable group of 100 postal employees, matched in terms of age, sex, grade, and years of experience. The Post Office counseling program was an open access one, which was located within the Occupational Health Service. By virtue of the fact that counseling was located in occupational health, there was a tendency for a large number of referrals to emanate from this quarter. However, ongoing publicity (letters to employees, articles in the "house" magazine and presentations to management groups) served to promote other avenues. A breakdown of sources of referral revealed that the groupings were from occupational health (40%), self-referral (31.5%) and welfare services (19%), the remaining 9.5% coming from managers, the human resource function and the trade unions.

Furthermore, the clients were spread across the entire organization, including cleaners, postmen and women, postal officers and executives, technical engineers and senior staff, in approximately the same proportion as they are represented in the whole organization.

Program

The stress counseling program consisted of Rogerian, client-centered counseling (Rogers, 1961), where postal employees attended individual sessions with trained clinical psychologists. (Details of the service can be found in Allison, Cooper, Reynolds, 1989.)

Within the first three years, a major cluster of counseling had been on mental health issues, which represent 46% of the caseload. These clients normally presented with anxiety and/or depression symptoms. Of the remaining 54% of referrals, the second largest group was that of "relationship" problems (24%), with the majority of these experiencing marital difficulties. Other areas included alcoholism and other addictions, bereavement, assault, physical illness or disability, social problems and panic attacks.

Measures

To assess the effectiveness of stress counseling within the Post Office, before and after data were collected on sickness absence, mental health, self-esteem, organizational commitment and changes in health behaviors.

Absence Data. Sickness absence data were collected from the organization's personnel files, which are detailed records of all absences from work. Data were collected for the six months immediately preceding a client's first interview, and for the six months following his/her final interview. The following information was recorded: number of unauthorized sickness events; number of work days lost as a result of these events; number of authorized absence events; number of work days lost due to authorized leave and number of warnings received. Similar data were collected for the control group, where the counseling period was left unrecorded for each client, a period of one month was left unrecorded for the controls. At the time of this analysis, pre- and post-counseling sickness absence data were available for 188 employees/clients; and for 100 matched controls.

Mental Health. Three subscales of the widely used and validated mental health questionnaire, the Crown Crisp Experiential Index (Crown & Crisp, 1979), formerly known as the Middlesex Hospital Questionnaire, were used to assess free floating anxiety, somatic anxiety and depression. These scales consist of 24 items and can be summated for a total mental health score. Reliability and validation data can be found in Crown and Crisp (1979).

Job Satisfaction. Job satisfaction was assessed by the 15-item Job Satisfaction Scale (Warr, Cooke, & Wall, 1979), which is a well validated and reliable measure.

Self-Esteem. This was measured by Rosenberg's Self-Esteem scales which consist of 10 Likert-type items and is well validated (Crandall, 1973).

Organization Commitment. Porter & Lawler's Organizational Commitment scale was used in this study. It consists of 15 items and has been reliably (and validly) utilized in many other studies (Griffin & Bateman, 1986).

Health Behaviors. The Health Behaviors Questionnaire consisted of 21 items, assessing smoking behavior, caffeine intake, alcohol consumption, exercise and other coping strategies (Cooper, Cooper, & Eaker, 1988). This scale was only given to employees counseled, to assess health behavior changes.

Questionnaire data for the self-report measures were collected on 135 counseled employees at the pre-counseling period, dropping to 113 at post-counseling. Control data were collected for 74 demographically matched postal employees at the pre-counseling period, dropping to 37 at the post-measure period. These data were collected before the first counseling session and three months after the last session. For controls, the data were collected twice, with a four-month gap between questionnaire instrumentation (a period roughly equal to the counseled group). The alpha reliability coefficients for all measures were above .70.

RESULTS

Questionnaire Data

To determine whether the scores obtained by the client group changed after counseling, t-tests were conducted, the results of which are shown in Table 1.

Results presented in Table 1 indicate that clients' psychological well-being improves at the completion of counseling. Clients are less anxious, less depressed, suffer from fewer psychosomatic symptoms of stress and have a higher level of self-esteem. Clients also show significant changes on a number of behavioral items. After counseling, the client group use less of the following health behaviors: drink coffee/coke or eat frequently; smoke; and use alcohol to cope with events. Respondents also indicate that they feel less guilty about drinking. On the other hand, they use more of the following measures to relax at work: relaxation techniques (e.g., meditation or yoga); informal relaxation techniques (e.g., deep breathing or imagining pleasant scenes); exercise; leaving work area and going somewhere (e.g., lunching away from the organization, taking time out); and using humor. Clients also find more time to relax and "wind down" after work. To determine whether the changes shown in Table 1 were due to practices within the Post Office unrelated to counseling, pre- and post-questionnaire scores from the controls were compared. Table 2 indicates no significant differences between pre- and post-questionnaire scores for the control groups.

Where average (or mean) scores are calculated for groups, a high standard deviation indicates that individual member scores differ to quite an extent from the mean. Such deviation diminishes the representativeness of the mean as a score for

TABLE 1   Differences in Pre- and Post-Test Scores for Counselled Group

Variable	Mean	SD	T-value	2-Tail Prob.
Anxiety				
Pre	10.44	3.51	9.01	< .001
Post	7.47	4.17		
Somatic anxiety				
Pre	8.81	3.93	8.85	< .001
Post	5.29	3.58		
Depression				
Pre	8.41	3.03	7.04	< .001
Post	5.81	3.50		
Self-esteem				
Pre	27.51	4.99	-4.98	< .001
Post	29.64	5.14		
Job satisfaction				
Pre	60.68	15.10	-1.09	NS
Post	61.99	15.80		
Organisational commitment				
Pre	64.24	17.86	.40	NS
Post	63.72	18.06		
Drink coffee, coke or eat frequently				
Pre	3.54	1.09	3.74	< .001
Post	3.19	1.19		
Smoke				
Pre	2.40	1.62	2.03	< .05
Post	2.29	1.58		
Use relaxation techniques				
Pre	1.46	.86	-5.06	< .001
Post	1.97	1.13		
Use informal relaxation techniques				
Pre	1.96	1.05	-3.68	< .001
Post	2.43	1.17		
Exercise				
Pre	2.53	1.16	-2.21	< .05
Post	2.79	1.14		
Leave work area and go somewhere				
Pre	2.72	1.18	-1.96	.05
Post	2.94	1.15		
Use humour				
Pre	3.41	.94	-2.55	.01
Post	3.61	.93		
Use alcohol to cope with events				
Pre	1.61	.74	4.00	< .001
Post	1.35	.59		
Feel guilty about drinking				
Pre	1.47	.67	1.97	.05
Post	1.35	.59		
Relax after work				
Pre	2.81	.99	-5.52	< .001
Post	3.42	1.00		

TABLE 2   Differences in Pre- and Post-Test Scores for Control Group

Variable	Mean	SD	T-value	2-Tail Prob.
Anxiety				
Pre	3.11	3.34	-.94	NS
Post	3.78	4.02		
Somatic anxiety				
Pre	3.63	3.20	.49	NS
Post	3.31	3.05		
Depression				
Pre	3.36	3.12	-.11	NS
Post	3.43	3.55		
Self-esteem				
Pre	33.20	4.38	-1.09	NS
Post	34.14	3.86		
Job satisfaction				
Pre	65.21	17.81	.34	NS
Post	63.92	20.23		
Organisational commitment				
Pre	68.99	19.86	.20	NS
Post	68.17	20.18		

individual group members, and, thereby, diminishes the utility of the t-test (which assesses changes or differences in group means). One way of overcoming this problem is to determine the number of people whose scores change significantly between two time periods. Table 1 shows that some of the standard deviations from the mean client group scores are high (e.g., pre-somatic anxiety, M = 8.81, SD = 3.93). Therefore, a test of the number of people whose scores changed significantly was conducted. The criterion set for a significant change was an increase or decrease by at least half a standard deviation. Results of this analysis are shown in Table 3.

Table 3 indicates that the mental health measures show the greatest change after counseling, with 62%, 61% and 60% of clients showing significant improvements in anxiety, somatic anxiety and depression respectively. Few people show a significant deterioration on these measures. Self-esteem shows the next highest improvement (39% of clients), followed by job satisfaction (24%) and organizational commitment (16%).

It can be seen from Tables 1 and 2 that the pre-test scores between the counseled and control groups, on all the questionnaires, were significantly different (p ≤ .05). This is not surprising when you consider that nearly 50% of the counseled group were suffering from clinical anxiety or depression (see Table 6), and were seeking psychotherapeutic help, particularly in a culture (the U.K.) where psychological counseling or therapy is not as acceptable as it would be in other

TABLE 3   Number of People Whose Scores Change Significantly Pre- and Post-Counselling

	Anxiety		Somatic anxiety		Depression		Self-esteem		Job satisfaction		Organiza-tional commit-ment	
	N	%	N	%	N	%	N	%	N	%	N	%
Improvers	69	62	69	61	67	60	42	39	25	24	17	16
Stayers	36	32	32	28	27	24	53	49	58	57	67	63
Deteriorators	7	6	12	11	18	16	13	12	19	19	22	21

societies (e.g., U.S.). It can also be seen that, although the client group significantly improved their health on a number of dimensions in Table 1, there was still a statistically significant difference between their post-test scores and those of the control group, with the exception of the job satisfaction and organizational commitment scales. This again should not be so surprising, given the poor levels of pre-test health behaviors of the treatment group, and the immediate instrumentation of the post-measures after counseling. There are those who would argue that the beneficial impact of stress counseling is likely to continue for months after the experience. Our measure of change, therefore, might be interpreted as a conservative measure of improvement.

Sickness Absence Data

To determine whether the level of sickness absence changed in the six months following counseling, compared to the six months immediately preceding counseling, a series of paired t-tests were conducted, as well as for controls (shown in Table 4).

Table 4 shows that there is a significant fall in number of absence events, days lost and warnings in the six months immediately following counseling. Pre- and post-scores for the controls of postal employees show no significant differences. The only result which is significant ($p < .05$) is the mean number of authorized absence events occurring in the two periods, which increases from .23 to .39.

The number of clients whose sickness absence fell in the post-interview six months, in comparison to the pre-interview six months, is shown in Table 5. A change of half a standard deviation was the criterion used to determine a significant change in scores. A similar analysis was conducted with the control group data, also shown in Table 5.

Table 5 indicates that 27% and 22% of the client group show significant improvements in unauthorized absence events and days respectively. A very small percentage of the sample show deterioration in scores. In terms of events, the control group show both quite high improvement (25%) and deterioration (33%).

TABLE 4   Differences in Pre- and Post-Sickness Absence Levels for Counselled and Control Groups

Variable	Counselled Group (N = 188)			Control Group (N = 100)		
	Mean	SD	T-value	Mean	SD	T-Value
Events						
Pre	2.26	3.46	4.75**	.77	.94	-1.26
Post	1.22	1.46		.93	.95	
Days						
Pre	27.61	59.16	3.33**	8.01	37.79	-.26
Post	11.14	33.87		8.82	26.06	
Authorised events						
Pre	.23	.62	-.64	.23	.62	-2.18*
Post	.27	.75		.39	.96	
Authorised days						
Pre	1.01	6.27	1.05	.46	1.70	-1.17
Post	.53	1.64		.78	2.27	
Warnings						
Pre	.18	.40	2.64**	.09	.29	.90
Post	.09	.31		.06	.24	

$**p \leq .001$; $*p \leq .05$, two-tailed probability.

Little change occurs with absence days. A 3 x 2 chi-squared test of improvers and deteriorators for the two groups showed significant results for events ($\chi^2(2) = 40.75$) and days ($\chi^2(2) = 16.26$), both significant ($p < .05$).

It can be seen in Table 4 that the pre-test sickness absence between the counseled and control groups, on the number of "sickness events" and "days lost," were significantly different ($p \leq .05$). As in the case of the job satisfaction and health measures, this is likely to be a function of self-selection. After counseling, however, the counseled group more closely resembled the control group on post-test sickness absence "events," but were still significantly higher than controls on "days lost," but with a substantial narrowing of the gap. The positive impact of counseling on the sickness absence data seems more substantive than on the questionnaire measures, which could lend support to the notion that the effectiveness of counseling is likely to extend over several months or even a year after treatment. Post-counseling sickness absence data, unlike the questionnaire data, were collected for the six month period after the last counseling session.

Details on the Counseling Service

The problems presented to counselors were classified under the three headings: life, organization, and job. Table 6 depicts the four most frequently presented

TABLE 5   Number of People Whose Sickness Absence Levels Change after
Counselling

	Counselled Group				Control Group			
	Events		Days		Events		Days	
	N	%	N	%	N	%	N	%
Improvers	51	27	41	22	25	25	3	3
Stayers	128	68	138	74	42	42	91	91
Deteriorators	9	5	8	4	33	33	6	6

problems in each category. The most common problem was poor mental health or
excessive stress (49% of sample). It was also found that referrals to the service
were through the occupational health department (40%), followed by self (31%)
and welfare department (19%). In most cases, counselors were able to see clients in
the same month as they were referred and, on average, a counseling relationship
lasted for one month with 70% of the sample having about 4 sessions. The severity
of some of the clients' problems is reflected in the finding that 25% of the present
sample admitted to the counselors that they had had some degree of suicidal
thoughts. These were classified by the counselors as vague (12%), active (7%) and
high (6%).

DISCUSSION

The present project involved an empirical evaluation of an in-house stress
counseling service provided for postal employees in the U.K. Pre- and post-
counseling sickness absence data were presented for 188 experimental subjects
and 100 random sample controls. Paired t-tests showed a number of significant
changes in experimental pre- and post-scores: absence events, days and warnings
decline. No significant differences were observed for the control group on absence
during a comparable pre- and post-period.

Calculations of the number of subjects whose unauthorized absence scores
changed in either direction (with a change of at least half a standard deviation being
set as the criterion for a significant change) showed that 27% of experimentals
showed an improvement in absence events and 22% in days lost (compared to 5%
and 4% showing a significant deterioration on these measures).

Controls showed a high number of improvers, stayers and deteriorators for
events (25%, 42% and 33%, respectively). This is probably because of the small
initial standard deviation in scores (SD = .94), which set the change criterion at a
low level.

Another indicator of the mental ill health of this group is the finding that 25%

TABLE 6   Frequencies of Client 'Presenting' Problem

Problem	No. of Clients	% of Total Clients (N = 249)
Life		
Mental health/anxiety and depression	121	49
Marital/established relationship	44	18
Work issue	26	10
Family problems	18	7
Organisation		
Lack of appreciation	16	6
Excessive demand	9	4
Role conflict	6	2
Lack of career development	5	2
Job		
Long work hours	47	19
Conformity	39	16
Relationship/conflict	25	10
Environmental	7	3

of the sample had had some level of suicidal thoughts. The effectiveness of the counseling service is shown by the changes between pre- and post-scores obtained by this group. Anxiety, somatic anxiety and depression all fall significantly and self-esteem rises significantly.

One criticism of the control group used in the study is that, to make the two groups comparable, it is necessary to have pre-scores set at the same levels and then to observe subsequent changes. Due to the fact that sickness absence records within the Post Office are not kept on a computerised system, it was not possible to sample a sufficient number of records to identify employees with high absence levels and not visiting the counseling service. Furthermore, employees with high levels of absence are subject to an internal disciplinary procedure. Therefore, employees involved in the procedure could not have been viewed as a control group but as a second treatment group.

Counselors can also serve as a teaching input, creating greater awareness of potentially problematic lifestyles and indicating ways in which individuals can change to help themselves (largely through the use of stress management techniques). The results of this study indicate that the counselors may also have served such a function (some lifestyle changes or health behaviors are indicated).

Analysis of the number of people whose questionnaire scores improved, deteriorated or remain unchanged was conducted for the experimental group. Improvement in anxiety, somatic anxiety, depression and self-esteem was found for 62%, 61%, 60%, and 39% of the sample, respectively. Few people show a deterioration in scores on these measures: 6%, 11%, 16%, and 12%, respectively.

Work attitudes (i.e., job satisfaction and organizational commitment) show much less of an improvement: 24% improve and 19% deteriorate on job satisfaction; 16% improve and 21% deteriorate on organizational commitment. A similar pattern is reflected in the results obtained from the paired t-tests. Neither job satisfaction nor organizational commitment show significant pre-post changes. This might be explained by the fact that counseling is an intervention targeted at the individual, and has its greatest impact at this level. It is, perhaps, somewhat ambitious to increase a person's coping capabilities, and expect this to act as a panacea for all organizational problems. At times, it may be necessary to effect changes at work (i.e., adopt primary stressor reduction strategies) to achieve more favorable attitudes towards the organization. The present findings warrant further research on the effect of implementing person-targeted stress interventions at work on work-related attitudes. It may be of value to note that, while work attitudes may not change after counseling, work behavior is likely to change, in terms not only of reduced absence (evidence of which is found in the present study), but also improved productivity and better relationships with superiors and subordinates.

Further research is needed on the impact of stressor reduction strategies. Furthermore, to be able to advise organizations on how to proceed, comparative research is required, addressing questions like the relative effectiveness of in-house and external counseling services (i.e., EAPs). It is likely that different situations and different organizations will need alternative services, and it would be of value to identify the relative advantages of each system. A second important research issue is the success of intervention programes that combine primary, secondary and tertiary interventions in comparison with programes targeting only one of these.

REFERENCES

Allison, T., Cooper, C.L., & Reynolds, P. (1989, September). Stress counseling in the workplace. *The Psychologist*, 384–388.

Beehr, T.A., & Newman, J.E. (1978). Job stress, employee health, and organizational effectiveness: A facet analysis, model, and literature review. *Personnel Psychology, 31*, 665–699.

Cooper, C.L. (1987). Stress management interventions at work [Special issue]. *Journal of Managerial Psychology, 2*(1), 2–30.

Cooper, C.L., Cooper, R.D., & Eaker, L.H. (1988). *Living with stress.* Middlesex: Penguin.

Cooper, C.L., & Payne, R. (1988). *Causes, coping and consequences of stress at work.* New York & London: John Wiley & Sons.

Crandall, R. (1973). The measurement of self-esteem and related constructs. In J.P. Robinson, & P.R. Shaver (Eds.), *Measures of social psychological attitudes* (Rev. ed., pp. 45–167). Ann Arbor, MI: Institute for Social Research, University of Michigan.

Crown, S., & Crisp, A.H. (1979). *Manual of the Crown-Crisp Experiential Index.* Kent: Hodder and Stoughton Educational.

Defrank, R.S., & Cooper, C.L. (1987). Worksite stress management interventions: Their effectiveness and conceptualization. *Journal of Managerial Psychology, 2*, 4–10.

Griffin, R.W., & Bateman, T. (1986). Job satisfaction and organizational commitment. In C.L. Cooper & I.T. Robertson (Eds.), *International review of industrial and organizational psychology* (pp. 157–189). New York: John Wiley & Sons.

Highley, C., & Cooper, C.L. (1993). Evaluating employee assistance/counseling programes. *Employee Counseling Today, 5*(5), 13–18.

Ivancevich, J.M. (1987). Job stress: From theory to suggestion [Special issue]. *Journal of Organizational Behavior Management, 2*(1), 1–80.

Marshall, J.R., & Cooper, C.L. (1979). Work experience of middle and senior managers: The pressure and satisfaction. *International Management Review, 19,* 81–96.

Matteson, M.T., & Ivancevich, J.M. (1987). Individual stress management interventions: Evaluation of techniques. *Journal of Managerial Psychology, 2,* 24–30.

McLean, A.A. (1974). *Occupational stress.* IL: Charles C. Thomas.

Murphy, L.R. (1984). Occupational stress management: A review and appraisal. *Journal of Occupational Psychology, 57,* 1–15.

Murphy, L.R. (1988). Workplace interventions for stress reduction and prevention. In C.L. Cooper & R. Payne (Eds.), *Causes, coping and consequences of stress at work.* New York: John Wiley & Sons.

Orlans, V. (1991). *Evaluating the benefits of EAPs.* Paper presented at the First European EAP Conference, London.

Porter, L.W., Steers, R.M., Mowday, R.T., & Boulian, P.V. (1974). Organizational commitment, job satisfaction and turnover among psychiatric technicians. *Journal of Applied Psychology, 59,* 603–609.

Rogers, C.R. (1961). *On becoming a person.* Boston: Houghton Mifflin.

Sauter, S.L., Hurrell, J.J., Jr., & Cooper C.L. (1989). *Job control and worker health.* Chichester and New York: John Wiley & Sons.

Shapiro, O., Cheesman, M., & Wall, T.D., (1993). *Secondary prevention—review of counseling and EAPs.* Royal College of Physicians Conference, London.

Singer, J.A., Neale, M.S., & Schwartz, G.E. (1987). The nuts and bolts of assessing occupational stress: A collaborative effort with labor. In L.R. Murphy & T.F. Schaenborn (Eds.), *Stress management in work settings.* (National Institute for Occupational Safety and Health Publication No. 87–111). Washington, DC: U.S. Department of Health and Human Services.

Swanson, N.G., & Murphy, L.R. (1991) Mental health counseling in industry. In C.L. Cooper & I.T. Robertson (Eds.), *International review of industrial and organizational psychology* (pp. 265–282). New York & London: John Wiley & Sons.

Walsh, D.C., & Hingson, R.W. (1985). Where to refer employees for treatment of drinking problems. *Journal of Occupational Medicine, 27,* 745–752.

Warr, P., Cooke, J., & Wall, T.D. (1979). Scales for the measurement of some work attitudes and aspects of psychological well-being. *Journal of Occupational Psychology, 52,* 129–148.

Relations Between Exercise and Employee Responses to Work Stressors: A Summary of Two Studies

Steve M. Jex
Paul E. Spector
David M. Gudanowski
Ronald A. Newman

Given the cost of healthcare, organizations are understandably concerned about improving employee health. A popular method of enhancing employee health and reducing stress management has been physical activity or exercise. In fact, as of 1983, 14% of American employers offered some type of fitness program to their employees (Fielding & Breslow, 1983). This number has risen considerably since that time, both in the United States and abroad (Gebhardt & Crump, 1990). It is also more common today to see fitness programs as a component of broad-based organizational health promotion programs (cf. Erfurt, Foote, & Heirich, 1992).

One reason for this trend is that consistent exercise has been shown to be associated with physical benefits such as lowered resting heart rate and faster heart rate recovery in response to stressors (Cox, Evans, & Jamieson, 1978). It has also been hypothesized that exercisers are less likely than non-exercisers to suffer from coronary heart disease or become ill when confronted with life stressors (American Heart Association, 1991; Cooper, 1968; Roth & Holmes, 1985). Thus, exercise would appear to be a viable means of improving the physical health of employees.

Authors' Notes: Study 1 was based on the first author's doctoral dissertation completed at the University of South Florida under the direction of the second author. The authors thank Mike Gasser and Gail Nagy for assistance in data collection.

In addition to these physical benefits, it is widely believed that exercise provides psychological benefits such as elevated mood, resistance to psychological strain in response to life stressors, and reduced anxiety levels in the presence of a stressor (Folkins & Sime, 1981). As a result, it is believed that exercise can be used as a treatment for the psychological impact of work-related stressors. Unfortunately, little research has examined the effects of exercise on employee responses to the work environment. This is also true of other stress management treatments such as relation training, cognitive behavior modification, and other forms of behavior change (Beehr, Jex, & Ghosh, in press).

The little research that has been done on the effects of exercise in organizations can be classified into two types: (1) studies which examine the effects of worksite fitness programs, and (2) studies specifically designed to examine the effects of exercise, independent of the evaluation of any worksite fitness program. Evaluation studies have consistently found that fitness programs improve the physical condition of participating employees (Bruning & Frew, 1987; Cox, Shephard, & Corey, 1981; Harrison & Liska, 1994; Pauley, Palmer, Wright, & Pfeiffer, 1982), and may be associated with reductions in medical care costs in organizations (Erfurt et al., 1992; Shephard, Corey, Renzland, & Cox, 1982). Fitness programs have also been associated with improved self-concept, as well as reductions in anxiety, depression, absenteeism and turnover (Belles, Norvell, & Slater, 1988; Cox et al., 1981; Linden, 1969; Pauley et al., 1982; Tucker, 1990).

Although evaluation studies provide consistent evidence for the physical benefits of fitness programs, they are more difficult to evaluate with regard to psychological and behavioral benefits. Since employees typically self-select into these programs, preexisting differences, rather than exercise, may be responsible for the psychological and behavioral effects reported. Steinhardt and Young (1992), for example, found that fitness program participants reported higher levels of both attitudinal commitment to exercise and self-motivation than non-participants. In addition, because these programs are offered by organizations, it is difficult to determine whether the effects are due to *exercise* or merely the fact that the organization has offered another benefit.

Other studies have more directly examined the psychological effects of exercise in organizations. Ranney (1981), for example, found that lower levels of physical fitness combined additively with organizational stressors, life stressors, and Type A personality to predict anxiety, exhaustion, and diastolic blood pressure among male white-collar employees. In a similar study, Csanadi (1982) found that psychological disorders were significantly related to both life and work stressors only among law enforcement officers who did not perform regular physical activity of an aerobic nature. Other studies have found the relation between exercise and psychological strain to be quite small or non-significant (Dryson, 1986; Hendrix, Ovalle, & Troxler, 1985; Kobasa, 1982).

Though not plagued by problems of self-selection, it is also difficult to draw firm conclusions from studies directly examining the effects of exercise. For example, in both the Ranney (1981) and Csanadi (1982) studies, it is possible that uncontrolled differences between exercisers and non-exercisers were responsible for observed differences in psychological strain. This is analogous to the self-selection problem which plagues evaluation studies. In addition, since these studies used different measures of both exercise and psychological outcomes, it is difficult to compare their results. Furthermore, as pointed out by Jex and Heinisch (in press), self-report measures of exercise vary considerably and, in many cases, have unknown construct validity.

PSYCHOLOGICAL BENEFITS OF EXERCISE:
A DISPOSITIONAL APPROACH

In all exercise studies reviewed, uncontrolled variables may have been responsible for both differences in exercise adherence and psychological strain. Thus, it is possible that exercise is *not* really associated with positive psychological effects in work settings. Based on this assumption, a dispositional model of the relation between exercise and work-related strain is plausible. The basic idea of such a dispositional model is that personality or dispositional traits are associated with higher levels of exercise adherence *and* lower levels of psychological strain. Therefore, it is possible that relations between exercise and psychological strain may be due to the fact that both of these are associated with the common personality traits.

Although this model has not been tested, theoretical and empirical evidence supports relations between certain personality traits and both exercise adherence and psychological strain. Self-motivation, for example, has been shown to predict exercise adherence (Dishman, Ickes, & Morgan, 1980). It is also plausible that highly self-motivated individuals, or those who have a dispositional tendency to persist at a task, experience lower levels of psychological strain because they use more direct, problem-focused coping (Coyne, Aldwin, & Lazarus, 1981). Indirect evidence of this is shown by the positive relationship between "conscientiousness," a personality trait very much like self-motivation, and job performance (Barrick & Mount, 1991).

Dispositional optimism, the tendency to expect positive outcomes, has been empirically shown to be negatively related to psychological strain (Jex & Spector, in press; Scheier & Carver, 1985). It is also plausible that highly optimistic individuals exercise consistently since they will anticipate positive outcomes and be more committed to achieving exercise-related goals (Harrison & Liska, 1994). In addition, since optimists tend to be healthier than pessimists, exercise adherence may represent an overall pattern of better health (Leiker & Hailey, 1987; Scheier & Carver, 1985; Scheier et al., 1986).

Negative affectivity, the tendency to experience negative emotion, has been shown to be strongly related to psychological strain in work settings (Brief, Burke, George, Robinson, & Webster, 1988; Chen & Spector, 1991; Jex & Spector, in press; Tombaugh & White, 1989). Negative affectivity is also positively related to both reports of physical symptoms and poor health habits (Brief et al., 1988; Costa & McCrae, 1985, 1987). Thus, unlike those exhibiting high levels of optimism, it is proposed that those exhibiting high levels of negative affectivity will *not* exercise consistently because they possess *poor* health habits.

Locus of control represents beliefs regarding control over external reinforcements (Rotter, 1966). Organizational research suggests that an internal locus of control is associated with positive outcomes such as high motivation and job satisfaction (Spector, 1982). Externals have also been shown to have more negative reactions than internals when faced with the same stressors (Cvetanovski & Jex, in press; Storms & Spector, 1987). Wallston, Wallston, and DeVellis (1978) proposed that those with an internal *health* locus of control feel that they have control over the state of their health and therefore will engage in positive health behaviors. Externals, not feeling as much in control, will not be as likely to engage in positive health behaviors. Empirical evidence supports this proposition. Internals have been shown to adhere more readily to an exercise program and seek out more health-related information than externals (Carter, Lee, & Greenlocke, 1987; Trice & Price-Greathouse, 1987). It is important to note, in the present context, that individuals with an external locus of control tend to generalize these beliefs to other domains such as work (Rotter, 1975).

The Present Research

Two field studies were conducted to partially test the idea that exercise and psychological strain are related because they are correlated with the same personality traits. The first examined whether relations between exercise adherence and measures of psychological strain would be affected by controlling for dispositional optimism and self-motivation. Based on the literature reviewed and the proposed dispositional model, it was hypothesized that self-motivation and optimism would be positively correlated with exercise adherence and negatively correlated with psychological strain. It was also predicted that exercise would be negatively correlated with both physical (symptoms, sick days and doctor visits) and psychological strain (dissatisfaction, anxiety, frustration, turnover intent). A final prediction was that controlling both self-motivation and dispositional optimism would substantially reduce the relation between exercise adherence and psychological strain.

STUDY 1

Method

Subjects

Subjects were obtained from a random sample of 380 female clerical employees at the University of South Florida. Since the study required heart rate measurement, the original sample of 380 were contacted and asked if they would be willing to participate in the study. If they agreed, a questionnaire was administered when the heart rate measurement was obtained. Of the original 380, 250 (66%) agreed to participate, although nine were not eligible because of physical conditions which would have affected the resting heart rate measure. Of the 241 able to participate, 214 (89%) returned usable questionnaires. These employees ranged in age from 19 to 74 years, with a mean age of 42 years. Tenure with the university ranged from three to 336 months, with a mean of 92 months.

Measures

Personality Variables. Self-Motivation was measured with a shortened version of the Self-Motivation Inventory (SMI) (Dishman et al., 1980). The original SMI consisted of 40 items assessing the general tendency to persevere at a task. For the present study, 20 items were randomly selected. For each of the statements on the SMI, responses ranged from 1 ("extremely uncharacteristic of me") to 5 ("extremely characteristic of me"). Dishman et al. (1980) reported acceptable internal consistency (.86) and test-retest reliability (.92) for the SMI. Internal consistency of the shortened version was found to be acceptable (.79). As evidence of validity, Dishman et al. (1980) reported two studies in which SMI scores reliably predicted both adherence to a 20-week exercise program and a six-month physical training program designed to prepare female athletes for collegiate rowing competition.

Dispositional Optimism was measured using the Life Orientation Test (LOT; Scheier & Carver, 1985). The LOT consists of 12 items, with 8 actually contributing to the scale and four being filler items. For each of the 12 items, responses ranged from 0 ("strongly disagree") to 4 ("strongly agree"). Scheier and Carver (1985) reported an internal consistency of .76 and test-retest reliability of .79 for the LOT. To assess validity, Scheier and Carver (1985) reported that LOT scores were positively related to self-esteem and internal locus of control and negatively related to depression, hopelessness, perceived stress and alienation. In addition, those individuals identified as optimists by the LOT have been found to use more effective coping strategies, report fewer physical symptoms, and recover more quickly from coronary bypass surgery (Scheier et al., 1986) than pessimists.

Exercise Adherence. One measure of exercise adherence was a four-item scale from Kobasa, Maddi, and Puccetti (1982) which assessed, in general, the extent to which subjects participated in sports activities, non-sports physical activities, how strenuous these activities were, and how many hours were spent in both sports and non-sports activities. Epidemiological research has shown that those who score higher on this scale (indicating regular exercise) show a decreased likelihood of heart attacks (Paffenberger & Hale, 1975). Additionally, Kobasa et al. (1982) found that individuals who were classified as regular exercisers by this scale reported fewer illnesses than those who were not. Taken together, this evidence suggests that this is a valid measure of regular exercise adherence. Alpha in this study was .82.

Since heart rate is a good indicator of physical fitness level (Aunola, Nykyri, & Rusko, 1978) and many of the studies reviewed have used this measure, Heart Rate was used as a second indicator of exercise adherence. Since physiological measures tend to be highly unreliable (Fried, Rowland, & Ferris, 1984), five separate resting heart rate measures were taken from each subject and averaged.

Physical Strain. Physical strain was measured by self-reported health symptoms (Spector, Dwyer, & Jex, 1988), number of visits to the doctor and sick days. Symptoms were measured by presenting subjects with a list of 21 specific health symptoms (e.g., nausea, backache, eyestrain, and fatigue) preceded by the question "During the past 30 days, did you have...?" Subjects indicated whether they had experienced the symptom or not and whether the condition required a visit to a doctor. The validity of self-reported symptoms has been shown by correlations between these measures and more objective measures of health such as visits to a physician, taking medications, and physician diagnoses (Caplan, Cobb, French, Van Harrison, & Pinneau, 1975; Kobasa, Maddi, & Courington, 1981). Subjects were also asked to indicate both the number of times they had visited a doctor and been absent due to illness during this period.

Psychological Strain. The four measures of psychological strain were overall Job (Dis)Satisfaction, Frustration, Anxiety, and Intent to Quit. Overall job (dis)satisfaction was assessed by the overall job satisfaction scale of the Michigan Organizational Assessment Scale (Cammann, Fichman, Jenkins, & Klesh, 1979). This is a three-item scale which assessed overall satisfaction with one's job. Responses to each of the three items ranged from 1 ("strongly disagree") to 6 ("strongly agree"). Internal consistency has been estimated to be .90 for this scale (Spector et al., 1988) and was .79 here.

Frustration was measured by the three-item Peters and O'Connor (1980) frustration scale. Three statements pertaining to frustration experienced in the workplace were rated on a scale from 1 ("strongly disagree") to 6 ("strongly agree"). Peters and O'Connor estimated the internal consistency of this scale to be .81 and it was .77 here.

A modified version of the 10-item state scale of Spielberger's State Trait Personality Inventory (STPI; Spielberger, 1979) was used to measure anxiety. Specifically, subjects were asked how they felt at work during the past 30 days. Some of the descriptions were feeling calm, tense, nervous, worried, and frightened. Each item was answered in terms of the degree to which the respondent felt that way. Responses ranged from 1 ("not at all") to 4 ("very much so"). The internal consistency of this scale has been estimated to be .90 (Spector et al., 1988) and was .81 here.

Turnover intent was measured by a single item asking the respondent to indicate how often she had considered quitting her present job. Responses ranged from "never" to "extremely often" on a six-point scale.

The validity of psychological strain measures is evidenced by the fact that they have been used extensively in the work stress literature, have been shown to be reliably related to a variety of work-related stressors, and intercorrelate well (cf. Jex & Beehr, 1991).

Procedure

For the 241 employees agreeing to participate in the study, a time was scheduled during the workday to obtain heart rate measures and deliver the questionnaire. When the experimenter arrived at the worksite, each subject was asked to sit quietly while the purpose of the study was reviewed. After the purpose of the study and requirements of participation were explained, a series of five, 10-second resting heart rate measures were taken, with each measure separated by a pause of 10 seconds. Since heart rate measurement is affected by a number of factors (Fried et al., 1984), subjects were asked to refrain from smoking or drinking caffeinated beverages during the hour prior to heart rate measurement. An effort was also made to keep the time of measurement relatively constant.

Following heart rate measurement, subjects were given the questionnaire booklet and asked to complete it at their own convenience. Completed questionnaires were returned to the senior author through campus mail. Data collection took place from March through May of 1988.

Results

Descriptive Statistics

Descriptive statistics on all study variables are presented in Table 1. Included are sample sizes, means, standard deviations, and reliabilities (coefficient alphas where appropriate). As can be seen, coefficient alphas were all reasonably high, ranging from .77 to .97. The lowest was .77 for frustration, while the highest was .97 for the five heart rate measures. For the exercise scale, where coefficient alpha was not appropriate, average intercorrelation among items was .39. Although not presented in Table 1, it was found that observed ranges from most of the variables covered the entire possible range; thus, restriction of range did not appear to be a problem. An exception was self-motivation, which ranged from 50 to 100 out of a possible range of 20 to 100.

Correlations

Correlations among all variables are also presented in Table 1. As can be seen, the correlation between Optimism and Self-Motivation was strong (.41). However, self-reported Exercise Adherence and resting Heart Rate were not significantly related, suggesting that perhaps resting heart rate was *not* a good indicator of exercise adherence. Measures of physical strain also did not intercorrelate well. Symptoms Requiring a Doctor's Attention was weakly correlated with Symptoms Not Requiring a Doctor's Attention (-.13), moderately related to Doctor Visits (.36), but unrelated to Sick Days. In addition, Doctor Visits was moderately

TABLE 1 Intercorrelations for All Study 1 Variables

	1	2	3	4	5	6	7	8	9	10	11	12
1. Optimism												
2. Self-Motivation	.41*											
3. Exercise	.06	.09										
4. Heart Rate	.02	.08	-.12									
5. Symptoms (Doc)	-.05	.13	-.09	.05								
6. Symptoms (No Doc)	-.11	.01	.13*	-.06	-.13							
7. Doctor Visits	-.06	-.02	-.15*	-.08	.36*	.01						
8. Sick Days	.03	-.03	-.16*	-.06	.09	.05	.37*					
9. Satisfaction	.16*	.07	-.13*	.02	.00	-.17*	.05	.09				
10. Frustration	-.15*	-.07	.00	-.06	.06	.38*	.06	.03	-.33*			
11. Anxiety	-.34*	-.12	.03	-.05	.12	.45*	.10	.01	-.36*	.50*		
12. Intent	-.03	-.01	.16*	-.13	.03	.28*	.06	.01	-.57*	.35*	.34*	
N	214.0	213.0	214.0	214.0	214.0	214.0	214.0	213.0	213.0	214.0	214.0	213.0
Mean	28.5	79.7	2.4	74.6	0.6	5.1	1.1	2.6	14.1	10.1	19.3	2.8
Standard Deviation	5.5	9.5	1.8	8.8	1.5	3.4	1.9	3.6	3.1	3.7	6.2	1.5
Coefficient Alpha	0.82	0.79	0.39[a]	0.97	NA	NA	NA	NA	0.79	0.77	0.81	NA

*p < .05
[a]Average interitem correlation among items using r to z transformation.

correlated with Sick Days (.37). Measures of psychological strain were all significantly intercorrelated, ranging from a high of -.57 (Satisfaction–Turnover Intent) to -.33 (Satisfaction–Frustration).

In terms of the dispositional model proposed earlier, neither Optimism nor Self-Motivation was significantly related toExercise. Optimism, however, was weakly related to both Job Satisfaction and Frustration (.16 and -.15, respectively) and moderately related to Anxiety (-.34). Self-Motivation was not related to any of the psychological strain measures.

Exercise Adherence was weakly correlated with Symptoms (No Doctor) (.13), Doctor Visits (-.15), and Sick Days (-.16). However, the correlation between exercise and symptoms was in the opposite direction than predicted. Exercise was also significantly correlated with two psychological strain measures: Job Satisfaction andTurnover Intent (-.13 and -.16, respectively). These correlations were not only quite small but in the opposite direction than predicted. Heart Rate was not significantly related to any of the strain measures.

Discussion

Study 1 assessed relations between exercise adherence, physical and psychological strain, and personality. An attempt was also made to discover whether the relation between exercise and psychological strain could be reduced by controlling for personality traits related to both.

As predicted, exercise adherence was negatively related to number of doctor visits and sick days. These correlations were weak, but consistent with previous research on the relations between exercise and physical outcomes (e.g., American Heart Association, 1991; Pauley et al., 1982). Somewhat surprisingly, exercise was *positively* related to physical symptoms, although this correlation was quite low. It is possible that those who exercise frequently may be healthier overall but have more *minor* physical symptoms than non-exercisers—perhaps injuries caused by the exercise. Neither personality characteristic was correlated with exercise adherence, and only optimism was correlated with outcomes (satisfaction, anxiety and frustration). Exercise was also weakly related to both job satisfaction and turnover intentions in the opposite direction than predicted. These results, however, are inconsistent with studies suggesting that exercise provides positive psychological benefits (Pauley et al., 1982; Ranney, 1981). On the other hand, this finding may signify that those who were exercising were those most in need. Kohler and Swim (1991), for example, found that employees experiencing high levels of job-related tension were most willing to participate in an exercise program on their own time.

Since exercise was so weakly related to measures of psychological strain, it was difficult to test the effect of controlling for personality traits hypothesized to be associated with both. As a result, little support was provided for the dispositional model of the relation between exercise adherence and psychological strain.

Overall, the results of this study suggest that exercise is only slightly associated with positive physical outcomes. Exercise, however, was not associated with positive psychological outcomes or personality. As a result, the effect of uncontrolled personality differences could not be assessed adequately. Because of limitations of the present study, however, these results are difficult to interpret. One limitation was that the distribution for exercise was skewed considerably, with 26% of respondents indicating that they do not engage in any form of exercise, and many reporting very low levels. Given that the rate of participation in company-sponsored fitness programs is typically less than 20% (Gebhardt & Crump, 1990), this is not surprising. It is also possible that the exercise scale had questionable construct validity, despite being used successfully elsewhere (Kobasa et al., 1982; Paffenberger & Hale, 1975). Questions on this scale were quite general and gave respondents considerable freedom in deciding what was and was not exercise. It is possible, therefore, that the scale did not clearly distinguish between those who consistently exercised and those who did not (Jex & Heinisch, in press).

Another limitation was the fact that the sample was quite homogeneous, consisting entirely of female clerical employees. Perhaps this sample, which exhibited a limited range of exercise participation, was not a good one to use in testing the effects of exercise. Since many of the women in this sample had parental and household responsibilities, time restrictions may have prevented regular exercise participation.

STUDY 2

Based on the limitations identified in Study 1, a second study was conducted to assess whether the results of Study 1 were due to the measurement of exercise or the uniqueness of the sample. Thus, a more extensive measure of exercise was used and both males and females representing a variety of job types were included in the sample. Another change from Study 1 was the selection of additional personality traits to examine. In addition to dispositional optimism, negative affectivity and health locus of control were included. Self-motivation was not included since it was unrelated to both exercise and psychological strain in Study 1.

Method

Subjects

Subjects were 154 male and female non-faculty employees of both the University of South Florida and Central Michigan University. These employees represented a variety of jobs. In all, 500 employees from both universities were selected. The overall response rate was 30%. Employees ranged in age from 23 to 68 years of age, with a mean of 41.2 years.

Measures

Personality Variables. Dispositional Optimism was again measured with the LOT (Scheier & Carver, 1985). The psychometric properties of the LOT are given in Study 1.

Negative Affectivity was measured with the negative affect subscale of the Positive and Negative Affect Scales (PANAS; Watson, Clark, & Tellegen, 1988). This consists of 10 adjectives reflecting negative affective states (e.g., "distressed," "upset," "hostile"). Subjects were asked to indicate the degree to which each of these adjectives characterized how they feel in general, with responses ranging from 1 ("very slightly or not at all") to 5 ("extremely"). According to Watson et al. (1988), if subjects use the PANAS items to describe how they generally feel, these scores demonstrate stability and reflect the underlying trait of negative affectivity. Internal consistency of the NA scale has been estimated to be .87 (Watson et al., 1988). Watson et al. (1988) provided validity evidence for the NA scale by showing its relation to established measures of negative affect (e.g., trait anxiety) and psychopathology, e.g., the Hopkins Symptom Checklist (HSCL; Derogatis, Lipman, Rickels, Uhlenhuth, & Covi, 1974), Beck Depression Inventory (BDI; Beck, Ward, Mendelson, Mock, & Erbaugh, 1961), and the A-State Scale of the State-Trait Personality Inventory (STAI; Spielberger, 1979).

Health Locus of Control was measured with the self-care (SC) subscale of the Lau-Ware Health Locus of Control Scales (Lau & Ware, 1981). The SC subscale consists of eight items reflecting the degree to which an individual believes he or she has control over his/her health. Subjects were asked to indicate the degree to which they agreed with each statement. Responses ranged from 1 ("strongly disagree") to 5 ("strongly agree"). Internal consistency of the SC subscale has been estimated to be .65 (Lau & Ware, 1981). Validity evidence has been shown by the fact that those with an internal health locus of control, as measured by the SC subscale, are more likely to perform self-care health behaviors (e.g., eating proper foods, controlling one's weight) and have regular medical checkups than externals (Lau, 1982).

Exercise Adherence. The primary measure of Exercise Adherence was a 21-item scale developed by Pasman and Thompson (1988). Each item was a statement about exercise behavior (i.e., "If I miss a planned workout, I attempt to make up for it the next day"). Subjects were asked to indicate how often they could make each statement. Responses ranged from 1 ("never") to 4 ("always"). The internal consistency of this scale was estimated to be .89 for the present study. Pasman and Thompson (1988) provided validity evidence by showing that scores on this scale reliably distinguished consistent runners and weight lifters from sedentary individuals, although items were not specific to either sport. It was felt that this scale more effectively captured the underlying construct of "exercise adherence" than simply asking subjects whether or not they exercise.

In addition to the Pasman and Thompson (1988) scale, subjects were asked to indicate the number of hours per week they engaged in the following activities:

Jogging, Weight Lifting, and Aerobic Dance. These activities were selected because of their popularity and the fact that many health clubs provide opportunities for participation. In addition, both males and females participate in all three activities.

Physical Strain. Physical strain was measured by asking subjects the number of times they needed to see a doctor in the past three months.

Psychological Strain. Measures of psychological strain (Job (Dis)Satisfaction, Frustration, Anxiety, and Turnover Intent) were the same as those used in Study 1, and details of each can be found there.

Procedure

At each of the two universities, a random sample of both male and female non-faculty employees performing a variety of jobs was taken from the university directory. Once selected, employees were sent a questionnaire along with a letter describing the purpose of the study and asking for their participation. Completed questionnaires were mailed to the senior author. Data were collected between August and December, 1989.

Results

Descriptive Statistics

Descriptive statistics for all variables in Study 2 are presented in Table 2. Included are sample sizes, means, standard deviations, and coefficient alphas (where appropriate). As can be seen, reliability estimates were acceptable for most scales, the lowest being .64 for the health locus of control scale. As in Study 1, restriction of range was not a severe problem. Several variables, however, did not cover the entire possible range. This was particularly true for exercise, with the highest observed value of 66 out of 84, and work anxiety, with the highest observed value of 34 out of 48.

Correlations

Correlations among all variables are also presented in Table 2. As can be seen, Optimism and Negative Affectivity were negatively correlated, as would be expected (-.33). Health Locus of Control, however, was not correlated with either Optimism or Negative Affectivity. Exercise Adherence, as measured by the self-report scale, was significantly related to reports of participation in Jogging, Weight Lifting, and Aerobic Dance (.47, .36, and .21, respectively). The three activities were not highly intercorrelated, however. Intercorrelations among psychological strain measures ranged from -.69 (Satisfaction–Turnover Intent) to -.30 (Satisfaction–Anxiety).

In terms of the proposed dispositional model, none of the personality measures were significantly correlated with Exercise Adherence, although Optimism and Health Locus of Control were weakly related to Jogging (.14 and .19, respectively).

Optimism was also correlated with all measures of psychological strain as predicted. These correlations ranged from -.30 with Anxiety to -.20 with Turnover Intent. Correlations between Negative Affectivity and psychological strain were somewhat stronger, ranging from .51 with Anxiety to -.22 with Satisfaction. Thus, optimism was the only personality trait related to measures of both exercise and psychological strain.

Exercise Adherence was weakly correlated with Doctor Visits as was predicted (-.14), but unrelated to measures of psychological strain. None of the three exercise activities were related to Doctor Visits. Jogging, however, was weakly correlated with Anxiety as expected (-.16), while Aerobics was weakly correlated with Frustration (.15) in the opposite direction than expected.

Gender Comparison

Since the results of Study 1 may have been due to the fact that the sample was all female, correlations between exercise measures and all other variables were examined separately for males and females. Optimism was moderately associated with Jogging among females (.34) but unrelated to this activity among males (-.04). In addition, Health Locus of Control was related to Weight Lifting among males (.23) but unrelated among females (-.13). However, since, out of 32 gender comparisons, these were the only two that were significant beyond the .05 level, it was concluded that they may have been due to chance. Thus, it did not appear as though gender differences were important.

Regression

The focus of the regression analysis was the prediction of both anxiety and frustration from optimism and exercise activities. In both cases, the objective was to find out whether exercise had any predictive power after the effects of optimism were controlled. From the correlations presented in Table 2, it was shown that jogging predicted a small amount of variance in anxiety if entered into the regression equation alone. Table 3, however, shows that the contribution of jogging to the prediction of anxiety became non-significant both in terms of the beta weight (-.16 to -.12) and R^2 change when entered after optimism. Aerobics also predicted a small amount of variance in frustration when entered alone. Controlling for optimism reduced the magnitude of its beta weight slightly but appeared to have little effect on the R^2 change index.

Discussion

Study 2 was conducted to test the same hypotheses as Study 1, with some modification in the measurement of exercise, the personality traits examined, and the sample. Specifically, Study 2 employed a more extensive measure of exercise adherence along with reports of specific exercise activities. In addition, negative

TABLE 2 Descriptive Statistics and Intercorrelations for Study 2 Variables

	1	2	3	4	5	6	7	8	9	10	11	12
1. Optimism												
2. Negative Affectivity	-.33*											
3. Health Locus of Control	.11	-.03										
4. Exercise	.11	-.04	.12									
5. Jogging	.14*	-.07	.19*	.47*								
6. Weights	-.01	-.09	-.02	.36*	-.09							
7. Aerobics	-.08	-.02	.07	.21*	.01	.17*						
8. Satisfaction	.21*	-.22*	-.04	-.04	-.03	-.06	-.08					
9. Frustration	-.27*	.32*	.11	-.06	-.12	-.04	.15*	-.32*				
10. Anxiety	-.30*	.51*	-.06	-.07	-.16*	-.06	.07	-.30*	.47*			
11. Intent	-.20*	.25*	.06	.07	-.05	.13	.08	-.69*	.50*	.32*		
12. Doctor Visits	-.05	.08	.03	-.14*	-.06	-.08	-.02	-.13*	.05	.27*	.04	
N	150.0	144.0	150.0	137.0	154.0	154.0	154.0	152.0	154.0	145.0	148.0	154.0
Mean	27.8	16.8	29.9	38.4	0.5	0.4	0.5	14.2	11.1	19.4	2.7	0.5
Standard Deviation	3.5	5.6	3.6	9.3	1.1	1.1	1.4	3.3	3.7	4.7	1.4	0.9
Coefficient Alpha	0.82	0.87	0.64	0.89	NA	NA	NA	.87	0.84	0.80	NA	NA

*p < .05

affectivity and health locus of control, two traits which could be associated with both exercise adherence and psychological strain, were examined. Finally, the sample was more heterogeneous, including males and females performing a variety of jobs.

Results indicated that none of the three personality traits were related to overall exercise adherence, although optimism and health locus of control were related to the number of hours subjects engaged in jogging. Exercise adherence was weakly related to physical strain (doctor visits), but unrelated to measures of psychological strain. Although none of the exercise activities were related to physical strain, jogging was negatively related to anxiety, and hours of aerobic dance was positively related to frustration. Both of these correlations were quite small, however, and the correlation between aerobics and frustration was in the opposite direction than predicted. However, as in Study 1, the relation between aerobics and frustration may simply reflect the fact that individuals were engaging in this activity to "work out their frustrations." The relation between jogging and anxiety was reduced somewhat by controlling for optimism.

SUMMARY

Taken together, the results of Studies 1 and 2 are consistent with previous studies which have found the relation between exercise and psychological strain to be small or non-existent (Dryson, 1986; Hendrix et al., 1985; Kobasa, 1982). Furthermore, the results of Study 2 showed that relations between jogging and anxiety were reduced by controlling for dispositional optimism. This provides very limited support for the idea that relations between exercise and outcomes are due to common personality traits. Perhaps personality traits not assessed here, or other individual differences, will be shown in future research to be important.

A good place to start in this regard would be with self-efficacy (Bandura, 1986). Self-efficacy reflects a person's belief that he or she can successfully complete a task or some other course of action. Harrison and Liska (1994) found that individuals who believed that they could achieve their exercise goals were more committed to them and, in fact, more likely to achieve them. Compared to personality, self-efficacy is probably more malleable. Thus, focusing on self-efficacy, as opposed to personality traits, may be more helpful to organizations in guiding their efforts to enhance fitness program participation.

Beyond investigating the effects of exercise, these studies also highlight important methodological issues. For example, in both studies, the low correlations between exercise and other variables may have been due to the *measurement* of exercise. This suggests to the present authors that the variable "exercise" is quite complex and multi-faceted. Considerably more thought and effort should be put into conceptualizing and measuring this construct in the future.

A second methodological issue to be considered is the skewed nature of the

TABLE 3   Summary of Hierarchical Regression Analyses

Variable	Beta Anxiety	R^2	ΔR^2
Step 1: Optimism	-.30*	.09*	
Step 2: Optimism	-.28*		
Jogging	-.12	.10*	.01

Variable	Frustration	R^2	ΔR^2
Step 1: Optimism	-.27*	.08*	
Step 2: Optimism	-.26*		
Aerobics	.13	.09*	.01

*$p < .05$

distributions of exercise activities. For example, in Study 2, the percentage of subjects engaging in jogging, weight lifting, and aerobics was only 20, 14, and 14, respectively. This is consistent with past research on fitness program participation (Gebhardt & Crump, 1990). Unfortunately, such skewness attenuates correlations since they are based on the assumption of normally distributed variables (Pedhazur, 1982). One solution to this dilemma might be to seek a diverse sample so that there is adequate variance in exercise. Another option is to preselect samples so that everyone is engaged in *some* minimal level of exercise activity. Whichever solution is chosen, researchers investigating correlates of exercise need to be sensitive to this issue.

In conclusion, the two studies presented in this paper are inconsistent with the notion that exercise participation greatly enhances the psychological functioning of employees in organizational settings. These studies, however, were cross-sectional and dealt only with *self-reported* exercise behavior. Such limitations suggest that the present results must be interpreted cautiously and that more research is needed. Further research examining the effects of actual exercise behavior over time while controlling for differences between exercisers and non-exercisers would be most beneficial. By designing quality research, organizations will be given a better idea of what to expect from exercise and fitness programs and will establish them for realistic reasons. It is hoped that this research will also provide organizations with ways to increase participation in such programs. At present, there is evidence that participation is lowest among those most at risk for health problems (Harrison & Liska, 1994).

REFERENCES

American Heart Association. (1991). *Exercise and your heart.* TX: Author.

Aunola, S., Nykyri, R., & Rusko, H. (1978). Strain of employees in the machine industry in Finland. *Ergonomics, 21,* 509–519.

Bandura, A. (1986). *Social foundations of thought and action: A social cognitive theory.* Englewood Cliffs, NJ: Prentice-Hall.

Barrick, M.R., & Mount, M.K. (1991). The Big Five personality dimensions and job performance. *Personnel Psychology, 44,* 1–26.

Beck, A.T., Ward, C.H., Mendelson, M., Mock, J., & Erbaugh, J. (1961). An inventory for measuring depression. *Archives of General Psychiatry, 4,* 561–571.

Beehr, T.A., Jex, S.M., & Ghosh, P. (in press). The management of occupational stress. *Handbook of Performance Management.*

Belles, D., Norvell, N., & Slater, S. (1988). *The psychological benefits of Nautilus weight training in law enforcement personnel.* Unpublished manuscript, University of Florida, Gainesville.

Brief, A.P., Burke, M.J., George, J.M., Robinson, B.S., & Webster, J. (1988). Should negative affectivity remain an unmeasured variable in the study of job stress? *Journal of Applied Psychology, 73,* 199–207.

Bruning, N.S., & Frew, D.R. (1987). Effects of exercise, relaxation, and management skills training on physiological stress indicators. *Journal of Applied Psychology, 72,* 515–521.

Cammann, C., Fichman, M., Jenkins, D., & Klesh, J. (1979). *Michigan Organizational Assessment Questionnaire.* Unpublished manuscript, University of Michigan, Ann Arbor.

Caplan, R.D., Cobb, S., French, J.R.P., Jr., Van Harrison, R., & Pinneau, S.R., Jr. (1975). *Job demands and worker health: Main effects and occupational differences* (HEW NIOSH No. 75–160). Washington, DC: U.S. Government Printing Office.

Carter, J.A., Lee, A.M., & Greenlocke, K.M. (1987). Locus of control, fitness values, success expectations and performance in a fitness class. *Perceptual and Motor Skills, 65,* 777–778.

Chen, P.Y., & Spector, P.E. (1991). Negative affectivity as the underlying cause of correlations between stressors and strains. *Journal of Applied Psychology, 76,* 398–407.

Cooper, K.H. (1968). *Aerobics.* Philadelphia: Lippincott.

Costa, P.T., Jr., & McCrae, R.R. (1985). Concurrent validation after 20 years: Implications of personality stability for its assessment. In J.N. Butcher & C.D. Spielberger (Eds.), *Advances in personality assessment* (Vol. 4, pp. 31–54). Hillsdale, NJ: Erlbaum.

Costa, P.T., Jr., & McCrae, R.R. (1987). Neuroticism, somatic complaints, and disease: Is the bark worse than the bite? *Journal of Personality, 55,* 299–316.

Cox, J.P., Evans, J.F., & Jamieson, J.L. (1979). Aerobic power and tonic heart rate responses to psychosocial stressors. *Personality and Social Psychology Bulletin, 5,* 160–163.

Cox, M., Shephard, R.J., & Corey, P. (1981). Influence of an employee fitness program upon fitness, productivity, and absenteeism. *Ergonomics, 24,* 795–806.

Coyne, J.C., Aldwin, C.M., & Lazarus, R.S. (1981). Depression and coping in stressful episodes. *Journal of Abnormal Psychology, 90,* 439–447.

Csanadi, S.B. (1982). *Physical activity and stressor-strain relationships in law enforcement.* Unpublished doctoral dissertation, University of South Florida, Tampa.

Cvetanovski, J., & Jex, S.M. (in press). Locus of control of unemployed people and its relationship to psychological and physical well-being. *Work and Stress.*

Derogatis, L.R., Lipman, R.S., Rickels, K., Uhlenhuth, E.H., & Covi, L. (1974). The Hopkins Symptom Checklist (HSCL): A self-report symptom inventory. *Behavioral Science, 19,* 1–15.

Dishman, R.K., Ickes, W., & Morgan, W.P. (1980). Self-motivation and adherence to habitual physical activity. *Journal of Applied Social Psychology, 2,* 115–132.

Dryson, E.W. (1986). Stress and some associated factors in a representative sample of the New Zealand workforce. *New Zealand Medical Journal, 10,* 668–670.

Erfurt, J.C., Foote, A., & Heirich, M.A. (1992). The cost effectiveness of worksite wellness programs for hypertension control, weight loss, smoking cessation, and exercise. *Personnel Psychology, 45,* 5–27.

Fielding, J.E., & Breslow, L. (1983). Health promotion programs sponsored by California employers. *American Journal of Public Health, 73,* 538–542.

Folkins, C.H., & Sime, W.W. (1981). Physical fitness training and mental health. *American Psychologist, 36,* 373–389.

Fried, Y., Rowland, K.M., & Ferris, G.R. (1984). The physiological measurement of work stress: A critique. *Personnel Psychology, 37,* 583–615.

Gebhardt, D.L., & Crump, C.E. (1990). Employee fitness and wellness programs in the workplace. *American Psychologist, 45*(2), 262–272.

Harrison, D.A., & Liska, L.Z. (1994). Promoting regular exercise in organizational fitness programs: Health-related differences in motivational building blocks. *Personnel Psychology, 47,* 47–72.

Hendrix, W.H., Ovalle, N.K., & Troxler, R.G. (1985). Behavioral and physiological consequences of stress and its antecedent factors. *Journal of Applied Psychology, 70*(1), 188–201.

Jex, S.M., & Beehr, T.A. (1991). Emerging theoretical and methodological issues in the study of work-related stress. In K. Rowland & G. Ferris (Eds.), *Research in personnel and human resources management* (Vol. 9, pp. 311–365). Greenwich, CT: JAI.

Jex, S.M., & Heinisch, D.A. (in press). Assessing the relationship between exercise and employee mental health: Some methodological concerns. *Workplace health: Employee fitness and exercise.*

Jex, S.M., & Spector, P.E. (in press). The impact of negative affectivity on stressor–strain relations: A replication and extension. *Work and Stress.*

Kobasa, S.C. (1982). Commitment and coping in stress resistance among lawyers. *Journal of Personality and Social Psychology, 42,* 707–717.

Kobasa, S.C., Maddi, S.R., & Courington, S. (1981). Personality and constitution as mediators in the stress–illness relationship. *Journal of Health and Social Behavior, 22,* 368–378.

Kobasa, S.C., Maddi, S.R., & Puccetti, M.C. (1982). Personality and exercise as buffers in the stress–illness relationship. *Journal of Behavioral Medicine, 5,* 391–404.

Kohler, S., & Swim, J. (1991). *Health promotion exercise program: The role of work-related variables.* Unpublished manuscript, Pennsylvania State University.

Lau, R.R. (1982). Origins of health locus of control beliefs. *Journal of Personality and Social Psychology, 42,* 322–334.

Lau, R.R., & Ware, J.F. (1981). Refinements in the measurement of health-specific locus-of-control beliefs. *Medical Care, 29,* 1147–1158.

Leiker, M., & Hailey, B.J. (1987). *The link between hostility and disease: Poor health habits?* Paper presented at the Southeastern Psychological Association Convention, Atlanta, GA.

Linden, V. (1969). Absence from work and physical fitness. *British Journal of Industrial Medicine, 26,* 47–53.

Paffenberger, R.J., Jr., & Hale, W.E. (1975). Work activity and coronary heart mortality. *New England Journal of Medicine, 292,* 545.

Pasman, L.N., & Thompson, J.K. (1988). Body image and eating disturbance in obligatory runners and obligatory weight lifters and sedentary individuals. *International Journal of Eating Disorders, 7,* 759–768.

Pauley, J.T., Palmer, J.A., Wright, C.C., & Pfeiffer, G.J. (1982). The effect of a 14-week employee fitness program on selected physiological and psychological parameters. *Journal of Occupational Medicine, 24,* 457–463.

Pedhazur, E.J. (1982). *Multiple regression in behavioral research.* New York: Holt, Rinehart, and Winston.

Peters, L.H., & O'Connor, E.J. (1980). The behavioral and affective consequences of performance-related situational variables. *Organizational Behavior and Human Performance, 25,* 79–96.

Ranney, J.M. (1981). A study on the relationship between physical fitness and occupational strain among white collar workers. *Dissertation Abstracts International, 42,* 1769A.

Roth, D.L., & Holmes, D.S. (1985). Influence of physical fitness in determining the impact of stressful life events on physical and psychologic health. *Psychosomatic Medicine, 47,* 164–173.

Rotter, J.B. (1966). Generalized expectancies for internal versus external control of reinforcement. *Psychological Monographs, 80*(1), Whole No. 609.

Rotter, J.B. (1975). Some problems and misconceptions related to the construct of internal versus external locus of control. *Journal of Consulting and Clinical Psychology, 43,* 56–67.

Scheier, M.F., & Carver, C.S. (1985). Optimism, coping, and health: Assessment and implications of generalized outcome expectancies. *Health Psychology, 4,* 219–247.

Scheier, M.F., Matthews, K.A., Owens, J., Abbott, A., Lebfevre, C., & Carver, C.S. (1986). *Optimism and recovery from coronary artery bypass surgery.* Unpublished manuscript, Carnegie-Mellon University, Pittsburgh, PA.

Shephard, R.J., Corey, P., Renzland, P., & Cox, M. (1982). The influence of an employee fitness and life modification program upon medical care costs. *Canadian Journal of Public Health, 73,* 259–263.

Spector, P.E. (1982). Behavior in organizations as a function of locus of control. *Psychological Bulletin, 91,* 482–497.

Spector, P.E., Dwyer, D.J., & Jex, S.M. (1988). Relations of job stressors to affective, health, and performance outcomes: A comparison of multiple data sources. *Journal of Applied Psychology, 73,* 11–19.

Spielberger, C.D. (1979). *Preliminary manual for the State-Trait Personality Inventory (STPI).* Unpublished paper, University of South Florida, Tampa.

Steinhardt, M.A., & Young, D.R. (1992). Psychological attributes of participants and nonparticipants in a worksite health and fitness center. *Behavioral Medicine, 18,* 40–46.

Storms, P.L., & Spector, P.E. (1987). Relationships of organizational frustration with reported withdrawal reactions: The moderating effect of locus of control. *Journal of Occupational Psychology, 60,* 227–234.

Tombaugh, J.R., & White, L.P. (1989). *The effects of organizationally based social support on survivors' perceived stress and work attitudes.* Paper presented at the Society for Industrial and Organizational Psychology Convention, Boston.

Trice, A.D., & Price-Greathouse, J. (1987). Locus of control and AIDS information seeking in college women. *Psychological Reports, 60,* 665–666.

Tucker, L.A. (1990). Physical fitness and psychological distress. *International Journal of Sport Psychology, 21,* 185–201.

Wallston, K.A., Wallston, B.S., & DeVellis, R. (1978). Development of the multidimensional health locus of control (MHLC) scales. *Health Education Monographs, 6,* 161–170.

Watson, D., Clark, L.A., & Tellegen, A. (1988). Development and validation of brief measures of positive and negative affect. *Journal of Personality and Social Psychology, 54,* 1063–1070.

Index